The VNR Dictionary of Environmental Health and Safety

Edited by Frank S. Lisella, Ph.D., M.P.H.

Library of Congress Catalog Card Number 93-25838
ISBN 0-442-00508-3

I(T)P Van Nostrand Reinhold is an International Thomson Publishing company. ITP logo is a trademark under license.

Printed in the United States of America

Van Nostrand Reinhold
115 Fifth Avenue
New York, NY 10003

International Thomson Publishing
Berkshire House, 168-173
High Holborn, London WC1V 7AA
England

Thomas Nelson Australia
102 Dodds Street
South Melbourne 3205
Victoria, Australia

Nelson Canada
1120 Birchmount Road
Scarborough, Ontario
M1K 5G4, Canada

International Thomson Publishing GmbH
Königswinterer Str. 518
5300 Bonn 3
Germany

International Thomson Publishing Asia
38 Kim Tian Rd., #0105
Kim Tian Plaza
Singapore 0316

International Thomson Publishing Japan
Kyowa Building, 3F
2-2-1 Hirakawacho
Chiyada-Ku, Tokyo 102
Japan

16 15 14 13 12 11 10 9 8 7 6 5 4 3 2 1

Library of Congress Cataloging in Publication Data

Dictionary of environmental health and safety/edited by Frank S. Lisella.
p. cm.
Includes bibliographical references.
ISBN 0-442-00508-3
1. Environmental health—Dictionaries. 2. Environmental protection—Dictionaries. 3. Industrial safety—Dictionaries.
I. Lisella, Frank S.
RA566.D53 1994
363'.03—dc20 93-25838
CIP

The VNR Dictionary of Environmental Health and Safety

To the memory of my parents, Bertha Mano Lisella and Frank J. Lisella. To Audrey and Richard. And to friends and faculty at Millersville University, Tulane University, and the University of Iowa.

Contents

Preface

Among the worst nightmares an editor of a book of this type could have is be reminded by a colleague that a certain key definition or series of definitions was omitted from the published text. Please be assured that the contributors and I have made every effort to prevent this eventuality. However, given the dynamics of the field we tried to embrace, we recognize that the capture of every single term was an impossibility. We all did strive, however, to include what we believe are among the most relevant environmental health and safety terms and those that will have the most utility to our readership. Each of the contributors provided terms from his or her specialty area, with the hope that the final product would be a holistic representation of this challenging field. We apologize in advance for any terms that may have inadvertently been omitted and welcome readers' suggestions on terms that should be included in revised versions. These suggestions should be sent to the publisher.

I would like to thank personally each of the contributors and numerous other individuals who helped with the preparation of this dictionary. Their experience, patience, time, and effort contributed immeasurably to the completion of the book.

Special thanks are also due to my wife, Lynn, who gave so unselfishly of her own time in the creation and management of the data base of terms and in the editing of the manuscript. Without her guidance, encouragement, and dedication to the project, its completion would not have been possible.

Frank S. Lisella
Atlanta, GA

Introduction

This dictionary, which contains well over 7000 terms, is intended to be a ready reference for environmental health and safety practitioners as well as for those in other professions who may need concise yet comprehensive definitions of particular terms in the various disciplines that constitute this field. Technical specialists may find the dictionary useful for searching out terms not ordinarily used in their everyday professional activities. Those in other professions who need to use a reference book for clarification of terminology in environmental health and safety will find an extensive listing of the core terms in this area.

Recognizing the number of disciplines represented and the breadth of the field of environmental health and safety, the contributors have attempted to include those terms that, in their experience, would be most helpful to users. The terms included have been selected from the fields of agronomy, biosafety, biostatistics, ecology, environmental law, epidemiology, general sanitation (food and water as well as refuse and sewage disposal), hazardous-material control, industrial hygiene, microbiology, radiation, and safety. Because it covers this broad range of topics, the dictionary should be useful to persons in federal, state, and local governments, acedemia, and the voluntary and corporate sectors.

How to Use This Book

Many of the terms listed in this dictionary have multiple definitions—e.g., the term "absorption" has slightly different meanings in the fields of chemistry, industrial hygiene, toxicology, and radiation, among others. When possible, each relevant definition for any particular term has been included. Additionally, many of the terms are cross-referenced. The word "*see*" is used to refer the reader to a directly related term that is defined elsewhere in the text. The words "*see also*" following any definition help readers locate indirectly related terms.

The book contains two appendices. Appendix I is an alphabetical listing of key acronyms directly related to the environmental health and safety field. Appendix II is a listing, in alphabetical order, of those chemicals defined in the text, along with the respective Chemical Abstracts Service Registry Numbers (CAS). This will aid readers who wish to compare data on a particular chemical regardless of the synonym under which the material may be published. The final portion of the book contains a list of references consulted in the preparation of this dictionary. Readers can consult these references for additional, in-depth information on particular terms.

Contributors

(HMB) **Harold M. Barnhart, Ph.D.**
Head, Environmental Health Program
College of Agriculture
Department of Food Science and Technology
University of Georgia
Athens, Georgia 30602

(RMB) **Robert M. Boyd, B.S.**
Radiological Safety Officer
Georgia State University
Atlanta, Georgia 30303

(SMC) **Susan M. Carstensen, M.S.**
Ecologist
Biology Department
Emory University
Atlanta, Georgia 30322

(MAG) **Michael A. Gladle, B.S., C.I.H.**
Asbestos Program Manager
Environmental Health and Safety Office
Emory University
Atlanta, Georgia 30322

(DEJ) **David E. Jacobs, M.S., C.I.H.**
National Center for Lead-Safe Housing
205 American City Bldg.
Columbia, MD 21044

(FSL) **Frank S. Lisella, Ph.D., M.P.H.**
Director
Environmental Health and Safety Office
Emory University
Atlanta, Georgia 30322

(SAN) **Shirley A. Ness, M.S., C.I.H., C.S.P.**
Amoco Oil Company
Chicago, Illinois 60601

(VWP) **Virginia W. Parks, M.A.T.**
Assistant Professor
Mathematics Department
DeKalb College
Clarkston, Georgia 30021-2396

(JHR) **John H. Richardson, D.V.M., M.P.H.**
Biosafety Consultant
2982 Danbyshire Court
Atlanta, Georgia 30345

(SWT) **Scott W. Thomaston, M.S.**
Chemical Safety Officer
Environmental Health and Safety Office
Emory University
Atlanta, Georgia 30322

(DEW) ***Deborah E. Weimer, J.D.***
Attorney–Advisor
Centers for Disease Control and Prevention, PHS
Department of Health and Human Services
Atlanta, Georgia 30333

The VNR Dictionary of Environmental Health and Safety

A

Abatement The removal of a nuisance; elimination or reduction of the degree or intensity of pollution. *See also* Nuisance. (DEW)

Abel test A colorimetric test that involves the use of moist potassium iodide paper, which turns violet or blue in the presence of gases evolved from nitroglycerin, nitrocellulose, and nitroglycol. This test is used primarily for quality control purposes in the manufacture of explosives. (FSL)

Abiocen The nonliving components of the ecosystem. (FSL)

Abiotic Indicating the absence of life; nonbiological. (FSL)

Ablation Removal of a layer, particularly of surface ice or loose surface materials, by the wind. (FSL)

Abrasive Any of the many types of commercially available materials used to scrub, scour, polish, or smooth surfaces. These materials give scouring and polishing properties to hand soaps, cleaners, and soap pads. They may be made of natural or synthetic substances. Naturally occurring abrasives include sand, pumice, quartz, and feldspar ground to fine particle size. (HMB)

Abrasive blasting A process used for cleaning surfaces that incorporates materials such as alumina, sand, and steel grit in a high-pressure air stream. (FSL)

Abscisic acid ($C_{15}H_{20}O_4$) A plant hormone, or auxin, that stimulates leaf drop and dormancy in buds and seeds. The auxins probably inhibit the synthesis of nucleic acid and proteins. (FSL)

Abscissa In a coordinate system, the distance of a point from the vertical axis as measured along a line parallel to the horizontal axis. *See also* Ordinate. (VWP)

Absolute humidity The percentage of water vapor present in a unit mass of air. The mass of water vapor present in a unit volume of the atmosphere, measured either as grams per cubic meter or actual partial pressure. (FSL, DEJ)

Absolute pressure Pressure that has been measured relative to zero pressure. (FSL)

Absorbed dose For any ionizing radiation, the energy imparted to matter by ionizing particles per unit mass of irradiated material at the place of interest. *See also* RAD. (RMB)

Absorption **(1)** (Toxicology) The ability of a substance to penetrate the body of another; the movement of a chemical from the site of exposure (oral, dermal, respiratory) within the biologic system across a biologic barrier and into the bloodstream or lymphatic system. **(2)** (Chemistry) The process by which one material is pulled into and retains another to form a blended or homogeneous solution. The absorption of gases, for example, is dependent upon temperature, humidity, and various physical principles. **(3)** (Physiology) The process whereby porous tissues such as the skin and intestinal walls permit the passage of liquids and gases into the bloodstream. **(4)** (Radiation) The process by which the number of particles or quanta in a beam of radiation is reduced or degraded in energy as it passes through some medium. The absorbed radiation may be transformed into mass, other radiation, or energy by interaction with the electrons or nuclei of the atoms upon which it impinges. (FSL, RMB)

Absorption coefficient Fractional decrease in the intensity of a beam of radiation per unit thickness (linear absorption coefficient), per unit mass (mass absorption coefficient), or per atom (atom absorption coefficient) of absorber. For ordinary use in shielding problems, etc., the mass absorption coefficient is preferable. This is equal to the linear absorption coefficient divided by the density of the absorbing material and has units of cm^2 gram. (RMB)

Absorption tower A structure used in the production of sulfuric acid from sulfur dioxide and water. Towers of this type are usually found in chemical manufacturing plants. (FSL)

Absorption trench A component of a subsurface sewage disposal system that consists of a trench not over 36 inches in width underlain with a minimum of 12 inches of clean, coarse aggregate, and covered with a minimum of 12 inches of soil, and that contains a distribution pipe. *See also* Sand filter trench. (FSL)

Acacia Leguminous tree found in tropical and subtropical areas that serves as a vital source of commercial products such as dyes and perfumes. (FSL)

Acanthite An ore of silver found in hydrothermal deposits. In addition to its phototrophic uses, it has considerable application in industrial situations because of its high reflective and conductive potential. (FSL)

Acaricide Chemical used to kill ticks and mites. *See also* Insecticides. (FSL)

Acceleration The shortening of the time in which the performance of an obligation is due or an expected interest is realized. Can occur by contract or by operation of law. (DEW)

Accelerator (Radiation) A device for imparting a large amount of kinetic energy to charged particles, such as electrons, protons, deuterons, and helium ions. Common types of accelerators are the cyclotron, synchrotron, synchrocyclotron, betatron, linear accelerator, and Van de Graaff electrostatic generator. (RMB)

Acceptor A charge of explosives or a blasting agent that receives an impulse from an exploding donor charge. (FSL)

Accessory mineral A mineral that occurs in such insignificant proportions in a rock that it does not figure into the classification of that rock. (FSL)

Accident An unplanned, unforeseen event that results in physical damage to property and/or the illness, injury, or death of an individual or group of individuals in the home, workplace, or leisure environment. *See also* Accident investigation; Incident; Near miss. (SAN)

Accident analysis A periodic review and study of records related to accidents, near misses, and incidents, performed to discern trends, discover causes, and identify preventive and corrective actions. *See also* Accident; Incident; Near miss. (SAN)

Accident investigation A detailed review of an individual accident or near miss conducted to identify the causes and develop control measures to prevent a reoccurrence. *See also* Incident; Near miss. (SAN)

Accident site The location of an unexpected occurrence, failure, or loss, such as a release of hazardous materials, either at a plant or along a transportation route. (FSL)

Acclimation The process by which humans develop a tolerance to various environmental or occupational stressors. For example, newly hired individuals will not be as tolerant to a hot working environment as will conditioned workers. Acclimatization periods for heat, cold, high altitude, and other stressors have been defined. (DEJ)

Accountability In order for a safety and health program to be effective, managers, supervisors, and employees must be held responsible for meeting their assigned responsibilities. The contractual relationship between company management and all contractors must bind contractors and hold them responsible for providing contract workers with safe and healthful working conditions. Specific performance tasks and clear measurements of performance relative to safety and health responsibilities must be a critical part of managers' and supervisors' performance standards. Performance by managers, supervisors, and employees in meeting safety and health responsibilities must be evaluated, rewarded, and/or modified as appropriate. *See also* Safety management systems. (SAN)

Accretion The addition of soil to land by gradual, natural deposits. *See also* River wash. (FSL)

Acer The genus name for maple trees and shrubs, which occur in temperate regions of the World. They represent a valuable source of timber for furniture manufacture and maple sugar products, and are an important cash crop in the northeastern United States. (FSL)

Acetic Acid (CH_3COOH) The acid contained in vinegar and that also obtained from wood, acetylene, or alcohol. The most common method of production is to react methanol, a catalyst, and carbon monoxide. It can also be obtained by the fermentative oxidation of ethanol. Glacial acetic acid is the purest generally available form of the compound (99.8%). In 1985, this chemical ranked 33rd in terms of volume produced in the United States. Pure acetic acid is a minimal fire risk and is moderately toxic by ingestion and inhalation. (FSL)

Acetone (CH_3COCH_3) (dimethyl ketone; 2-propanone) A colorless, volatile solvent that has a sweet odor, has a flash point of 15°F, and is miscible with water, chloroform, ether, and many oils. This chemical, which is moderately toxic by ingestion, is flammable and a dangerous fire hazard. Explosive limits in air range from 2.6 to 12.8%. The threshold limit value (TLV) is 750 ppm in air. (FSL)

Acetylene ($HC{\equiv}CH$) A colorless gas typically prepared: by reacting water with calcium carbide; by cracking petroleum hydrocarbons with steam; from the partial oxidation of natural gas; or from fuel oil by the modified arc process. This chemical, which is used in the manufacture of vinyl chloride, vinyl acetate, and acetaldehyde, is highly flammable and a dangerous fire risk, and forms explosive compounds with silver, mercury, and copper. (FSL)

Acid-cleaning compound Any of a class of cleaning compounds frequently used in food-service facilities and in food processing plants. As with alkaline cleaners, pH is used to describe the nature of the cleaning compound. A pH from 0 to 7 characterizes acidic substances. As pH decreases from 7 to 0, the acidity increases. These chemicals may be divided into subclasses—e.g., strong acid cleaners, used for cleaning operations in food processing and including agents such as muriatic, sulfuric, and phosphoric acids. These compounds are highly corrosive to most surfaces and must be used with great care. They are used in cleaning operations for the removal of encrusted surface material and mineral deposits; mild acid cleaners are mildly corrosive, but may irritate the skin and eyes. Examples are acetic acid, hydroxyacetic acid, and gluconic acid. They are used primarily to remove mineral buildup. (HMB)

Acid deposition A complex chemical and atmospheric phenomenon that occurs when emissions of sulfur and nitrogen compounds and other substances are transformed by chemical processes in the atmosphere, often far from the original sources, and then deposited on earth in either wet or dry form. The wet form, popularly called "acid rain," can fall as rain, snow, or fog. The dry form includes acidic gases and particulates. (FSL)

Acid food Food with its natural or inherent acidity below that required by most spoilage organisms. Microorganisms grow best at pH levels around 7.0 (6.5–7.5). Some organisms, e.g., molds and yeasts are more tolerant of acidic pH, and for that reason acid foods tend to undergo spoilage by molds and yeasts. Examples of acidic foods include fruits, vegetables, soft drinks, vinegar, and wine. (HMB)

Acid-forming bacteria These organisms ferment the lactose in milk and result in production of acid, mainly lactic acid, which combines with a protein, calcium caseinogenate, liberating casein, which precipitates in the form of a curd. *Streptococcus lactis* is an important acid-producing bacterium in milk that causes souring, thereby affecting the quality of the product. (HMB)

Acid mantle The lipid (oily) outside layer of the skin structure, composed of oil and sweat and easily removed by washing. The acid mantle has a pH less than 7.0. (DEJ)

Acid rain Precipitation contaminated with sulfur dioxide, nitrous oxide, and other chemicals from power plants and industrial sites. Acid rain has been shown to affect aquatic life, trees, and vegetable crops adversely and to contribute to soil erosion and corrosion of buildings. *See also* Acid deposition. (FSL)

Acid soil A soil with a pH of less than 7, but usually less than 6.6 (as a more practical reference). This term normally refers to the upper soil horizons or root zones. (SWT)

Acquired immunodeficiency syndrome (AIDS) A severe, life-threatening disease first recognized as a distinct syndrome in 1981. This syndrome represents the late clinical stage of infection with the human immunodeficiency virus (HIV), which most often results in progressive damage to the immune system and various organ systems, especially the central nervous system. The $CD4^+$, or T-helper lymphocytes are the primary target cells for HIV infection. A decrease in the number of these cells correlates with the risk and severity of HIV-related clinical conditions. Studies of the natural history of HIV infection have documented a wide spectrum of disease manifestations, ranging from asymptomatic infection to life-threatening conditions characterized by severe immunodeficiency, serious opportunistic infections, and cancer. Approximately 253,448 cases of AIDS were reported in the United States as of December 1992. Virtually all countries of the world have cases of the disease; cases have been noted among all races, ages, and socioeconomic classes. In the United States, populations that have been primarily considered to be at high risk for AIDS are men who have sex with men, users of injected drugs, heterosexual contacts of infected partners, children of infected mothers, and patients with hemophilia. Community or social (casual) contact with an HIV-infected person presents no risk of infection. Sexual exposure and contact with blood or tissues places an individual at risk. HIV has been recovered from body fluids other than blood, such as tears, saliva, urine, and bronchial secretions, but transmission after contact with these substances has not been reported. (FSL)

Acre-foot The volume of water (43,560 cubic feet) required to cover one acre of surface to a depth of one foot. (FSL)

Acremite A mixture of approximately 94% ammonium nitrate and 6% fuel oil, initially developed as a cheap explosive for open-pit mining. (FSL)

Acroosteolysis A condition reported in workers exposed to vinyl chloride and manifested by ulcerating lesions on the hands and feet. (FSL)

Acrophobia The fear of high places. (FSL)

Actinomyces A genus of bacteria pathogenic for humans and animals. These organisms are nonmotile, nonspore-forming, and anaerobic to facultative anaerobic, and stain Gram-positive. (FSL)

Action level **(1)** An exposure limit usually set at 50% of the permissible exposure limit (PEL) in the Occupational Safety and Health Administration (OSHA) standards. Exposures exceeding the action level typically require implementation of medical surveillance, training, and monitoring programs, but not necessarily further measures (e.g., engineering controls) aimed at reducing exposures. The action level concept was originally recommended to OSHA by the National Institute for Occupational Safety and Health, and was predicated on the idea that keeping exposures below 50% of the PEL increased the likelihood that random exposures on any given day would also be below the PEL. An important exception to the 50% criterion is lead, which has a permissible exposure limit of 50 $\mu g/m^3$ and an action level of 30 $\mu g/m^3$ (both 8-hour time-weighted averages). **(2)** A point above which adverse health effects are possible, based upon the level or concentration of a chemical in feeds or food. The concentration of a specific chemical is calculated as an 8-hour time-weighted average. This serves as the reference point for the initiation of legal action by regulatory authorities and initiates certain required activities, such as exposure monitoring and medical surveillance. *See also* Tolerances. (DEJ) (FSL)

Activated alumina A granular, porous form of aluminum oxide capable of absorbing other substances from gases or liquids. This chemical is used in chromatographic analyses and as a catalyst. (FSL)

Activated carbon (charcoal) A form of carbon with a high potential for the absorption of gases, vapors, and colloidal particles. This material is frequently incorporated into filters in heating and air conditioning systems for the purpose of removing odors from ambient air. It is also an effective component of most municipal water treatment systems, where it is added in the mixing basins for taste and odor control. (FSL)

Activated sludge An aerobic biological process used in the secondary treatment of wastewater. Naturally occurring, biologically active microorganisms (activated sludge) are brought into contact with organic wastes in sewage in the presence of mechanically introduced air. The activated sludge process involves the same phenomenon as aerobic stabilization in a stream, except it is artificially accelerated. When mixed with the appropriate amount of biologically active sludge, sewage that usually has received primary treatment will undergo rapid clarification. Mechanically introduced air accelerates this process, in which dissolved nutrients and finely suspended solids are transported to the surface of sludge floc, where large numbers of microorganisms exist. A continuous supply of activated sludge must be mixed with the incoming wastewater, which necessitates the return of sludge (return activated sludge) from the secondary clarifier to the incoming sewage. Mechanical aeration is typically supplied through diffusers located throughout the aeration basin. After passage through the aeration basin, sewage is allowed to settle in secondary clarifiers, and, after appropriate detention time, clarified effluent may be discharged, some may be wasted, and a portion (return activated sludge) is cycled to mix with the incoming sewage. Several different types, which are modifications of the conventional activated sludge system, may be used and include extended aeration systems and contact stabilization systems. The activated sludge system has the advantages of lower capital cost, flexibility in its action, and, generally, production of a good effluent. Its disadvantages include high operating costs, the large volume of sludge for disposal, and the need for skilled operators. *See also* Primary treatment; Secondary treatment. (HMB)

Activation analysis A method of chemical analysis, especially for small traces of material, based on the detection of characteristic radionuclides. (RMB)

Activation energy The energy necessary to cause a particular reaction to begin. Nuclear activation energy is the amount of outside energy that must be added to a nucleus before a particular nuclear reaction will begin. Chemical activation energy is the amount of outside energy necessary to activate an atom or molecule so as to cause it to react chemically. (RMB)

Active ingredient In any pesticide product, the component that kills, or otherwise controls, target pests. Pesticides are regulated primarily on the basis of active ingredients. *See also* Pesticide. (FSL)

Active material Fissionable material, such as plutonium-239, uranium-235, or the thorium-derived uranium-233, that is capable of supporting a chain reaction. In the military field of atomic energy, the term refers to the nuclear components of atomic weapons exclusive of the natural uranium parts. (RMB)

Activity The rate of decay of radioactive material expressed as the number of nuclear disintegrations per second. *See also* Curie. (RMB)

Activity/Production Index A measure of changes in production activity, economics, and/or other factors that affect the quantity of hazardous waste generated in a given year, compared to the preceding year. The index is used to distinguish interyear hazardous-waste generation quantity changes resulting from waste minimization activity and from production, activity, economics, or other factors. (FSL)

Act of God Generally, an act caused by a force of nature with no human intervention. For purposes of CERCLA, an unanticipated grave natural disaster or other natural phenomenon of an exceptional, inevitable, and irresistible character, the effects of which could not have been prevented or avoided by the exercise of due care or foresight (CERCLA §9601(1)). *See also* Force majeur. (DEW)

Acute exposure **(1)** (Chemical) A sudden, short, rapid association with a chemical compound. **(2)** (Radiation) Radiation exposure of short duration. Generally taken to be the total dose absorbed within 24 hours. **(3)** (Biologic) A brief encounter with a pathogenic or nonpathogenic microorganism. *See also* Chronic exposure. (MAG, RMB)

Acute hazardous waste Any hazardous waste with an EPA Waste Code beginning with the letter *P*, or any of the following *F* codes: F020, F021, F022, F023, F026, and F027. These wastes are subject to stringent quantity standards for accumulation and generation. (FSL)

Acute toxicity The ability of a substance to cause poisonous effects resulting in severe biological harm or death soon after a single exposure or dose. Also, any severe poisonous effect resulting from a single short-term exposure to a toxic substance. *See also* Chronic toxicity; Toxicity. (FSL)

Adaptation A change in an organism's structure or habit that helps it adjust to its surroundings. (FSL)

Additive In foods, the term characterizes a substance in terms of use rather than the nature of the substance. The FDA defines food additives as substances "the intended use of which results or may reasonably be expected to result, directly or indirectly, either in their becoming a component of food or otherwise affecting the characteristics of food. A material used in the production of containers and packages is subject to the definition if it may reasonably be expected to become a component, or to affect the characteristics, directly or indirectly, of food packed in the container." "Affecting the characteristics of food" does not include having such physical effects as protecting the contents of packages, preserving shape, or preventing moisture loss. If there is no migration of a packaging component from the package to the food, the packaging component does not become a component of the food and thus is not considered a food additive. A substance that does not become a component of food, but that is used, for example, as an ingredient of the food to give a different flavor, texture, or other characteristic to the food, may be a food additive. Additives may be classed as either intentional or incidental. Intentional additives are those of known composition that have been purposely added to foods to achieve specific effects. Incidental additives are those substances that have no planned function in food, but that have become a part of it accidentally or indirectly during some phase of production, processing, packaging, or storage. Additives are intentionally used in the food industry for one or more of the following purposes: to maintain or improve nutritional value, to maintain freshness, to help in processing or preparation, or to make food more appealing. *See also* Incidental; Intentional additive. (HMB)

Add-on control device An air pollution control device, such as a carbon absorber or incinerator, that reduces pollution in an exhaust gas. The control device usually does not affect the process being controlled and thus is "add-on" technology as opposed to a pollution-reducing alteration to the basic process. (FSL)

Adenoma A neoplasm, usually benign and noninvasive, composed of epithelial tissues in which the tumor cells form glands or glandlike structures. (MAG)

Adhesion Molecular attraction that holds the surfaces of two substances in contact. (FSL)

Adiabatic **(1)** Refers to a reaction that occurs without a gain or loss of heat. **(2)** Of, pertaining to, or designating a reversible thermodynamic process executed at a constant energy; taking place without a gain or loss of heat energy (without heat entering or leaving the system). *See also* Adiabatic precipitation; Adiabatic temperature. (SMC)

Adiabatic precipitation Precipitation that occurs as a result of adiabatic cooling or contact with moist air. When warm air rises, it expands and cools adiabatically (without loss of heat energy in the system) because of decreasing atmospheric pressure. The cooler air is able to hold less water vapor than is the warmer air, so the water vapor condenses and forms precipitation. Adiabatic precipitation is especially important in regions near the equator and on the windward side of mountain ranges, which are areas in which warm moist air rises. Conversely, the adiabatic warming of air as it descends and compresses creates a desiccating effect as the warmer, drier air takes up more moisture. This desiccating effect predominates near certain latitudes (about 30 degrees N and S) and on the leeward side of mountain ranges. On the leeward side of mountains this desiccating effect is also called rain shadow. *See also* Adiabatic. (SMC)

Adiabatic temperature The temperature attained by a reaction undergoing a volume or pressure change in which no heat enters or leaves the system. *See also* Adiabatic. (FSL)

Adjudication Under the Administrative Procedure Act, agency process for the formulation of an order, which is defined as the whole or part of a final disposition, whether affirmative, negative, injunctive, or declaratory in form, of an agency in a matter other than rulemaking but including licensing. 5 USC §551(7). (DEW)

Adjustable barrier guard A barrier that requires a different position for each application. These guards are commonly used on die presses and must be adjusted for each die setup. *See also* Machine guarding. (SAN)

Adjuvant In immunology, a vehicle, such as a suspension of minerals, in which an antigen is absorbed for the purpose of increasing antigenicity; a substance added to a drug product that affects the action of an active ingredient in a predictable manner. (FSL)

Administrative control A measure initiated to reduce worker exposure to various stresses in the work environment—e.g., limitation of the amount of time employees can be exposed to hazardous chemicals, noise, dust, or fumes by rotating them to other locations within the industrial setting. (FSL)

Administrative law judge A hearing officer who presides over administrative procedure hearings of federal agencies under the Administrative Procedure Act. (DEW)

Administrative order A legal document directing an individual business or other entity to take corrective action or refrain from an activity. (FSL)

Administrative Procedure Act A federal law that sets out procedural requirements for federal agency decision making through rulemaking and adjudication. It also outlines the principles for judicial review of agency actions. 5 USC §551–559, 701–706 (additional sections found throughout 5 USC). (DEW)

Adrenal gland In mammals, a gland adjacent to the kidney that produces the hormone adrenaline, or epinephrine. This hormone influences the heartbeat rate, dilates blood vessels, increases blood sugar, and plays a major role in other physiologic activities of mammals. (FSL)

Adsorption **(1)** The condensation of gases, liquids, or dissolved substances on the surface of solids. **(2)** The attraction and retention of atoms, molecules, or ions on the surface of solids. Adsorption is due to hydrogen bonding, electrostatic bonding, or coordination reactions. Negatively charged particles such as those in certain clay minerals or other soil colloids are examples of particles that will adsorb cations. **(3)** (Radiation) The taking in of one substance by another substance. Tritium is often adsorbed by zirconium accelerator targets. (FSL) (SWT) (RMB)

Adsorption complex Any of the various organic and inorganic substances (humic acid or clay minerals) in soil that are capable of adsorbing molecules or ions. (SWT)

Adulterants Chemical impurities or substances that by law do not belong in a food or in a pesticide. (FSL)

Adulticide An insecticide used to destroy mature forms of arthropods. *See also* Insecticides. (FSL)

Advanced wastewater treatment Any treatment of sewage that goes beyond the secondary or biological treatment stage and includes the removal of nutrients such as phosphorus and nitrogen and a high percentage of suspended solids. *See also* Primary wastewater treatment; Secondary wastewater treatment. (FSL)

Advection Movement caused by the motion of heat, air, water, or another fluid. It specifically refers to the horizontal movement by wind currents of chemical pollutants, particulates, or heat. (FSL)

Aeolin deposit A sediment, often called eolian, that has been deposited after being transported by the wind. (FSL)

Aerate To promote the exchange of soil gas with atmospheric gases. *See also* Aeration, soil. (SWT)

Aeration The process of bringing about contact between air and a liquid such as water. Aeration may be used to oxidize iron or manganese and to remove odors from water such as those caused by hydrogen sulfide and algae. Aeration is also effective in increasing the oxygen content of water deficient in dissolved oxygen. In sewage treatment, aeration is the process whereby a material becomes saturated with air or another gas. Air is introduced by agitation or other means into raw sewage to oxidize that material as a preliminary step in the treatment process. (FSL)

Aeration, soil Exchange of soil air with atmospheric air. Aeration depends upon the volume and continuity of soil pores. Strongly aggregated soils tend to be well aerated. A well-aerated soil is of approximately the same composition as atmospheric air. *See also* Aggregate, soil. (SWT)

Aeration tank A chamber used to inject air into water. (FSL)

Aeroallergen Airborne material, such as particulates, pollen, dusts, and dander, that may precipitate an allergic response in susceptible persons. (FSL)

Aerobic Describes: an environment with molecular oxygen present; organisms that live or grow in the presence of molecular oxygen; and reactions that occur in the presence of molecular oxygen. *See also* An aerobic. (SWT)

Aerobic treatment Process by which microbes decompose complex organic compounds in the presence of oxygen and use the liberated energy for reproduction and growth. Types of aerobic processes include extended aeration, trickling filtration, and use of rotating biological contactors. (FSL)

Aerosol A dispersion of solid or liquid particles of microscopic size in a gaseous medium. Smokes, fogs, and mists are examples of aerosols. (DEJ)

Aerozine A mixture of 50% anhydrous hydrazine and 50% dimethylhydrazine used as a liquid fuel for rocket engines. (FSL)

Aflatoxins Some of the most commonly encountered and most toxic of the mycotoxins found as contaminants in foods. Mycotoxin is a term used to designate a broad class of secondary metabolites resulting from mold growth occurring on a specific substrate. The name *aflatoxin* is derived from the mold that produces it, *Aspergillus flavus*. There are four main aflatoxins: B^1, B^2, G^1, and G^2, of which B^1 is the most common and most toxic. Aflatoxins are highly stable and highly carcinogenic. The most limiting factor for the formation of mycotoxins may be moisture. High water activity and aerobic conditions are necessary for molds to grow, and therefore for mycotoxin production. Aflatoxins and other mycotoxins pose a threat to both human and animal health. Prevention and control of mycotoxin formation is essential. Measures to eliminate conditions that favor the growth of molds on suitable substrates are fundamental to the prevention of mycotoxin formation. (HMB)

Afterburner In incinerators, a burner located in such a way that the combustion gases are made to pass through its flame to remove smoke and odors. (FSL)

Aftershock Any of a series of small tremors following an earthquake of significant magnitude. These typically occur in close proximity to the area of the initial shock. (FSL)

Agar A gelationlike substance, derived from certain species of seaweed, used as a medium for culturing microorganisms and as a thickening agent in foods. (FSL)

Agar contact plate Refers to a method for estimating the sanitary quality of surfaces. Plastic disposable petri plates containing an appropriate bacteriological medium are filled sufficiently so that the meniscus of the agar rises above the rim of the plate, to give a slightly convex surface. When this agar surface is placed in contact with the surface to be evaluated, the agar will make proper contact. The contact plate method is recommended particularly when quantitative data are sought for flat, impervious surfaces; it should not be used for pervious, creviced, or irregular surfaces. When properly incubated, a total count can be obtained for the number of microorganisms recovered from the surface being tested. An example of the contact technique is the RODAC (replicate organism direct agar contact) plate method. *See also* RODAC plate. (HMB)

Agency for Toxic Substances and Disease Registry (ATSDR) Agency of the U.S. Public Health Service of the Department of Health and Human Services, created by the Comprehensive Environmental Response, Compensation, and Liability Act of 1980 as amended, to implement specific health-related authorities under that statute. Responsibilities include: the preparation of public health assessments of all National Priorities List sites, and, in response to petitions from the public, the preparation of toxicological profiles of priority hazardous substances, the creation of a registry of persons exposed to hazardous substances, epidemiologic studies of exposed persons, and the gathering and dissemination of related educational materials. 42 USC §9604(i), 42 CFR Part 90. (DEW)

Agent A biological, physical, or chemical entity capable of causing disease. (FSL)

Agent Orange A toxic herbicide and defoliant that was used in the Vietnam conflict. It contains 2,4,5-trichlorophenoxyacetic acid (2,4,5-T) and 2,4-dichlorophenoxyacetic acid (2,4-D), with trace amounts of dioxin. (FSL)

Agglomerate Rock composed of volcanic material, with the majority of individual pieces greater than 1 inch in diameter. (FSL)

Agglomeration The process by which precipitation particles grow larger by collision or contact with cloud particles or other precipitation particles. (FSL)

Agglutination The process of uniting solid particles coated with a thin layer of adhesive material. (FSL)

Aggradation The accumulation of deposited material on land and the associated raising of the elevation of the area. (FSL)

Aggregate Crushed stone used primarily as construction material and for underlaying of subsurface sewage disposal systems. Typical soil absorption systems have 6 inches of aggregate material under the individual tile lines and 2 inches above. (FSL)

Aggregate, soil Soil particles held in a single mass such as a clod, crumb, or block. Aggregation is influenced by chemical and biological activity such as that of plant roots, or microbial mucilage and mycelia. (SWT)

Agoraphobia A combination of phobias in which, in a situation involving open or closed places, the person feels cut off from escape to a safe place. (FSL)

Agric horizon An illuvial soil horizon at which the clay and humus from the cultivated layer above have accumulated. *See also* Illuvial. (SWT)

Agricultural pollution The liquid and solid wastes from farming, including runoff and leaching of pesticides and fertilizers, erosion materials and dust from plowing, animal manure and carcasses, and crop residues and debris. (FSL)

Agronomy A specialized agricultural science that involves the study and practice of field crop production and soil management; the scientific management of land. (SWT)

A-horizon The surface horizon of a mineral soil. This horizon usually has the maximum amount of biological activity, organic matter, and eluviation. (SWT)

AIDS *See* Acquired immunodeficiency syndrome.

Air and Water Management Association A national technical association, formerly called the Air Pollution Control Association, composed of professionals concerned with air pollution control and waste management issues. (FSL)

Air balance In laboratories, it is necessary that the air supply and exhaust systems be kept in balance. This involves fan tests, which include measurements of static pressure, fan and motor performance, air volume, current draw, and temperature rise, and resulting adjustments of dampers and sheaves. (FSL)

Airborne particulates Total suspended particulate matter found in the atmosphere as solid particles or liquid droplets. Chemical composition of particulates varies widely, depending on location and time of year. Airborne particulates include windblown dust, emissions from industrial processes, smoke from the burning of wood and coal, and the exhaust of motor vehicles. (FSL)

Airborne transmission The dissemination of microbial agents to a suitable portal of entry, usually the respiratory tract. (FSL)

Air changes Usually expressed as air changes per hour or air changes per minute, air changes are calculated by dividing the volumetric airflow rate in a space or room (in cubic feet per minute, for example) by the size of the room (in square feet for example), divided by unit time. (DEJ)

Air changes per hour (ACH) The movement of a volume of air in a given period of time; if a dwelling has one air change per hour, all the air in that structure will be replaced in a one-hour period. (FSL)

Air cleaners Systems that remove airborne contaminants (dusts, fumes, mists, vapors, gases, smoke, or odors) and are situated at air exhaust stacks or inlets. (DEJ)

Air curtain A method of containing oil spills. In this method, air bubbling through a perforated pipe causes an upward water flow that slows the spread of oil. The method can also be used to stop fish from entering polluted water. (FSL)

Air door A ventilation unit consisting of controlled, outgoing air that, when used around doors and windows, helps keep out flying insects. Air doors can be a valuable aid in a facilities pest control program. (HMB)

Air dose (Radiation) A dose of X-rays or gamma rays, expressed in roentgens, delivered at a point in free air. In radiological practice it consists of the radiation of the primary beam and that scattered from surrounding air. (RMB)

Air-dry soil Soil with a moisture content, at equilibrium, that is the same as that of the surrounding atmosphere. Actual moisture content depends on temperature and relative humidity. (SWT)

Air-equivalent ionization chamber *See* Air wall ionization chamber.

Airflow The volumetric rate at which air flows through a space, usually measured in cubic feet per minute (CFM) or cubic meters per second (CMS). Also, the speed at which air moves through a space, usually measured in feet per minute or meters per second. (DEJ)

Air gap An unobstructed, vertical distance through the free atmosphere that separates the outlet of a potable water supply from the flood-level rim of a potentially contaminated receptacle. (HMB)

Air locks Devices that are usually found at entrances and/or exits and that prevent air in one space from entering another space. The device usually consists of a chamber with at least two doors and a separate exhaust ventilation system. Air locks are used in areas where biological, radioactive, or chemical hazards exist. (DEJ)

Air, makeup Air that replaces other air exhausted from a space. Insufficient makeup air is one possible cause of insufficient exhaust airflow. *See also* Replacement air. (DEJ)

Air mass A widespread body of air that can gain certain meteorological or pollution characteristics—e.g., development of a heat inversion or smog—while set in one location. The characteristics can change as the mass moves away. (FSL)

Air Moving and Conditioning Association (AMCA) An organization, located at 30 West University Drive, Arlington Heights, Illinois 60004, that develops standards for special-use fans—e.g., those used in laboratory ventilation systems. (FSL)

Air pollutant Any substance in air that could, in high enough concentration, harm people, animals, or vegetation, or damage nonliving material. Such pollutants may be in the form of solid particles, liquid droplets, gases, or combinations of these forms. (FSL)

Air porosity, soil The proportion of the bulk volume of soil that is filled with air at any given time or under a given condition, such as a specified moisture tension. (SWT)

Air-purifying respirator A device worn by an individual that filters the air to be breathed and returns it to an acceptable state. This type of respirator relies on the person's breathing to force air through the filter medium or utilizes a powered blower to provide breathing air (as in a powered air-purifying respirator, also known as a PAPR). Air-purifying respirators are not permitted in oxygen-deficient atmospheres or those containing contaminants with poor odor-warning properties. These respirators are also known as chemical cartridge respirators or powered air-purifying respirators. In the United States, all respirators must be formally approved by the National Institute for Occupational Safety and Health for specific work conditions. *See also* Respirator. (DEJ)

Air quality control region An area, designated by the federal government, in which communities share a common air pollution problem. Several states may be involved. (FSL)

Air quality standards The level of pollutants, prescribed by regulations, that may not be exceeded during a specified time in a defined area. (FSL)

Air shed A term used primarily in Europe in computing the air pollution burdens expected from designated pollution sources. Air shed calculations take into account air current movement and rates of chemical change. (FSL)

Air titration A field analytical method involving the use of an impinger or bubble to draw air through a liquid reagent that changes color in direct proportion to the concentration of the contaminant in the air. Air titration usually lacks the accuracy and precision of laboratory methods. (DEJ)

Air wall ionization chamber An ionization chamber in which the materials of the wall and electrodes are selected so as to produce ionization essentially equivalent to that in a free air ionization chamber. This is possible only over limited ranges of photon energies. Such a chamber is more appropriately termed an air-equivalent ionization chamber. (RMB)

Aithen counter A device named for the English inventor John Aithen that is used to count the number of condensation nuclei present in a unit volume. (RMB)

Alachlor A herbicide, marketed under the trade name Lasso, used mainly to control weeds in corn and soybean fields. (FSL)

Alamogordo cattle A small herd of Hereford cattle that was exposed to fallout from the first atomic bomb test at Alamogordo, New Mexico, in July, 1945. The animals were later moved to Oak Ridge, Tennessee, where they were studied in detail to determine the health effects associated with their exposure to radiation. (FSL)

Alar Trade name for daminozide (succinic acid-2,2-dimethylhydrazide), a pesticide that makes apples redder, firmer, and less likely to drop off trees before harvest. It is also used to a lesser extent on peanuts, tart cherries, concord grapes, and other fruits. (FSL)

Alarms, audible Horns, buzzers, whistles, and other noise-making devices that annunciate under a given set of circumstances, or the signals from them. These are installed in sensing instruments, in emergency alerting systems, and on heavy equipment. The noise level of the alarm should be higher than that of the operations around the device, so that the alarm can be heard. Many alarms associated with instruments or as stand-alone systems provide a combination of audible and visual signals. *See also* Alarms, visual. (SAN)

Alarms, visual Devices that produce signals generally composed of steady, turning, or flashing lights (often red, yellow, or blue) placed in a strategic location to warn personnel of an emergency or dangerous condition; or the signals themselves. Many alarms associated with instruments or as stand-alone systems provide a combination of audible and visual signals. *See also* Alarms, audible. (SAN)

Alban A cutan that is composed of materials that have been strongly reduced. *See* Cutan. (SWT)

Albedo The reflectance of a surface; the amount of solar radiation reflected by a body in relation to the amount incident upon it, usually expressed as a percentage (the albedo of the earth is 34%). The reflectance of the earth and different surfaces on the earth (e.g., snow, parking lots, tree tops) have a strong influence on climate. (SMC, SWT)

Albic horizon A light-colored surface or lower horizon from which clay and free iron oxides have been removed or so segregated as to permit the color to be determined by the primary sand and silt particles. (SWT)

Aldehydes A group of volatile, colorless chemical compounds containing the CHO group that are intermediate between acids and alcohols. (FSL)

Aldicarb An insecticide, sold under the trade name Temik, that is made from ethyl isocyanate. (FSL)

Aldrin ($C_{12}H_8Cl_6$) A chlorinated hydrocarbon insecticide used for the control of a wide range of insects. Since the chemical is persistent in soil, it has been banned for agricultural use in the United States. (FSL)

Alfisols Mineral soils that have no mollic epipedon, or oxic or spodic horizon, but do have an argillic or natric horizon that is at least 35% base saturated. Usually formed in humid areas under deciduous forests. (SWT)

Algae Photosynthetic eukaryotic organisms in the Kingdom Protista. Algae are usually aquatic and include kelp and many other "seaweeds." So-called blue-green algae, or cyanobacteria, are photosynthetic bacteria, not true algae. True algae are simple rootless plants that grow in sunlit waters in relative proportion to the amounts of nutrients available. They can affect water quality adversely by lowering the dissolved oxygen in the water. They provide food for fish and other aquatic animals. (SMC) (FSL)

Algal bloom A massive increase in aquatic algae populations. Algal blooms often indicate a rapid influx of some nutrients—for example, nitrogen and phosphorus from fertilizer runoff. These blooms can block sunlight needed by underwater plants, which eventually die and decompose. This decomposition depletes the dissolved oxygen in the water, causing the death of fish and other aquatic animals. (SMC)

Algicide An agent such as copper sulfate that will kill algae. *See also* Pesticide. (FSL)

ALGOL Algorithmic language or data-processing language utilizing algebraic symbols to express problem-solving formulae for computer solutions. (DEJ)

Aliphatic hydrocarbon One of the major groups of organic compounds characterized by a straight- or branched-chain arrangement of carbon atoms. This group is composed of three subgroups: alkanes (paraffins), which are saturated and relatively unreactive; alkenes (olefins) and alkadienes, which contain double bonds and are reactive; and, alkynes (acetylenes), which contain triple bonds and are highly reactive. (MAG)

Alkali-forming bacteria Organisms that make milk alkaline, presumably by acting on proteins. They may also saponify some fat through the action of a lipase produced, resulting in a yellow, translucent, whey-like fluid. Although their activity is mostly overshadowed by that of the acid-producing bacteria and is seldom noticed, their control is essential in order to ensure high-quality dairy products. (HMB)

Alkaline cleaning compound A class of cleaning compound frequently used in food-service establishments and processing plants. The pH, which is a logarithmic measurement of hydrogen ion concentration is used to describe the nature of the cleaning compound. As pH increases from 7 to 14, alkalinity increases. These cleaners may be divided into subclasses—e.g., strong alkaline cleaners, which have good dissolving properties, are very corrosive, and may burn or ulcerate skin. An example is sodium hydroxide. These strong alkaline compounds are used to remove heavy soil, such as in an oven, and have little effect on mineral deposits. Heavy-duty alkaline cleaners are slightly corrosive to noncorrosive, may contain sodium pyrophosphate, sodium hexametaphosphate, and trisodium phosphate, and are used in CIP or other mechanized systems; they are excellent for removing fat deposits, but have little effect on minerals. Mild alkaline cleaners, which are used to clean lightly soiled surfaces, and may contain sodium carbonate, tetrasodium pyrophosphates, or alkyl aryl sulfonates (surfactants), are good water softeners but are of little value in the removal of mineral deposits. (HMB)

Alkaline soil A soil with a pH of 8.5 or higher, or with a high exchangeable sodium content (15% or more). (SWT)

Alkaloid A complex chemical constituent containing nitrogen, and that is found in certain plants. Most alkaloids are physiologically active in vertebrates, and many have a bitter taste and are poisonous. These chemicals—e.g., nicotine from tobacco leaves, strychnine, morphine, quinine, and caffeine—are known for their medicinal or poisonous qualities. (FSL) (SMC)

Allele Any one of a series of two or more different genes that may occupy the same portion or locus on a specific chromosome. (FSL)

Allelopathy The inhibition of growth and development of plants by chemicals produced by a nearby plant of the same or different species. (SMC)

Allergen An allergy may be defined as any unusual or exaggerated response to a particular substance in a person sensitive to that substance. Any of a wide variety of substances or environmental conditions that may provoke an allergic reaction are called allergens. Common allergens are foods, beverages, dust, pollen, drugs, vaccines, cosmetics, and sunlight. People are most commonly subjected to possible allergens in foods because they consume foods daily. Good sanitary practices are important in the control of unwanted or undesirable substances in foods that may elicit an allergic response in some individuals. (HMB)

Allochthonous Refers to materials transported into a system from outside that system—e.g., materials and organic matter transported into streams and lakes by runoff. (SMC)

Allopatric Occurring in separate geographic areas; refers especially to species or populations whose ranges do not overlap. *See also* Sympatric. (SMC)

Alloy A metallic substance prepared by the addition of other metal ingredients to base metals to obtain a desired structural quality (e.g., strength) as in the case of solder or brass. (FSL)

Alluvial soil A soil developing from recently deposited alluvium and showing essentially no horizon development. (SWT)

Alluvion A deposit of soil on a shore or stream bank as a result of the force of the water, as by a current or by waves, in a process so gradual that the amount of material added in a short period of time is too low to measure. (DEW)

Alluvium General term for detrital material, including clay, sand, gravel, deposited by or in transit through streams. (SWT)

Alopecia The loss of hair; often called fox mange. (FSL)

Alpha particle A specific particle, consisting of two protons and two neutrons, (a helium nucleus) ejected spontaneously from the nuclei of some radioactive elements. It has low penetrating power and short range. Even the most energetic alpha particles will generally fail to penetrate the skin. The danger arises when matter containing alpha-emitting isotopes is introduced into the lungs or intestinal tract. The mass of an alpha particle is 4.00277 atomic mass units. Symbol: α. (RMB)

Alpha radiation A stream of alpha particles. (RMB)

Alpha ray A strongly ionizing and weakly penetrating radiation stream of fast-moving helium nuclei. (FSL)

Alpine The area of a mountain above the tree line, but below the permanent snow line. (FSL)

Alternative Remedial Contracting System (ARCS) A turnkey contracting and subcontracting program administered by the Environmental Protection Agency and funded under the Comprehensive Environmental Response, Compensation, and Liability Act (CERCLA-Superfund Act of 1980) to investigate and remediate hazardous-waste sites. (FSL)

Alumino-silicates Compounds containing aluminum, silicon, and oxygen as main constituents. (SWT)

Aluminum (Al) A light, ductile metallic element found naturally as bauxite. (FSL)

Alveoli Numerous small, terminal air sacs in the lungs where pulmonary capillary blood is in close juxtaposition to the alveolar gas, permitting the rapid exchange of carbon dioxide and oxygen in the lungs. There are about 300 million alveoli situated at the ends of small air passageways in the lungs. Each resembles a tiny bunch of grapes in structure. The average adult male has an alveolar surface area of about 70 square meters. Alveoli are the main deposition site of respirable dust particles (1–10 microns in diameter) or respirable fibers (e.g., asbestos fibers) that can result in various respiratory diseases, such as silicosis and asbestosis. (FSL, DEJ)

Amalgam A mixture or alloy of mercury combined with other metals, such as tin, zinc, gold, and silver, or alloys. Dental amalgams are alloys of mercury, silver, and tin. (FSL)

Ambient air For purposes of the Clean Air Act, that portion of the atmosphere, external to buildings, to which the general public has access. 40 CFR §50.1(e). *See also* Clean Air Act. (DEW)

Amended water *See* Wetting agent. (MAG)

American Academy of Industrial Hygiene (AAIH) An organization composed of board-certified industrial hygienists. *See also* American Board of Industrial Hygiene. (DEJ)

American Association for Accreditation of Laboratory Animal Care (AAALAC) A nonprofit corporation established in 1945 as a scientific and educational organization to promote high-quality animal care and use through a voluntary accreditation program. (FSL)

American Association for Laboratory Animal Science (AALAS) An organization composed of individuals and institutions professionally concerned with the production, care, and use of laboratory animals. (FSL)

American Association of Occupational Health Nurses, Inc. (AAOHN) An organization dedicated to promoting the field of occupational health nursing, and formerly named the American Association of Industrial Nurses. (DEJ)

American Board for Occupational Health Nurses An organization that administers a certification program for registered nurses working in occupational health. To be certified, nurses must: satisfactorily complete an accredited basic nursing program, have worked in occupational health for at least five years, have the required number of continuing education units, and satisfactorily complete an examination. Upon completion of these requirements, a nurse may use the identifying initials *COHN* (certified occupational health nurse) after his or her name. (DEJ)

American Board of Industrial Hygiene (ABIH) An organization responsible for certifying industrial hygienists. The organization offers comprehensive and specialized tests to qualified applicants, and is responsible for determining whether individuals are qualified to practice industrial hygiene. Currently, the Board requires one to have at least 5 years' experience (or 4 years plus a Masters of Science degree) and to complete satisfactorily a two-day examination in order to qualify as a certified industrial hygienist. *See also* American Academy of Industrial Hygiene. (DEJ)

American College of Laboratory Animal Medicine (ACLAM) A specialty board founded in 1957 to encourage education, training, and research, to establish standards of training and experience, and to certify qualified laboratory animal specialists or diplomates. (FSL)

American Conference of Governmental Industrial Hygienists (ACGIH) A nongovernmental organization composed of industrial hygienists employed in government and academia, and founded in 1938. The ACGIH establishes threshold limit values (TLV) of certain chemicals, and co-sponsors (with the American Industrial Hygiene Association) the annual American Industrial Hygiene Conference. (DEJ)

American Industrial Hygiene Association (AIHA) A nongovernmental organization founded in 1939 that is composed of industrial hygienists employed in private industry, government, and academia. The Association is dedicated to the prevention of workplace-related illnesses or injuries affecting the health and well-being of workers or the community. The AIHA establishes workplace environmental exposure limits and operates a national laboratory accreditation program. (DEJ)

American Industrial Hygiene Foundation An organization that operates under the auspices of the American Industrial Hygiene Association and that provides fellowships to graduate industrial hygiene students, promotes the development of graduate schools of industrial hygiene, and encourages qualified individuals to enter the field. (DEJ)

American National Standards Institute (ANSI) An organization responsible for developing consensus standards on a wide range of matters, including occupational and environmental health and safety. ANSI has established standards for respiratory protection, exposure limits to chemicals and physical agents, and standards for safety harnesses, safety eyewear, hardhats, footwear, and other items. Some of these standards have been incorporated into those established by OSHA. (DEJ)

American Public Health Association (APHA) An organization founded in 1872 and dedicated to protecting and promoting personal and environmental health by exercising leadership in the development and dissemination of health policy. The organization represents all disciplines and specialties in public health, and is the largest public health association in the world, with a membership of 50,000. (DEJ)

American Society for Testing and Materials (ASTM) An organization dedicated to the development of consensus standards for materials characterization and use. Standards are typically developed with the input of specialists from government, private industry, and academia. (FSL)

American Society of Laboratory Animal Practitioners (ASLAP) An organization, formed in 1966, that has as its purpose the dissemination of ideas, experiences, and knowledge among veterinarians engaged in laboratory animal practice. (FSL)

American Society of Safety Engineers (ASSE) A not-for-profit, multidisciplinary professional organization of trained practitioners responsible for protecting people, property, and the environment. Members must be employed in the safety profession or one of its specialties, performing at least two of the following functions of the professional safety position: **(1)** identification and appraisal of hazardous conditions and practices and evaluation of the severity of accident or loss problems; **(2)** development of hazard control methods, procedures, and programs; **(3)** communication of hazard control information to those directly involved; **(4)** measurement and evaluation of the effectiveness of hazard control systems. (SAN)

American Veterinary Medical Association (AVMA) The most important national organization of veterinarians; its objective is to advance the science and art of veterinary medicine, including its relationship to public health and agriculture. (FSL)

Ames test A test that bears the name of its developer and that is used to determine the carcinogenicity of chemicals. It is often known as the *Salmonella* test. In the test, mutant strains of *Salmonella typhimurium* are cultured on a medium deficient in histidine while being exposed at the same time to a potential carcinogen and liver extracts. Mutagenic bacteria will back-mutate to contain a functional histidine gene, permitting bacterial growth. The level of mutagenicity can be determined by the number of colonies that develop. (FSL)

Amicus curaie Literally, *friend of the court*. A brief filed with the court by a nonparty in support of a particular position or outcome. (DEW)

Amine A class of organic compounds of nitrogen derived from ammonia (NH_3) by replacing one or more of the hydrogen atoms with alkyl groups. Amines are basic in nature and usually combine with hydrochloric or other strong acids to form salts. (FSL)

Amino acid An organic acid that is one of the building blocks in the formation of proteins. (FSL)

Ammonia (NH_3) A colorless, combustible gas (or liquid) that is lighter than air and has an extremely irritating odor. It is obtained by steam retorting a mixture of carbon monoxide, hydrogen, carbon dioxide, and nitrogen or by the partial combustion of natural gas. Ammonia is used as a fertilizer alone or in combination with other compounds, such as nitrates. It is the third-highest-volume chemical produced in the United States and is used in the production of hydrogen cyanide and nitric acid, and as a refrigerant. *See also* Ammonium nitrate. (FSL)

Ammonium fixation Absorbtion or adsorbtion of ammonium ions by a soil organic or mineral fraction in a way that makes them relatively insoluble in water and relatively nonexchangeable by usual methods of cation exchange. (SWT)

Ammonium nitrate (NH_4NO_3) A chemical compound used as a fertilizer and as a constituent of many explosives. This material is often included in "bore holes" used for the placement of explosives and is combined with dynamite to enhance the explosive potential of the latter. Ammonium nitrate runoff from agricultural areas represents a significant source of pollution and is responsible for algal blooms and eutrophication in many bodies of water. *See also* Ammonia. (FSL)

Ampere Practical unit of current; the flow of 1 coulomb per second. Abbreviation: amp or A. (RMB)

Amplification As related to radiation-detection instruments, the process (either gas, electronic, or both) by which ionization effects are magnified to a degree suitable for their measurement. (RMB)

Anadromous fish Fish that live most of their lives in oceans or lakes but that ascend streams or rivers to reproduce (spawn)—e.g., salmon. (SMC)

Anaerobes Organisms unable to multiply in any environment that contains oxygen. Anaerobic microorganisms, such as many bacteria and protozoa, have oxygen-sensitive enzymes and cannot function in the presence of molecular oxygen. Anaerobes obtain their energy from oxygen-independent metabolism. Some may be more aerotolerant than others. Those severely affected by the presence of oxygen are called strict anaerobes, or obligate anaerobes. (HMB)

Anaerobic **(1)** Without oxygen; **(2)** refers to cells or organisms that can live without oxygen or processes that occur in the absence of oxygen. Strict anaerobes cannot live in the presence of oxygen. *See also* Respiration. (SMC)

Anaerobic digestion The degradation of organic matter through the action of microorganisms in the absence of oxygen. This process is utilized for the stabilization of some of the more recalcitrant fractions of municipal wastewater—e.g., oil, grease, and sludge. Anaerobic digestion proceeds in two phases. The first is the acid phase, characterized by the growth of aerobic and facultative anaerobes, which, in turn, degrade most types of organic material, producing organic acids (mostly acetic acid) and lowering the pH and oxygen concentration of the mixture. The second is the methane phase, characterized by the growth of methanogenic bacteria, which convert the products of the acid-producers into a mixture of methane and carbon dioxide. After digestion, the stabilized sludge is dewatered and typically disposed of in a sanitary landfill or by other acceptable disposal methods. (HMB)

Analog A chemical compound that resembles another in structure but is not necessarily an isomer of it; in different species of plants or animals one of two organs or parts that differ in structure or development but are similar in function. (FSL)

Analysis of variance (ANOVA) **(1)** One-way: a method of comparing the means of a number of groups that are normally distributed. Information about the variability of the data within groups can be determined in this way. **(2)** Two-way: a method of comparing two different variables that classifies groups that are normally distributed. The effects of each variable are analyzed controlling for the effects of the other variables. (VWP)

Anamnestic response The accelerated rise in antibody titer in a person who has previously developed a primary immune response to a particular antigen. (FSL)

Anaphalaxis A condition of hypersensitivity in humans or animals to protein or other substances, caused by previous exposure to the substance and resulting in shock or other physical reactions. (FSL)

Anatomic Relating to the science of the morphology or structure of organisms. (FSL)

Anchorage A secure point of attachment for lifelines, lanyards or deceleration devices that is independent of the means of supporting or suspending an employee during hazardous work. *See also* Fall protection. (SAN)

Ancient Soil The layer of soil beneath the surface that is weathered as a result of past processes and climatic changes. (FSL)

Anechoic room A room with no boundaries to reflect sound energy generated inside; a "free field." (DEJ)

Anemia A disorder of blood as a whole; deficiency in the number of red corpuscles or of hemoglobin. (FSL)

Anemometer An instrument used to measure the motion of wind or air and that employs a pitot tube directed by a vane or rotor, or a pressure plate deflected against a spring or gravity. (FSL)

Anesthetic A characteristic or an ability of an agent to produce a loss of feeling or sensation, especially pain. (MAG)

Angiosarcoma A rare malignant tumor, believed to originate from the endothelial cells of blood vessels, that has been demonstrated to occur in persons known to have had chronic exposure to vinyl chloride or thorium oxide. (FSL)

Angle of repose The greatest angle above the horizontal plane at which the material (soil) on the walls of excavations will lie without sliding. *See also* Excavations, sloping. (SAN)

Ångstrom A unit of length, used chiefly in expressing short wavelengths. It equals 10^{-10} meter or 10^{-8} centimeter. Symbol: Å. (RMB)

Anhydrous Describes a substance free of water; when applied to chemicals such as oxides, indicates the absence of water or water in combination. (FSL)

Animal products Any products derived wholly or in part from animals. Such products include beef, pork, poultry, eggs, milk, and fish. Sanitary measures in food production, food processing, and food service are most critical when dealing with products of animal origin, because these products are likely to be contaminated and subsequently act as vehicles for the transmission of human pathogens. Sanitary surveys should focus on, among other factors, the appropriate sources, handling, and preparation of foods of animal origin. (HMB)

Anion Negatively charged ion. (FSL)

Anion exchange capacity The sum of exchangeable anions that can be adsorbed by a soil. Usually soils carry negative charges, but under certain conditions soil colloids may carry positive charges; hence, anion exchange occurs, as in phosphate fixation and retention. *See also* Cation exchange capacity. (SWT)

Anionic wetting agent One of several synthetic compounds having the property of a wetting agent and also considered synthetic detergents. Their addition to a solution lowers the surface tension. Synthetic detergents can be divided into different categories. For example, anionic wetting agents have a negatively charged ion when in solution. They have particularly good wetting properties and are compatible with alkaline cleaning agents, and therefore they are the most common wetting agents used in cleaning compounds. *See also* Surface-active agent, wetting agent. (HMB)

Annihilation radiation Photons produced when an electron and a positron unite and cease to exist. The annihilation of a positron-electron pair results in the production of two photons, each of which has at least 0.51 MeV energy. (RMB)

Annual Plant that completes its life cycle from seed to seed in a single year. (FSL)

Anode Positive electrode; electrode to which negative ions are attracted; electrode at which oxidation occurs. (FSL)

Anoplura The sucking lice, wingless parasitic insects that have flattened bodies and live on the blood of mammals. The human head louse, *Pediculus humanus corpus*, for example, is frequently detected among school children and may be found in epidemic proportions in those populations. (FSL)

Anoxia The absence of, or a diminished amount of, oxygen in blood, tissues, or a body of water. The deficiency of oxygen in organisms often results in an increased rate and depth of breathing. Anoxia in humans is often accompanied by dizziness, rapid heartbeat, and headache, and can cause death. *See also* Anoxic. (FSL) (DEJ)

Anoxic Without oxygen. *See also* Anoxia. (SMC)

ANPR Advanced notice of proposed rulemaking under the Administrative Procedure Act. (DEW)

Answer In a civil action, the pleading filed in response to a complaint. Sets out the defendant's defenses to the allegations in the complaint. (DEW)

Antabuse A chemical product (tetraethylthiuram disulfide, $[(C_2H_5)_2NCS]_2S_2$) used in the treatment of alcoholism. Antabuse creates toxic symptoms, such as nausea and vomiting, when ingested in association with alcohol. This material is also used as a fungicide. (FSL)

Antagonism The interaction of two chemicals that have an opposing, or neutralizing effect on each other, or (given some specific biological effect) a chemical interaction that appears to have an opposing or neutralizing effect beyond what might otherwise be expected. Mutual resistance, denoting mutual opposition in action between structure, agents, diseases, or physiologic processes. (FSL)

Antarctic "ozone hole" Refers to the seasonal depletion of ozone in a large area over Antarctica. (FSL)

Anthelmenthics Drugs used to treat infestations of parasite worms, such as tapeworms, roundworms, hookworms, and whipworms. (FSL)

Anthesis The time of flowering or full bloom of a flower. (SMC)

Antho- Pertaining to flowers. (SMC)

Anthracene $[C_6H_4(CH)_2C_6H_4]$ An aromatic hydrocarbon obtained during the distillation of coal. The material is yellow and crystalline, with blue fluorescence, and is soluble in alcohol and ether and insoluble in water. It is a carcinogen used as an intermediate in the production of dyes and in organic semiconductor research. (FSL)

Anthracite Clean, dense, hard coal. This type of coal burns with a minimum of smoke and has low sulfur content and high luster. This is so-called hard coal, and is typically found in northeastern Pennsylvania, and sections of Kentucky, West Virginia, and other states. *See also* Anthracosis; bituminous; black lung disease; pneumonoconiosis. (FSL)

Anthracosis A condition known as Collier's disease, Shaver's disease, miner's lung, and black lung; a pneumonoconiosis resulting from the accumulation of carbon from inhaled smoke or coal dust in the lungs. *See also* Anthracite; bituminous; black lung disease; pneumonoconiosis; Shaver's disease. (FSL)

Anthrax An acute bacterial disease usually affecting the skin. Also known as woolsorters' disease, ragpickers' disease, or malignant edema, this disease is caused by the organism *Bacillus anthracis*. The disease is transmitted by contact with tissues of infected animals (cattle, sheep, goats, horses, and others) or contaminated hair, wool, or hides. (FSL)

Anthropic epipedon A thick dark surface horizon formed under long-continued cultivation in which large amounts of fertilizer and organic matter have been added. The anthropic epipedon is more than 50% base saturated, has a narrow carbon : nitrogen ratio, and more than 250 ppm phosphorus soluble in citric acid. (SWT)

Anthropometry The measurement of the human body, including body dimensions, range of motion of body members, and strength, including both static and dynamic measurements. The branch of anthropology that deals with the comparative measurements of the human body. *See also* Ergonomics. (DEJ)

Antibiogram A list of those antibiotics to which a particular microorganism is sensitive or resistant. (FSL)

Antibiosis An antagonistic situation in which one organism produces a substance that is harmful to other organisms. (FSL)

Antibiotic A chemical substance produced by living organisms that inhibits the growth of or kills other organisms. Penicillin, which is produced by the mold *Penicillium notatum*, is an important antibiotic in common use. Numerous synthetic antibiotics are commercially produced as well. Commercially prepared antibiotics are substances produced by moldlike organisms, and purified and prepared as concentrates. Antibiotics exhibit varying degrees of selective antimicrobial activity. Some may be active against specific organisms or against numerous Gram-positive bacteria, whereas others exhibit a ''broad spectrum'' of activity against many organisms. Thousands of antibiotics have been isolated, and the search continues to add to the number available. Antibiotics may occur in foods naturally—e.g., nisin in milk—or as additives. Undesirable side effects resulting from use of antibiotics in foods include: development of microbial resistance, allergic reactions, diarrhea, nausea, and rashes. (HMB)

Antibiotic residues Contamination of foods with detectable levels of antibiotics, usually as a result of inappropriate use of the chemicals for the treatment of disease or as growth stimulators. Regulations prohibit or severely limit the level of antibiotics that may be present in foods because of the public health risk involved. *See also* Antibiotic. (HMB)

Antibiotic sensitivity testing The identification of the concentration of various antibiotics to which a particular microorganism is sensitive. (FSL)

Antibody A globulin found in tissue fluids and blood serum. Antibodies are produced in response to the stimulus of a specific antigen, and are capable of combining with that antigen to neutralize or destroy it. Globulins are often referred to as immune substances. *See also* Antigen. (FSL)

Anticline An upward fold in the structure of rocks. (FSL)

Anticoagulant Hemorrhagic agents such as warfarin and dicumarol (3,3′-methylenebis[4-hydroxy-2H-1-benzopyran-2-one]) that antagonize the action of vitamin K and reduce the activity of vitamin K-dependent clotting factors. Used as medication in humans and as rodenticides. *See also* Warfarin. (FSL)

Anti-Degradation Clause Part of federal air quality and water quality requirements prohibiting deterioration in cases in which pollution levels are above the legal limit. (FSL)

Anti-Drug Abuse Act Public Law (PL) 100-690. Section 2405 of this act established the Joint Federal Task Force to address the issue of hazardous-waste contamination by clandestine drug laboratories. (FSL)

Antifoaming agents Substances that suppress or inhibit the formation of foam in systems where it may interfere with processing. Foaming may be due to proteins, gases, nitrogenous materials, or certain microorganisms—e.g., Actinomyces. Examples of antifoaming agents are organic phosphates, sulfonated oils, silicone fluids, and octanol. (HMB)

Antigen That portion or product of a biologic agent capable of stimulating the formation of specific antibodies. *See also* Antibody. (FSL)

Antigenic character The chemical arrangement of the antigenic components of an agent; the uniqueness of the arrangement and of the components to each species or strain of agents; the specificity of immunity resulting from infection with that agent. (FSL)

Antigenicity Ability of an agent or its products to stimulate the formation of antibodies. *See also* Antibodies. (FSL)

Antigibberellin A chemical, such as maleic hydrazide, that can cause stunted growth in plants. (FSL)

Anti-knock compounds Chemicals that have been added to gasoline to prevent knocking or ''pinging'' in internal combustion engines. The most frequently used chemical has been tetraethyl lead, which, because of its health impacts on people and the environment, has largely been replaced by platinum and other chemicals. *See also* lead; lead poisoning. (FSL)

Antimony (Sb) An element with an atomic weight of 121.73 whose compounds are often incorporated in metals used in dyestuffs and lead acid batteries. As an occupational hazard, excessive exposure to antimony has been shown to be a sufficient source of dermatitis among workers. (FSL)

Anti-natalist An individual whose viewpoint or policies promote population control measures and the reduction of birth rates. (FSL)

Antirepeat safeguards Mechanisms incorporated into a machine, such as an industrial powered press, that prevent the machine from cycling more than one time. These are typically incorporated into operations that require the operator to reach into the machine, usually during a cycle; thus the safeguard is necessary to make certain the machine does not recycle while the operator's hands are vulnerable to injury by the machine. (SAN)

Antisepsis The prevention of sepsis by inhibition or destruction of the causative organisms. (FSL)

Antiseptic Chemical compounds that are capable of reducing the number of microorganisms on body surfaces. They are relatively mild substances used primarily on humans and animals, in contrast with disinfectants, which are used primarily on nonliving surfaces. It is possible to use a very dilute solution of disinfectant as an antiseptic—for example, tincture of iodine. *See also* Disinfectant. (HMB)

Antithrombin Any substance that inhibits or prevents the clotting of blood. *See also* Heparin. (FSL)

Antitoxin An antibody to the toxin of a microorganism, usually a bacterial exotoxin. Antitoxins combine with a specific toxin, in vivo and in vitro, with the consequent neutralization of toxicity. (FSL)

Apatite Any of a group of minerals in igneous rocks (metamorphosed limestone), apatites are the main source of phosphate. (FSL)

Aphagia The failure to eat, which results in illness and, subsequently, death of the affected animal or person. *See also* Hyperphagia. (FSL)

Ap horizon The surface layer of a soil that is disturbed by cultivation. (SWT)

Apical meristem The growing point at the tip of a root or shoot in a vascular plant. The apical meristem is composed of meristematic tissue and gives rise to the primary tissues. (SMC)

Apocrine Sweat glands that open into hair follicles. Apocrine sweat glands are limited to a few regions of the body—notably, the underarm and genital areas. (DEJ)

Applicable or relevant and appropriate requirements (ARARs) Under the Comprehensive Environmental Response, Compensation, and Liability Act, requirement that remedial actions carried out under the statute must attain a degree of cleanup measured by applicable or relevant and appropriate federal and state environmental laws. *See also* Comprehensive Environmental Response, Compensation, and Liability Act; remedial action. 42 USC §9621(d). (DEW)

Aquaculture **(1)** The agricultural management of aquatic plants or other organisms for food or other commodities; **(2)** term sometimes used as equivalent to *hydroponics*. *See also* Hydroponics. (SMC)

Aquatic Of or pertaining to water; especially, environments characterized by the continuous presence of water (oceans, lakes, rivers, ponds, streams, etc.) or organisms that live in such environments (in or on water). (SMC)

Aqueous film-forming foam (AFFF) A fluorinated surfactant, used as a fire suppressant, that has a foam stabilizer that is diluted with water to act as a temporary barrier to exclude air so as to prevent it from mixing with the fuel vapor. The AFFF develops a film on the fuel surface and suppresses the generation of fuel vapors. It is effective on both Class A and Class B fires. It cools and penetrates to reduce temperatures below the ignition level. *See also* Chemical foam. (FSL)

Aquic A reducing soil moisture regime with little dissolved oxygen due to saturation by ground water. (SWT)

Aquifer An underground geological formation, or group of formations, containing usable amounts of ground water that can yield significant quantities to wells and springs. (FSL)

Arachnology The study of arachnids (Class Arachnida, Phylum Arthropoda), which include spiders, ticks, and mites. (SMC)

Araneology *See* Arachnology. (SMC)

Arbitration A method of dispute resolution, entered into by agreement or law, in which the dispute is settled by an impartial third party. In binding arbitration, parties agree to accept findings of an arbitrator. In nonbinding arbitration, there is no obligation to accept the decision, and dissatisfied parties may resort to litigation or appeal as appropriate. (DEW)

Arboreal Of or pertaining to trees; especially, organisms that live on or in trees. (SMC)

Arboviruses Viruses that are transmitted from one host to another by one or more kinds of arthropods. (FSL)

Arctic smoke A fog/mist combination produced as the result of the passage of frigid air over warm water; a condition that occurs frequently off the coast of Norway. (FSL)

Area air monitoring A method of air monitoring in which the air-sampling equipment is in a fixed position within and/or at the perimeter of a work area. The height at which the sampling medium or instrument is placed is based on variables relating to the approximate height of the worker's breathing zone, chemical properties of the substance(s), and environmental conditions. The information derived from this type of monitoring is useful as an index of general contamination and is usually not a good indicator of personal exposures. (MAG)

Argillic Horizon An illuvial subsurface horizon characterized by an accumulation of silicate clays. Generally a B horizon. (SWT)

Argon (Ar) A nonmetallic element with the atomic number 18 that comprises nearly 1% of the atmosphere by weight. Argon is used as an inert gas shield in arc welding, furnace brazing, electric light bulbs (neon, fluorescent, sodium vapor), and in Geiger-counting tubes and lasers. (FSL)

Arithmetic mean ($\bar{x}$) A measure of central tendency for data. The sum of a set of n values divided by n. Formula:

$$\bar{x} = \frac{x_1 + x_2 + \cdots x_n}{n} = \frac{\sum_{i=1}^{n} x_i}{n} \qquad \text{(VWP)}$$

Aromatic hydrocarbon A major group of unsaturated cyclic hydrocarbons containing one or more rings made up of 6 carbon atoms. This group—most notably benzene—is chiefly derived from petroleum and coal tar. The name is due to the strong and not unpleasant odor characteristic of substances within this group. (MAG)

Arraignment Point in a criminal proceeding at which the defendant is formally charged with a crime and informed of his constitutional rights, including the opportunity to enter a plea. (DEW)

Arroyo A flat-floored stream channel with steep sides that is a typical landform in arid regions of the world. (FSL)

Arsenic (As) An element with an atomic weight of 74.9216 that occurs in three allotropic forms. It occurs in the elemental form, with sulfur as realgar or orpiment, with oxygen as white arsenic, and with some metals. Arsenic compounds are toxic and have been used extensively as a component of insecticides and, in some cases, as herbicides. Most of these uses have been discontinued, however, because of the availability of more efficient, less toxic materials. (FSL)

Arsenopyrite A mineral (FeAsS) that is a major ore of arsenic and that is found in hydrothermal deposits. Historically, arsenic compounds were used extensively for the control of insects such as the boll weevil. In their use as insecticides, however, these compounds have largely been replaced by chemicals, such as the organophosphates, that do not accumulate in the soil. Arsenic compounds are used as alloys for transistors and cable-sheathing coatings. (FSL)

Artery A blood vessel that conveys blood from the heart to any part of the body. *See also* Capillary; vein. (FSL)

Artesian spring A spring from which the flow of water depends on the difference in recharge and discharge elevations of the aquifer and on the size of the openings through which the water passes. This type of spring is more dependable than a gravity spring, but is very sensitive to the pumping of wells and possible drawdown from the same aquifer. (FSL)

Artesian well Well that penetrates aquifers in which the ground water is found under hydrostatic pressure. This occurs in an aquifer that is confined beneath an impermeable layer of material at an elevation lower than that of the intake area of the aquifer. Recharge areas or intake areas of confined aquifers are commonly at high-level surface outcrops of the formations. (FSL)

Arthropod Animals with hard segmented bodies bearing jointed appendices and representing approximately 740,000 species, including insects, crustaceans, and arachnids. *See also* Infestation. (FSL)

Articulate Jointed, joined. (SMC)

Artifact A product that exists because of an extraneous, especially human, action and does not occur in nature. (SMC)

Artificial recharge The practice of introducing surface water into an aquifer, usually by pumping, through recharge wells. (FSL)

Artificial reef A collection of spent materials, usually concrete and other solid items, that are deposited in the ocean to provide harborage for marine organisms. The practice of creating artificial reefs has come under scrutiny in recent years, because of the potential long-term impact of the deposited items on marine ecosystems. (FSL)

Asbestos A generic term used to describe a number of naturally occurring, fibrous, hydrated mineral silicates differing in chemical composition. They are white, gray, green, or brown, and are found in Vermont, Arizona, California, North Carolina, Africa, Mexico, and several Canadian provinces.

Asbestos fibers are characterized by high tensile strength, flexibility, heat and chemical resistance, and good frictional properties, making asbestos useful in a number of products for insulating and fireproofing purposes, such as insulation materials, brake linings, cement, gaskets, and special fabrics. Chrysotile, crocidolite, amosite, anthophyllite, tremolite, and actinolite are all forms of this material.

Exposure to asbestos fibers is known to cause a variety of diseases, including asbestosis (a diffuse, interstitial nonmalignant scarring of the lungs), bronchogenic carcinoma (a lung cancer), mesothelioma (a tumor of the lining of the chest cavity or lining of the abdomen), and cancer of the stomach, colon, and rectum. Asbestosis is evident on X-rays, and symptoms such as reduced or restrictive pulmonary function, "clubbing" of the fingers, and dyspnea may be present.

Abatement efforts to remove asbestos from schools, office buildings, and other structures are undertaken if the asbestos fibers are friable and therefore represent an inhalation hazard. (DEJ) (FSL)

Asbestos-containing waste Any solid waste containing more than 1%, by weight, of naturally occurring hydrated mineral silicates separable into commercially used fibers, specifically, the asbestiform varieties of serpentine, chrysatile, cummingtomite-grunerite, amosite, rieheckite, crocidolite, anthophyllite. *See also* Asbestos. (FSL)

Asbestosis Disease caused by prolonged exposure to asbestos fibers, which injure the lungs and diminish their oxygen absorbing properties, thus reducing lung function in the affected person. *See also* Asbestos. (FSL)

Ascorbic acid ($C_6H_8O_6$) A water-soluble vitamin (vitamin C) that is a constituent of the diet of humans and animals necessary for the formation of collagen and the maintenance of capillary walls and the healing process. A deficiency of vitamin C (scurvy) can be prevented by the consumption of oranges, grapefruit, and fresh vegetables. (FSL)

Asepsis The absence of infectious microorganisms. (FSL)

Aseptic technique Procedures designed to exclude infectious agents; laboratory or clinical techniques that do not result in the transfer of disease-producing microorganisms from one surface to another. (FSL) (DEJ)

Asexual reproduction Any reproductive process that does not include the union of gametes. Examples are budding (in yeast) and vegetative propagation by root sprouts (in aspen trees). (SMC)

Ash The remnants of a product after complete combustion. (FSL)

Asphalt A black solid or semisolid substance that is produced during the refining of some petroleums. This material is used as paving for streets, roads, driveways, etc. It contains several aromatic hydrocarbon compounds that have been linked to cancer in workers receiving prolonged exposure. (FSL)

Asphyxia The state of respiratory distress or suffocation due to the lack of oxygen. (FSL)

Aspirate To remove by suction a gas or body fluid from a body cavity, from an unusual accumulation, or from a container. (FSL)

Associated corpuscular emission The full complement of secondary charged particles (usually limited to electrons) associated with an X-ray or gamma-ray beam in its passage through air. The full complement of electrons is obtained after the radiation has traversed sufficient air to bring about equilibrium between the primary photons and secondary electrons. Electronic equilibrium with the secondary photons is intentionally excluded. (RMB)

Assumption of the risk A defense to a tort allegation that the plaintiff was voluntarily exposed to the risk that caused the injury. *See also* Contributory negligence. (DEW)

Asymptomatic Without symptoms. Diseases produce particular symptoms that are subjective evidence of infection. In certain cases individuals may harbor the causative agent of a disease without exhibiting any of the particular characteristic symptoms. This may occur at various times during the course of the illness, dependent on the individual and the disease. *See also* Carrier; inapparent infection. (HMB)

Atelectasis Incomplete expansion of a lung, caused by the partial collapse of the lung due to occlusion of a bronchus or by external compression. (DEJ)

Atmosphere **(1)** The whole mass of air surrounding the earth, and composed largely of oxygen and nitrogen. **(2)** A standard unit representing the pressure exerted by a 29.92-inch column of mercury at sea level at 45′ latitude and equal to 1000 grams per square centimeter. (FSL)

Atmospheric attenuation Reduction in intensity of radiant energy received at some distance after passage through the atmosphere, relative to that which would be received if the atmosphere were not present; a measure of the absorption of radiant energy in the air. The transmissivity is based upon the visibility (in miles) and the water vapor concentration in grams per cubic meter. The concept of transmissivity may also be applied to materials, such as glass and plastics. (RMB)

Atmospheric dispersion The dilution of air contaminants whereby the concentration of the pollutants is progressively decreased. Atmospheric dispersion is a most important mechanism for the distribution of contaminants and ultimately their removal from populated areas. Much remains unknown about this phenomenon and its understanding is becoming increasingly important because of the international implications of acid rain and other air pollution problems. (FSL)

Atmospheric pressure A value of normal or standard atmospheric pressure equivalent to the pressure exerted by a column of mercury 760 millimeters (29.92 inches) high. (FSL)

Atom The smallest particle of an element, which cannot be divided or broken up by chemical means. It consists of a central core called the nucleus, which contains protons and neutrons. Electrons move in orbitals and energy levels in the region surrounding the nucleus. (RMB)

Atomic absorption spectrophotometry A method commonly used for the analysis of heavy metals. An atomic vapor cloud is formed by dispersion of a solution containing the metal to be analyzed. The small droplets formed are passed through a flame, evaporating the solvent and decomposing the salt or oxide, and forming the atomic vapor cloud. A cathode lamp is used to produce radiation of a chosen wavelength. Because each metal absorbs radiation at a unique wavelength, different lamps are used for different metals. The amount of atomic absorbance is proportional to the concentration of the metal in the atomic vapor cloud. (DEJ)

Atomic energy Energy released in nuclear reactions. Of particular interest is the energy released when a neutron initiates the fission of an atom's nucleus into smaller pieces or when two nuclei are joined together, or undergo fusion, at millions of degrees of temperature. "Atomic energy" is really a popular misnomer. The energy involved is more correctly called "nuclear energy." (RMB)

Atomic Energy Act of 1954 An act of Congress regulating exposure to ionizing (i.e., alpha, beta, gamma, X-ray, neutron, etc.) radiation, and providing for licensing of holders of radioactive sources under authority of the Nuclear Regulatory Commission, designation of restricted areas, and regulation of effluent discharge into restricted areas. Codified at 42 USC 2011. (DEJ)

Atomic mass The weighted mean of the masses of the neutral atoms of an element expressed in atomic weight units. Unless otherwise specified, it refers to a naturally occurring form of the element. The atomic weight of any element is the relative weight of its atom compared with the weight of one atom of carbon taken as 12, hence a multiple of one-twelfth the weight of a carbon atom. (RMB)

Atomic mass unit Equals one-twelfth the mass of one neutral ^{12}C atom; abbreviation: amu. *See also* Atomic mass. (RMB)

Atomic number The number of electrons of the neutral atom of an element or the number of protons in the nucleus. (RMB)

Atomization The process of dividing a liquid into extremely small particles, thus increasing the surface area of a material. (FSL)

Atrophy Wasting away or diminution in the size of a cell, tissue, organ, or part, from defect, failure of nutrition, or disuse. (FSL)

ATSDR Agency for Toxic Substances and Disease Registry. (DEW)

Attack rate *See* Rate. (VWP)

Attainment area An area considered to have air quality as good as or better than the national ambient air quality standards as defined in the Clean Air Act. (FSL)

Attendant According to OSHA, an individual stationed outside a confined space who is trained in monitoring the number of persons inside the space, recognizing potential hazards, and observing activities inside and outside the space to determine whether it is safe for entrants to remain. This person maintains effective and continuous contact with personnel inside, has the authority to require entrants to evacuate the confined space immediately when a condition exists that is not allowed by the entry permit. The attendant detects behavioral effects of hazard exposure, any situation outside the space that could endanger the entrants, and any uncontrolled hazard within the permit space, and should know when to summon rescue. *See also* Confined space; entrants; permitting systems. (SAN)

Attenuation **(1)** Any of a number of processes for reducing the ability of a microorganism such as a virus to produce infection and disease (e.g., the fixing or adaption of the strain by passage in animal hosts, tissue culture systems, embryonated eggs, culture on artificial media). **(2)** (Radiation) The process by which primary quanta or particles are reduced in number following the passage through some medium. (FSL) (JHR)

Attenuation factor The ratio of the incident intensity of radiation to the transmitted intensity. (RMB)

Atterburg limits Limits applied to soil materials passing the number 40 sieve: **(1)** Shrinkage limit. Maximum water content at which a reduction in water content will not cause a decrease in soil mass volume. Defines the arbitrary limit between solid and semisolid states. **(2)** Plastic limit. Water content corresponding to an arbitrary limit between the plastic and semisolid states of soil consistency. **(3)** Liquid limit. Water content corresponding to the arbitrary limit between the liquid and plastic states of soil consistency. (SWT)

Attractant A chemical or other agent that lures insects or other pests by stimulating their sense of smell. (FSL)

Attractive nuisance Legal doctrine applicable to persons maintaining on their premises a condition, instrumentality, machine, or other agency that is dangerous to young children because of their inability to recognize peril. (DEW)

Attrition Wearing or grinding down of a substance by traction. A contributing factor in air pollution, as with dust. (FSL)

Audiogram A graphic recording of hearing levels referenced to a normal sound pressure level as a function of frequency. Audiograms are used in the diagnosis and treatment of hearing loss. *See also* Audiometer. (DEJ)

Audiometer A frequency-controlled audio signal generator that produces pure tones at various frequencies and intensities and that is used to measure hearing sensitivity or acuity. Measurement of hearing threshold (i.e., a person's ability to hear) results in an audiogram, a graphic presentation of the hearing threshold (measured in decibels) at selected frequencies (usually 125 to 8000 Hertz). *See also* Audiogram. (DEJ)

Audioradiograph A record of radiation from radioactive material in an object, made by placing its surface in close proximity to a photographic emulsion. (FSL)

Audubon Society A national organization dedicated to conservation, and named after James Audubon (1785–1851), an ornithologist noted for his paintings of birds and studies of their migratory patterns. (FSL)

Aufwuchs The community of organisms living on the surfaces of submerged plants and other objects. (SMC)

Aural insert protectors A form of hearing protector commonly known as earplugs. They are supplied in numerous configurations, consisting of foam, plastic, fine glass fiber, and wax-impregnated cotton. There are three types: formable, custom-molded, and premolded. (DEJ)

Autecology **(1)** The ecology of individual species in response to environmental conditions; **(2)** the study of organisms in relation to their physical (abiotic) environment. The emphasis in autecology is usually on the life history and natural history of a particular species and its physiology, development, and reproduction under certain environmental conditions. (SMC)

Authentication The requirements that for its admission in a legal proceeding, evidence be established to be what its proponent claims—for example, a certified copy of a document. Federal Rules of Evidence 901–903. (DEW)

Autochthonous Refers to materials produced and minerals cycled within a system. (SMC)

Autoclave A device that uses a combination of steam and pressure to bring about sterilization, i.e., the destruction of all microorganisms. The complete destruction of microorganisms by this method occurs at 15 pounds pressure, 120°C, for 15 minutes. (FSL)

Autoignition *See* Spontaneous combustion. (SAN)

Autoimmunity The condition whereby a person's tissues are subject to the deleterious effects of the person's own immune system. (FSL)

Autolysis The breakdown of animal and plant tissues that occurs after the death of the cells and is carried out by enzymes within these cells. (FSL)

Autopolyploid An organism with more than two haploid sets of chromosomes, all of which were derived from parents of the same species. (FSL)

Autosome A chromosome other than a sex chromosome. (FSL)

Autotroph An organism that converts inorganic substances into the nutritive substances and organic compounds it requires to produce energy for metabolic processes. Autotrophs require some external source of energy. Photosynthesizing plants and algae are autotrophs that use solar energy. Other autotrophs use chemical or heat energy. Autotrophs convert energy into forms usable by organisms for metabolic processes. Autotrophs, or producers, are the first trophic level in any food chain or food web. *See also* Heterotroph; producer. (SMC)

Available chlorine *See* Free chlorine. (HMB)

Available nutrients Nutrient ions or compounds (nitrogen, phosphorus, potassium, etc.) in forms that plants can absorb and utilize. Contents of designated "available" nutrients contained in fertilizers as determined by specified laboratory procedures. (SWT)

Available water **(1)** The water in soil that is available for absorption by plant roots. **(2)** The water in a microorganism's surroundings that is available—that is, not bound physically or chemically and therefore unavailable—for metabolic activity, which is carried out in aqueous environments. A dry environment usually results in the loss of the cell's water to the surroundings. The available water is sometimes expressed as water activity (A_w). *See also* Bound water; water activity. (HMB)

Available water capacity (available moisture capacity) The capacity of soils to hold water accessible for use by most plants. It is commonly defined as the difference between the amount of soil water at field moisture capacity and the amount at wilting point, commonly expressed as inches of water per inch of soil. (SWT)

Avalanche **(1)** A massive slide of rock debris and/or ice and snow. **(2)** (Radiation) The multiplicative process in which a single charged particle accelerated by a strong electric field produces additional charged particles through collision with neutral gas molecules. This cumulative increase of ions is also known as Townsend ionization or a Townsend avalanche. (RMB)

Average life (mean-life) The average of the individual lives of all the atoms of a particular radioactive substance. It is 1.443 times the radioactive half-life. (RMB)

Avian Of or pertaining to birds. (FSL)

Avicide An agent, such as strychnine, that is used to control bird populations. *See also* Pesticide. (FSL)

Avogadro's number 6.023×10^{23} atoms per gram-atomic weight or molecules per gram-molecular weight. The number of atoms in a gram-atomic weight of any element; also the number of molecules in the gram-molecular weight of any substance. This is one of the fundamental physical constants. (RMB)

Axon The essential conducting portion of a nerve fiber that is continuous with the cytoplasm of a neuron or nerve cell. (FSL)

Azobacter An important species of nitrogen-fixing bacteria that are aerobic and obtain their energy by breaking down carbohydrates in the soil. (FSL)

B

Bacillus The shape of a bacterial cell, also commonly referred to as a rod. Bacilli (pl.) are generally shaped like a cylinder. They may also be curved, spiral, or helical. Bacilli that are curved are designated as vibrios, spiral rods are designated as spirilla, and helical rods are designated spirochetes. (HMB)

***Bacillus cereus* food poisoning** An illness caused by the bacterium *Bacillus cereus*, which has been recognized as a foodborne pathogen since the 1950s. The genus *Bacillus* belongs to the family Bacillaceae, which is characterized as a sporulating Gram-positive bacillus found in the environment and in dried foods. Spores are not always killed by cooking temperatures, and, when the food cools, the organisms germinate. At least two clinically distinct forms of the illness exist. One, a diarrheal form, resembles *Clostridium perfringens* poisoning, producing primarily lower-intestinal-tract symptoms with an incubation period of 10–12 hours. The diarrheagenic toxin is easily inactivated by cooking temperatures. The second, an emetic form, more closely resembles staphylococcal intoxication with an incubation period of 2–4 hours, and produces upper-gastrointestinal symptoms.

Foods implicated with the diarrheagenic form are varied and include many meats and vegetable-containing dishes. The emetic form has most frequently been associated with rice that has been previously cooked and left unrefrigerated for long periods of time before being mixed with other items, and is then quickly stir-fried before serving. The emetic toxin is quite heat resistant, whereas the diarrheagenic toxin is easily destroyed by cooking. Although *Bacillus cereus* spores are quite heat resistant, there is no evidence that the vegetative cells are particularly resistant to conventional sanitation procedures. (HMB)

Backflow The flow of water or other substances into the distribution pipes of a potable water supply from any unintended source. Food service regulations generally carry the requirement that potable water supplies be installed to prevent the possibility of backflow. Regulations usually require that devices be installed to protect against backflow at all fixtures and equipment where an air gap at least twice the diameter of the water supply inlet is not provided. *See also* Air gap; back-siphonage. (HMB)

Backflow preventer Another name for a device designed to prevent backflow or back-siphonage in a piping system. These devices are designed by manufacturers for specific applications (e.g., double check valve with intermediate atmospheric vent for use on carbonated-beverage vending machine water supply lines, double check valve with atmospheric vent for use in faucets with gooseneck spout). See also Backflow; back-siphonage; cross connection; vacuum breaker. (HMB)

Background Level In air pollution control, the concentration of air pollutants in a definite area during a fixed period of time prior to the starting up or at the stoppage of a source of emission under control. (FSL)

Background Radiation Radiation arising from radioactive material other than that directly under consideration. Background radiation due to cosmic rays and natural radioactivity is always present. There also may be background radiation due to the presence of radioactive substances in other parts of a building, in the building material itself, etc. (RMB)

Back-pressure A condition whereby pressure higher than the supply pressure is created on the premises, causing reversal of flow into a water supply. Such conditions may be caused, for example, from pump action or thermal expansion from boilers. *See also* Backflow; back-siphonage. (HMB)

Back programs Activities in the occupational environment that aim at either healing injured backs or preventing back injuries through the use of safe lifting techniques and exercises to strengthen critical muscles. Training is the typical approach and most commonly involves teaching workers to lift properly. Some programs also train workers in strength and physical fitness. It is also possible to utilize testing to identify weak muscle groups that increase a worker's potential for back injury. Another approach involves redesigning jobs to fit the capabilities of the worker, using techniques such as job health and safety analysis. A third step is to select workers to fit the job by utilizing physical tests that simulate key job movements and/or incorporating certain medical criteria into the placement requirements. In the workplace the most common source of back injuries is overexertion during manual material handling. *See also* Job health and safety analysis; NIOSH lifting guidelines. (SAN)

Backscattering The process of scattering or reflecting into the sensitive volume of a measuring instrument radioactive radiations originally having no positive component of motion in that direction. It is dependent, therefore, on the nature of the mounting material, the nature of the sample, the type and energy of the radiations, and the particular geometrical arrangement. (RMB)

Back-siphonage A plumbing hazard. The reverse flow of water as a result of negative pressure in the potable water system. Back-siphonage can result from a break in a pipe, repair of a water main at an elevation lower than the service point, or reduced pressure from the suction side of booster pumps. The simplest method of preventing back-siphonage is to have an air gap between the free-flowing discharge end of a supply pipe and a source of contaminated water. Several kinds of mechanical backflow preventers are available as well. *See also* Air gap; backflow, backflow preventer. (HMB)

Bactericide A substance that kills bacteria. (HMB)

Bacteriophage A virus that lyses bacteria. (FSL)

Bacteriosis Pertains to any disease or abnormal condition caused by a bacterium. *See also* Bacteriostat; bacteriostasis. (HMB)

Bacteriostasis The arrest or inhibition of bacterial growth without killing of the bacterium. *See also* Bacteriostat. (HMB)

Bacteriostat A substance that inhibits or retards the growth of bacteria. An agent that inhibits the growth of bacteria in the presence of moisture but does not necessarily kill them. An agent that produces bacteriostasis. (HMB)

Badland A land area that is generally devoid of vegetation and broken by narrow ravines, sharp crests, and pinnacles resulting from serious erosion of soft geologic materials. Most commonly found in arid or semi-arid regions. (SWT)

Baffle A plate or grating used to block, hinder, or divert a flow or to change the direction of flow of a substance. (FSL)

Baffle chamber In incinerator design, a chamber designed to promote the settling of fly ash and coarse particulate matter by changing the direction and/or reducing the velocity of the gases produced by the combustion of the refuse or sludge. (FSL)

Bagasse The crushed remnants of sugar cane that remain as a fibrous residue after the sugar has been extracted. Dried airborne bagasse particles can irritate lung tissue and cause a respiratory condition known as bagassosis, which occurs among chronically exposed workers. *See also* Bagassosis; farmer's lung disease. (FSL)

Bagassosis Similar to farmer's lung disease, except caused by exposure to bagasse, the dried stalks of sugar cane. *See also* Bagasse; farmer's lung disease. (DEJ)

Baling Compacting solid waste into blocks to reduce volume and simplify handling. (FSL)

Ballistic separator A machine that sorts organic from inorganic matter for composting. (FSL)

Band application The spreading of chemicals over, or next to, each row of plants in a field. (FSL)

Barcelona Convention An international agreement, prepared in 1976 by the nations bordering the Mediterranean, that prohibits the discharge of a wide range of pollutants into the Mediterranean. (FSL)

Bar graph A pictorial representation of data using a horizontal scale and a vertical scale and rectangles to depict data size. (VWP)

Barium (Ba) A soft, silver-white element with an atomic weight of 137.43 that tarnishes rapidly when exposed to the air. Its compounds are used in the manufacture of paints, glass, and fireworks. (FSL)

Barrier cream A protective cream that may be used for preventing contact with harmful agents. Barrier creams and lotions should be used to supplement, but not replace, personal protective equipment. The types currently available include vanishing creams, which coat the skin; water-repellent types, which deposit a thin film of lanolin, petrolatum, or silicone on the skin; and solvent types, which contain ingredients that repel oils and solvents. (FSL)

Bar Screen In wastewater treatment, a device used to remove large solids. (FSL)

Basal application The application of a chemical on plant stems or tree trunks just above the soil line. (FSL)

Basal metabolic rate (BMR) Rate of heat production by the human body under neutral conditions. (DEJ)

Base level Lowest level to which a stream can erode its channel, or the level of the main stream into which the tributary flows. (FSL)

Base saturation percentage The amount that the adsorption complex of soil is saturated with exchangeable cations (other than aluminum or hydrogen). Expressed as a percentage of the total cation exchange capacity. Usually associated with the fertility of soil. (SWT)

Bauxite Aluminum ore, mainly hydroxides, that are used as fillers in plastics and rubber and as abrasives. (FSL)

Bayes' rule Let $E_1, E_2, E_3, \ldots E_n$ be n mutually exclusive events at least one of which is known to have occurred. Let A represent another event known to have occurred as a result of one of the events $E_1, E_2, E_3, \ldots E_n$. Then:

$$P(E_i|A) = \frac{P(A|E_i) \times P(E_i)}{\left[\sum_{i=1}^{n} P(A|E_i)\right][P(E_i)]}$$

$$i = 1, 2, 3 \ldots n$$

(VWP)

Bearing wall A building wall that supports vertical loads in addition to its own weight. (FSL)

Bedrock The tight, consolidated rocks below loose superficial material such as soil or alluvium. (FSL)

Behavior modification programs Safety programs that emphasize feelings and emotions of workers in order to improve their attitude and work practices. Behavioral programs often offer rewards for performing safe work practices. Many systems reward work groups or individual workers after a period of accident-free work. *See also* Safety campaigns. (SAN)

Benching In excavations, a method of protecting employees from cave-ins by excavating the sides of a site to form one or a series of horizontal levels or steps, usually with vertical or near-vertical surfaces between levels. *See also* Sloping; shoring. (SAN)

Benign Pertaining to the mild character of an illness or the nonmalignant character of a neoplasm. (FSL)

Benthic Of or referring to the bottom of a large aquatic environment. *See also* Benthos. (FSL)

Benthos **(1)** The bottom of a large aquatic environment (river, lake, sea, or ocean). **(2)** Organisms that dwell on or are fixed to the bottom of an aquatic environment anywhere between the depths and the high-water mark. (SMC)

Bentonite A mineral, largely composed of smectite clay, produced by the alteration of volcanic ash in situ. This material is used in water softeners and as a major component of drilling muds. (FSL, SWT)

Benzene (C_6H_6) A highly flammable narcotic liquid known to be carcinogenic. This aromatic hydrocarbon is found in coal tar and is used as an industrial solvent and for extraction purposes in laboratory procedures. It is also used in the manufacture of styrene and lacquers. (FSL)

Benzopyrene ($C_{20}H_{12}$) A carcinogen found in tars. It is produced in the manufacture of asphalt, tar, and other organic materials. (FSL)

Beri-beri A disease caused by vitamin B (thiamine) deficiency common among populations that survive on diets of polished rice. (FSL)

Bernoulli distribution Another name for the normal distribution, named after J. Bernoulli (1654–1705), who discovered it at the end of the 17th century. *See also* Normal distribution. (VWP)

Bernoulli trial The type of random variable that has the value 1 with probability p and the value 0 with probability $1 - p$. (VWP)

Berylliosis Chronic poisoning caused by exposure to the dust or fume of beryllium metal, beryllium oxide, or soluble beryllium compounds. The disease is manifested by loss of appetite and weight, weakness, cough, extreme difficulty in breathing, cyanosis, and cardiac failure. The disease can appear 5–20 years after exposure has ceased and is characterized by the occurrence of granulomatous fibers in the lungs. It is commonly progressive in severity, and can be accompanied by enlargement of the heart, liver, and spleen, kidney stones, and cyanosis. *See also* Beryllium. (DEJ)

Beryllium (Be) A white, hard, brittle, metallic element (atomic number 4, atomic weight 9.0121). Beryllium is a carcinogen (OSHA) and has a high toxicity, especially by inhalation of dust. The TLV is 0.002 mg/m^3 of air. *See also* Berylliosis. (FSL)

Best available control technology (BACT) An emission limitation based on the maximum degree of emission reduction achievable through application of production processes, available methods, systems, and techniques. (DEW)

Beta decay Radioactive change by emission of a beta particle. In beta decay, a neutron decays into a proton, with the emission of an electron, or beta particle; or a proton transforms into a neutron and emits a positron. In both cases, the charge of the nucleus is changed without changing the number of nucleons. (RMB)

Beta naphthylamine A chemical used in the production of aniline dyes and that is known to be a bladder carcinogen. (FSL)

Beta particle A small particle ejected spontaneously from the nucleus of a radioactive element. It has the mass of the electron, has a charge of negative one (-1) or positive one ($+1$), and has a mass of 1/1840 that of a proton or neutron. It has a low penetrating power and short range. The most energetic of beta particles will penetrate the skin and other tissues. The damage is manifested by skin burns. Abbreviation: β^- or β^+. (RMB)

Beta ray A stream of beta particles of nuclear origin more penetrating but less ionizing than alpha rays per unit length of travel; a stream of beta particles emitted in certain radioactive disintegrations. (RMB)

Betatron A circular electron accelerator providing a pulsed beam of high-energy electrons or X-rays by magnetic induction. (RMB)

Bhopal A community in India located in close proximity to a pesticide production facility responsible for massive release of methyl isothiocyanate (MIC) gas in 1984. This incident caused an estimated 1800–5000 deaths and injured up to 300,000 persons. MIC has poor warning properties, particularly as related to smell, and this factor was felt to be responsible, in part, for the high mortality and morbidity connected with this disaster. (FSL)

Biennial A plant that requires two growing seasons to complete its life cycle. Generally, only vegetative growth occurs in the first year and sexual reproduction occurs in the second year. (SMC)

Bile The yellowish-brown or green fluid secreted by the liver and discharged into the small intestine, where it aids in the emulsification of fats, increases peristalsis, and retards putrefaction. (FSL)

Bill *See* Statute. (DEW)

Bimodal distribution A set of data with two modes. (VWP)

Binder The resins in a mixture of materials that hold together the other ingredients. (RMB)

Binding energy The energy represented by the difference in mass between the sum of the component parts of a nucleus and the actual mass of the nucleus. It represents the amount of energy required to separate the individual nucleons. (RMB)

Binomial distribution The spread of the number of successes in n independent trials, where the probability of success on each trial is p and the probability of failure is $1 - p$. *See also* Binomial probability function. (VWP)

Binomial probability function If p is the probability that an event will happen in a single trial (the probability of a success), and $q = 1 - p$ is the probability that an event will fail to happen in any single trial (the probability of a failure), then the probability that an event will happen x times in n trials (i.e., x successes and $n - x$ failures will occur) is:

$$p(x) = \binom{n}{x} p^x q^{n-x}$$

See also Combinations. (VWP)

Binomial probability table Table in which individual probability values are provided for a range of values of trials (*n*) and successes (*x*), and that gives probability of a success (*p*). (VWP)

Bioaccumulative Substances that increase in concentration in living organisms (that are very slowly metabolized or excreted) as they breathe contaminated air, drink contaminated water, or eat contaminated food. *See also* Biological magnification. (FSL)

Bioassay Use of living organisms to measure the effect of a substance, factor, or condition by comparing pre- and postexposure data. (FSL)

Biochemical Describes an event or action involving the chemistry of living organisms and the chemical changes occurring therein. (FSL)

Biochemical oxygen demand (BOD) The amount of dissolved oxygen required to decompose organic matter in water; a numerical estimate of contamination in water, expressed in milligrams per liter (mg/L) of dissolved oxygen. The greater the BOD, the greater the degree of pollution. (FSL)

Biochemistry The study of the chemistry of organisms, their chemical constituents, and vital processes. (FSL)

Biocide An agent used to kill organisms. (FSL)

Bioclimatic Of or pertaining to the effects of climate on organisms. (FSL)

Biodegradable Capable of being broken down into individual components by the action of microorganisms. (FSL)

Biodegradation The process or series of processes by which organisms render chemicals less harmful to the environment. (FSL)

Biogas Methane (CH_4), also known as marsh gas, that has been produced as a result of the anaerobic decomposition of manure. Several small-scale pilot plants have been developed in the United States to produce this material and use it as a fuel source. Additional research is underway in several developing countries, including India, to determine the feasibility of biogas production as an emerging source of energy. *See also* Biomass. (FSL)

Biological amplification The concentration of a persistent substance, such as a chlorinated hydrocarbon insecticide, by the organisms in a food chain, such that, at each level, the amount of the substance is increased. For example, 0.001 ppm of DDT in earthworms may create levels as high as 1 ppm in robins and other birds consuming large numbers of earthworms. (FSL)

Biological clock An internal timing mechanism that governs the biological rhythms of some organisms. A biological clock can also synchronize an organism's internal rhythms with external environmental stimuli, such as changes in the dark/light (day/night) cycle. *See also* Circadian rhythm. (SMC)

Biological control The use of animals and organisms that eat or otherwise kill or outcompete pests. (FSL)

Biological exposure index (BEI) Set of reference values established by the American Conference of Governmental Industrial Hygienists as guidelines for the evaluation of potential health hazards in biological specimens collected from a healthy worker who has been exposed to chemicals to the same extent as a worker with inhalation exposure to the threshold limit value. The values apply to 8-hour exposures, 5 days per week. BEI values are not intended for diagnosis of occupational illness. Usually expressed as a concentration of a chemical or its metabolite in exhaled air, blood, and/or urine. (DEJ)

Biological half-life The time required for the body to eliminate one-half of an administered dose of any substance by regular processes of elimination. This time is approximately the same for both stable and radioactive isotopes of a particular element. (RMB)

Biological magnification The process whereby certain substances such as pesticides or heavy metals move up the food chain, work their way into a river or lake, and are eaten by aquatic organisms, such as fish, that in turn are eaten by large animals or humans. The substances become concentrated in tissues or internal organs as they move up the chain. (FSL)

Biological monitoring *See* Biomonitoring.

Biological oxidation The process by which living organisms in the presence of oxygen convert organic matter into a more stable or a mineral form. (HMB)

Biological product As defined in the regulations for the interstate transport of etiologic agents, a product prepared and manufactured in accordance with the provisions of 9 CFR Parts 102–104 and 21 CFR Parts 312 and 600–800, and that, in accordance with such provisions, may be shipped in interstate traffic. 12 CFR Part 72. *See also* Etiologic agents. (DEW)

Biologicals Preparations made from organisms and their products, including vaccines, cultures, etc., intended for use in diagnosing, immunizing, or treating humans or animals or in research pertaining thereto. (FSL)

Biological safety cabinets Primary and partial containment work enclosures typically 4 feet to 6 feet in length. They are used for manipulations of agents or materials that may produce infections or sensitization via the respiratory route. The velocity through the front opening of these cabinets is in the range of 75 to 125 linear feet per minute. There are six general classes or configurations of biological safety cabinets, designated as I, IIA, IIB^1, IIB^2, IIB^3, and III. Class I biological safety cabinets are ventilated work chambers in which air is drawn into the work chamber by an exhaust blower located external to the cabinet; the inflow air stream protects the worker and the immediate laboratory environment from droplets and aerosols generated within the work chamber. Room air entering the work opening is unfiltered. Exhaust air from the work chamber is discharged through a high-efficiency particulate air (HEPA) filter integral to the cabinet. Although there are differences in the configuration of the other classes of cabinets (Classes II and III), they all share a common feature: the incoming air as well as the exhaust air from the cabinet passes through a HEPA filter, thus providing protection to the worker and preventing contamination of the materials being worked with. *See also* High-efficiency particulate air filter. (JHR)

Biological treatment A treatment technology that uses bacteria to consume waste. This treatment breaks down organic materials. (FSL)

Biomass Weight of living material in all or part of an organism, population, or ecosystem. Biomass is often expressed as dry weight per unit area. Also called biota. (SMC)

Biome A major regional ecological community characterized by distinctive vegetation and climate. Biomes in North America include desert biome, grassland biome (tallgrass prairie and shortgrass prairie), temperate and boreal forest biomes, taiga, and tundra. (SMC)

Biomedical waste This term includes the following: **(1)** Pathological waste. All recognizable human tissues and body parts removed during surgery, obstetrical procedures, autopsy, and laboratory procedures. **(2)** Biological waste. Bulk blood and blood products, exudates, secretions, suctionings, and other bulk body fluids that cannot be or are not directly discarded into a municipal sewer system. **(3)** Cultures and stocks of infectious agents and associated biologicals, including cultures from medical and pathological laboratories, cultures and stocks of infectious agents from research and industrial laboratories, wastes from the production of biologicals, discarded live and attenuated vaccines, and culture dishes and devices used to transfer, inoculate, and mix cultures. **(4)** Contaminated animal carcasses, body parts, their bedding, and other wastes from animals that have been exposed to infectious agents, capable of causing disease in human beings during research, production of biologicals, or testing of pharmaceuticals. **(5)** Sharps. Any discarded articles that may cause punctures or cuts. Such waste includes, but is not limited to, items such as needles, IV tubing and syringes with needles attached, and scalpel blades. **(6)** Chemotherapy waste. Any disposable material that has come in contact with cytotoxic agents (agents toxic to cells) and/or antineoplastic agents (agents that inhibit or prevent the growth and spread of tumors or malignant cells) during the preparation, handling, and administration of such agents. Such waste includes, but is not limited to, masks, gloves, gowns, empty IV tubing bags and vials, and other contaminated materials. The waste must first be classified as empty, which means that it occurs in such quantity that it is not subject to other federal or state waste management regulations prior to being handled as biomedical waste. **(7)** Discarded medical equipment and parts, not including expendable supplies and materials, that were in contact with infectious agents and have not been decontaminated. (JHR)

Biomonitoring **(1)** The use of organisms to test the suitability of effluents for discharge into receiving waters and to test the quality of such waters downstream from the discharge. **(2)** The measurement of changes in composition of body fluid, tissues, or expired air in order to determine excessive absorption of a contaminant. (DEJ, FSL)

Biopsy The removal and examination of tissue from living mammals. (FSL)

Bioremediation The process whereby soil, sludge, water, or an air stream is managed in order to promote optimal microbiological activity and, therefore, degradation or transformation of contaminated material into an innocuous product. (FSL)

Biosafety levels There are four biosafety levels that consist of specifications on laboratory practices and techniques, safety equipment, and laboratory facilities appropriate for the operations performed and the hazard posed by the infectious agents.

Biosafety Level 1 practices, safety equipment, and facilities are appropriate for undergraduate and secondary educational training and teaching laboratories and for other facilities in which work is done with defined and characterized strains of viable microorganisms not known to cause disease in healthy adult humans. *Bacillus subtilis*, *Naegleria gruberi*, and infectious canine hepatitis virus are representative of microorganisms meeting these criteria. Many agents not ordinarily associated with disease processes in humans are, however, opportunistic pathogens and may cause infection in the young, the aged, and immunodeficient or immunosuppressed individuals. Vaccine strains that have undergone multiple *in vivo* passages should not be considered avirulent simply because they are vaccine strains.

Biosafety Level 2 practices, equipment, and facilities are applicable to clinical, diagnostic, teaching, and other facilities in which work is done with the broad spectrum of indigenous moderate-risk agents present in the community and associated with human disease of varying severity. With good microbiological techniques, these agents can be used safely in activities conducted on the open bench, provided that the potential for producing aerosols is low. Hepatitis B virus, the salmonellae, and *Toxoplasma spp.* are representative of microorganisms assigned to this containment level. Primary hazards to personnel working with these agents may include accidental autoinoculation, ingestion, and skin or mucous-membrane exposure to infectious materials. Procedures with high aerosol potential that may increase the risk of exposure of personnel must be conducted in primary containment equipment or devices.

Biosafety Level 3 practices, safety equipment, and facilities are applicable to clinical, diagnostic, teaching, research, or production facilities in which work is done with indigenous or exotic agents by which the potential for infection by aerosols is real and the disease may have serious or lethal consequences. Autoinoculation and ingestion also represent primary hazards to personnel working with these agents. Examples of such agents for which Biosafety Level 3 safeguards are generally recommended include *Mycobacterium tuberculosis*, St. Louis encephalitis virus, and *Coxiella burnetii*.

Biosafety Level 4 practices, safety equipment, and facilities are applicable to work with dangerous and exotic agents that pose a high individual risk of life-threatening disease. All manipulations of potentially infectious diagnostic materials, isolates, and naturally or experimentally infected animals pose a high risk of exposure and infection to laboratory personnel. Lassa fever virus is representative of the microorganisms assigned to Level 4. (JHR)

Biosequence (soils) Related soils that differ primarily because of variations in the kinds and number of soil organisms as a soil-forming factor. *See also* Catena. (SWT)

Biosphere The relatively thin zone at or near the surface of the earth in which all organisms exist. The biosphere is formed by the intersection of the lithosphere, hydrosphere, and atmosphere, which provide the conditions necessary for life. (SMC)

Biostabilizer A machine that converts solid waste into compost by grinding and aeration. *See also* Compost. (FSL)

Biostatistics The area of applied statistics that deals with the application of statistical methods to medical and biological data. (VWP)

Biota All the flora, fauna, and other organisms of a particular place (habitat, ecosystem, planet, etc.). *See also* Biomass. (SMC)

Biotechnology Techniques that use organisms or parts of organisms to produce a variety of products (from medicines to industrial enzymes), to improve plants or animals, or to develop the microorganisms for specific uses, such as removing toxic compounds from bodies of water, or as pesticides. (FSL)

Biotic Of or pertaining to life. (SMC)

Biotic community A naturally occurring assemblage of plants and animals that live in the same environment and are mutually sustaining and interdependent. (FSL)

Biotin One of the B group of vitamins that is a water-soluble nitrogenous acid and is involved in the formation of fats and use of carbon dioxide in animals. (FSL)

Biotoxicology The scientific study of toxins produced by organisms, their effects, and the treatment of conditions they produce. (FSL)

Bird fancier's disease Extrinsic allergic alveolitis observed in some individuals who have been exposed to birds. The condition is accompanied by breathlessness or tightness of the chest, coughing, and wheezing. Extensive fibrosis can be seen in the chronic form. The disease can be caused by exposure to avian feed and/or antigens. (DEJ)

Bird infestation Many species of birds—e.g., pigeons, sparrows, or starlings—may present problems for a food processing or food-service establishment. Their droppings are unsightly and may carry pathogenic microorganisms. They may also be carriers of mites and other parasites. Stopping bird infestations is a fundamental part of a food sanitation program. The proper management and sanitation practices can reduce or eliminate the problem. Sanitary practices that remove food from the facility will reduce bird attraction. Entry into buildings can be reduced through maintenance of screens on doors, windows, and ventilation openings. Various methods to repel birds have been attempted. Noisemakers, mild electric shocks, and irritants have been used to discourage roosting near facilities. Commercially available repellents, chemical poisons applied by professionals, biological controls (sterilants), and trapping have been variously successful control approaches. *See also* Bird repellents. (HMB)

Bird Repellents Chemicals such as pyridine that are used as coatings on grains, seeds, and other foodstuffs and that, because of their burning taste and nauseating odors, create an undesirable situation for pigeons and other roosting birds. Upon ingestion, the repellents cause alarm in the birds, which in turn issue panic or distress calls and alert other birds of the danger. Frequently, birds overindulge in the consumption of the poison-laced baits and sometimes receive a fatal dose of the material.

Other repellents involve the use of caulking compounds that are placed in the roosting areas. These adhesive-like compounds may be impregnated with bitter-tasting chemicals that are unpleasant for the birds, or with adhesives that may stick to their feet and cause them to attempt to remove the material by pecking. Such induced activities not only annoy the affected birds, but are interpreted by other birds as danger signals.

Mechanical means, such as the use of screening materials and electric shocking devices have also been employed to minimize the clustering of undesirable bird populations. *See also* Bird infestation. (FSL)

Bittern The residue that remains after the crystallization of sodium chloride from salt water and that represents a source of bromine, iodine, and magnesium. (FSL)

Bituminous So-called soft coal, which has a high sulfur content and gives off considerable smoke and soot, particularly when improperly fired. *See also* Anthracite; anthracosis; pneumonoconiosis. (FSL)

Bivariate data Values of two different response variables that are obtained from the same population element. (VWP)

Black body An ideal body that would absorb all (and reflect none) of the radiation falling upon it. The spectral energy distribution of a black body is described by the Planck radiation equation. The total rate of emission of radiant energy is proportional to the fourth power of the absolute temperature (Stefan-Boltzmann law). (RMB)

Black Death Term used to describe the disease plague that occurred as a pandemic throughout Europe and, eventually, China and India, and that killed millions of persons. The causative organism of the disease is *Yersinia pestis* (formerly called *Pasteurella pestis* or *Bacillus pestis*), a bacterium maintained in nature as an infection of small rodents and transmitted by fleas. (FSL)

Black light The region of the electromagnetic spectrum between 300 nm and 400 nm in the ultraviolet region. This region is responsible for causing added pigmentation of the skin (burning and tanning) following exposure to ultraviolet light. (DEJ)

Black lung disease A disease contracted by coal miners, and is marked by varying degrees of pulmonary impairment, including X-ray abnormalities, cough, breathlessness, massive progressive fibrosis, formation of nodules and scar tissue in the lungs. Also known as coal miners' pneumoconiosis, Collier's disease, and Shaver's disease. (DEJ)

Blackwater Water that contains animal, human, or food wastes. (FSL)

Blaster The person responsible for the loading and firing of a blast. (FSL)

Blast Gate A device that regulates airflow in ductwork, similar to a damper, but usually operated by positioning a sliding metal plate across a duct. Often used to balance airflow in duct branches. Proper sizing of branch ducts sometimes eliminates the need for blast gates. (DEJ)

Blasting gelatin A mixture of approximately 92% nitroglycerin and 6% soluble gun cotton used as a comparative explosive in determinations of the relative weight/strength of a blast. (FSL)

Blasting mat A mat of woven steel wire, rope, old tires, or other suitable material, used to cover blast holes and prevent flying missiles. (FSL)

Blastomogen A substance capable of causing cancer. The term is often used synonymously with *carcinogen*. (DEJ)

Blast wave A pressure wave that radiates from an explosion. When the blast wave hits an object, the object may be propelled over a distance, bent, or broken, depending on its size and weight and the energy of the blast wave. Blast waves are measured in pounds/inch^2. *See also* Explosion. (SAN)

Bleach A concentration of 2–3% sodium hypochlorite ($NaOCl \cdot 5H_2O$) used as an oxidizer of organic materials that removes stains and color. Bleaches are significant antimicrobial agents and are used in higher concentrations as disinfecting agents at sewage treatment plants, water treatment plants, and sewage microbion ponds. In these situations HTH (high-test hypochlorite), which is approximately 70% calcium hypochlorite [$Ca(OCl)_2$], is used as the oxidizer. (FSL)

BLEVE See boiling liquid-expanding vapor explosions. (SAN)

Blinding The absolute closure of a pipe, line, or duct by fastening across its bore a solid plate or cap that completely covers the bore; also called blanking. The closure extends at least to the outer edge of the flange to which it attaches and is capable of withstanding the maximum upstream pressure. *See also* Bleed; double block. (SAN)

Blood The fluid and its suspended formed elements that are circulated through the heart, arteries, veins, and capillaries. This is the means by which oxygen is transported to the tissues and carbon dioxide is removed for excretion. (FSL)

Bloodborne pathogens Pathogenic microorganisms that are present in human blood and can cause disease in humans. These pathogens include, but are not limited to, hepatitis B virus (HBV) and human immunodeficiency virus (HIV). (FSL)

Bloodborne pathogens standard Regulation promulgated by the Occupational Safety and Health Administration (OSHA) on December 6, 1991 (29CFR 1910.1030), to eliminate or minimize exposure to hepatitis B virus (HBV), human immunodeficiency virus (HIV), and other bloodborne pathogens. On the basis of existing information, OSHA has made a determination that employees face a significant health risk as a result of occupational exposure to blood and other potentially infectious materials because such materials may contain bloodborne pathogens.

The regulation, which became effective on March 6, 1992, requires the preparation of an exposure control plan and the initiation of engineering and work practice controls, the use of personal protective equipment and clothing, training, medical surveillance, HBV vaccinations, and proper signage in laboratories, hospitals, or other facilities. (JHR)

Blood dyscrasia Any persistent change from normal of one or more of the blood components. (FSL)

Blood products Any product derived from blood, including but not limited to blood plasma, platelets, red and white blood corpuscles, and other derived licensed products, such as interferon. (FSL)

Bloom As applied to plankton or algae, a state of flourishing or being in an exceptionally healthy condition. Blooms are accentuated by the discharge of nitrogen- or phosphorus-containing compounds, such as those found in fertilizers or domestic sewage and that are often related to pollution. (FSL)

Blower's cataract *See* Glass blower's cataract. (DEJ)

Blue-green algae Any unicellular or filamentous algae of the class *Myxopyceae*. The bluish-green color is from the presence of blue pigments as well as chlorophyll. (FSL)

Board of Certified Safety Professionals (BCSP) The administering body for certified safety professionals, located in Savoy, IL. The Board reviews and approves applicants for certification, administers exams, and tracks maintenance points. (SAN)

Boatswain's chair A seat supported by slings attached to a suspended rope or wire that is designed to accommodate one worker in a sitting position. These seats are used to lift or lower workers in situations in which scaffolds and other working platforms cannot be used. OSHA requires that boatswain's chairs meet certain construction requirements. Due to the dangers involved, workers requiring this type of device must be protected by a safety belt and lifeline. (SAN)

Body belt A device used as a restraint to keep a worker from falling. Also known as a safety belt, work belt, harness, lanyard, lifeline, and dropline. The devices are divided into Class I (body belts), Class II (chest harnesses), Class III (harnesses), and Class IV (suspension belts). (DEJ)

Body burden, maximum permissible **(1)** (Radiation) An amount of radioactive material in a critical organ such that the whole-body dose is 0.3 rem per week or less; in the case of an alpha or beta emitter that is deposited in the bone, body burden is derived from the long-established maximum permissible body burden of radium, 0.1 microcurie, adjusted for possible less uniform deposition. Abbreviation: *q*. **(2)** The total amount of a substance stored in the body following exposure. The body burden for a particular substance is a function of its biological half-life and its biochemical uptake and elimination rate. (DEJ, RMB)

Body fluids Liquid emanating or derived from humans and limited to: blood; dialysis, amniotic, cerebrospinal, synovial, pleural, peritoneal, and percardial fluids; and semen and vaginal secretions. (FSL)

Bog A wetland ecosystem created by poor drainage and characterized by peat formation, acidic conditions, and the dominance of sphagnum moss. Some bogs also support various characteristic herbs (e.g., pitcher plants) or shrubs (especially shrubs in the Ericaceae family). (SMC)

Boiling liquid–expanding vapor explosions (BLEVE) A type of explosion that occurs when a container holding a liquid at a temperature well above its boiling point ruptures. When this occurs, the liquid evaporates or boils rapidly into a vapor state. BLEVEs involving liquefied propane gas produce fireballs. (SAN)

Bole The trunk of a tree. (SMC)

Bona fide In good faith. A bona fide purchaser is a purchaser in good faith who pays valuable consideration, believing the seller has a right to sell, and has no notice of rights in the property of another or intent to take unfair advantage. (DEW)

Bonding A safety procedure that involves the permanent joining of metallic parts to form an electrically conductive path that will assure electrical continuity and the capacity to conduct safely any current likely to be imposed. Bonding conductors should have the capacity to conduct any fault current that is imposed on them. Bonding jumpers are commonly used to prevent the buildup of static electricity between two metal containers when a flammable liquid is poured from one container to another. *See also* Grounding. (MAG, SAN)

Bone The skeletal tissue of vertebrates, consisting of cells arranged in a matrix of collagen fibers and cells containing calcium and phosphate. (FSL)

Bone marrow Soft material that fills the cavity in most bones; it manufactures most of the formed elements of the blood. (FSL)

Bone seeker Any compound or ion in the body that migrates preferentially to the bone. (FSL)

Bootleg The portion of a drilled blast hole that remains when the explosion does not fracture the rock to the bottom of the hole. (FSL)

Bordeaux mixture A fungicide that was used extensively before the advent of synthetic products. This mixture, which consisted of copper sulfate, quick lime, and water, was used to control fungal infections on fruit trees and potatoes. (FSL)

Boreal forest Forest found south of the tundra but north of the temperate forest zone. Boreal forests consist primarily of coniferous (evergreen) trees. (SMC)

Bored wells Wells typically constructed with earth augers turned either manually or by power equipment. In suitable soil and rocks, bored wells are often cased with concrete pipe, steel casing, or other suitable material. (FSL)

Boric acid (BH_3O_3) Colorless, odorless, transparent crystals, granules, or powder that occurs in nature as the mineral sassolite. Ingestion or absorption may cause nausea, vomiting, and diarrhea. Irritation to the skin may occur following exposure to the dry form. Boric acid is used as an insecticide for the control of cockroaches and black carpet beetles, for weatherproofing wood and fireproofing fabrics, as a preservative, and for the control of fungi on citrus fruits (FDA tolerance: 8 ppm boron residue). (FSL)

Boron An element occurring as borax or boric acid that is used for hardening steel and in the production of glass. (FSL)

Botanical pesticide A pesticide whose active ingredient is a plant-produced chemical such as nicotine or strychnine. (FSL)

Botany The study of plants (Kingdom Plantae). (SMC)

Bottle bill Proposed or enacted legislation that requires a returnable deposit on beer or soda containers and applies to retail store or other redemption centers. (FSL)

Bottom land *See* Flood plain. (SWT)

Botulism A severe illness resulting from ingestion of the toxin from the organism *Clostridium botulinium*. The illness may result in blurred vision, sore throat, or other symptoms of a nervous system disorder. This bacterium is a Gram-positive, strictly anaerobic, spore-forming bacillus. Seven antigenically distinct types, designated A–G, have been identified, but foodborne botulism is generally caused by types A, B, or E. These toxins interrupt nerve impulse transmission and are therefore classed as neurotoxins.

In typical cases of botulism, the onset of symptoms varies from a few hours to several days, most typically 12 to 48 hours. Cases in which symptoms have the shortest onset generally involve more severe illness, because in such cases a larger amount of neurotoxin has likely been ingested. Initially, symptoms may mimic those of gastroenteritis, with neurologic symptoms following or occurring simultaneously. Frequently, however, ptosis, blurred or double vision, weakness, and dry, sore throat are frequently first complaints. Vomiting, diarrhea, or constipation may also occur. Gradually more severe neurological symptoms prevail. Fatalities in the past were at 60% or higher, but in recent years have been at 30% or less. Respiratory paralysis is usually the cause of death. Treatment is usually with polyvalent antitoxins and should be administered without delay. Full recovery is generally possible, though likely slow, occurring over a period of weeks or months, depending upon the severity of the case.

Most botulism cases in the U.S. are associated with home-canned vegetables or fruits. Some preliminary heat treatment of canning or smoking is involved. In European countries, meats—e.g., fish, sausages, and smoked or preserved meats—are more typically involved. Prevention is most effective through adequate heat treatment, because the neurotoxin is readily destroyed by boiling. (HMB)

Botyoidal Grapelike, or having a shape resembling that of a bunch of grapes. (FSL)

Bound water Water that is physically or chemically encumbered and unavailable to microorganisms for survival and growth. (HMB)

Bovine Pertaining to or derived from cattle. (FSL)

Box plot display A pictorial representation of a set of data on a scale that may be horizontal or vertical. The box depicts the middle half of the data, where the hinges and median break the ranked set of data into four subsets. *See also* Hinge. (VWP)

Brackish Describes water that contains salt in solution, but less than that normally found in sea water. (FSL)

Brass Alloy that consists of copper and zinc. (FSL)

Break A split or crevice in a rock formation that may make blasting dangerous in areas where there are pockets or concentrations of naturally occurring combustible gases. (FSL)

Breakwater An offshore wall built to protect a beach from wave action. (FSL)

Breathing zone sample An air sample, collected inside a one-foot radius of an exposed worker's head, that has the same concentration of the contaminant being measured as does the air the individual is breathing. Typically, a breathing zone sample of an air contaminant is collected by a sampling device worn by the person and consisting of a battery-powered air pump, tubing, and collection media (such as a filter housed in a cassette for dusts, or a sorbant tube housed in a glass tube for gases) attached to the shirt lapel. Breathing zone samples are distinguished from area samples, which are collected according to location, not personal exposure. Also known as personal breathing zone sample or personal sample. (DEJ)

Breeder A reactor that produces more atomic fuel than it consumes. In such a reactor, a nonfissionable isotope is bombarded by neutrons and transformed into a fissionable material, such as plutonium, which can be used as fuel. Scientists are attempting to develop breeder reactors in which all the material burned can be replaced through this process. *See also* Converter. (RMB)

Bremsstrahlung A German word for "braking radiation." The production of electromagnetic radiation by the deceleration of a charged particle, usually an electron, that is passing through matter. Examples are the continuous spectrum from an X-ray tube, and the electromagnetic radiation often noted from pure beta emitters, such as P-32 and Sr-90. *See also* Scattered radiation. (RMB)

Brief In a law suit, the document in which facts, arguments, and points of law are presented in order to persuade the court to make a particular finding. (DEW)

British thermal unit (BTU) The quantity of heat needed to raise the temperature of 1 pound of water by 1 degree Fahrenheit. (FSL)

Bromelain A group of proteolytic enzymes derived from the pineapple plant and used as meat tenderizers. (FSL)

Bromine (Br) A brownish-red, corrosive element that is liquid at room temperature. It is used in the manufacture of dyestuffs and pharmaceuticals and is an effective water disinfection agent. *See also* Disinfectants. (FSL)

Bromophor Bromine used alone or in combination with other compounds as a sanitizer for utensils or processing equipment. It has found wider application for water treatment, but is effective for food service equipment. The effectiveness of bromine as a disinfectant is enhanced by the addition of chlorine to the solution. Their combination is synergistic. *See also* Disinfectants; sanitizer. (HMB)

Bronze Generic term for alloys of copper and tin. (FSL)

Brownian movement **(1)** The random movement of small suspended particles because of impacts with molecules in the medium. **(2)** A continuous agitation of particles in a colloidal solution caused by unbalanced impacts with molecules of the surrounding medium. The motion can be seen with a microscope when a strong beam of light is used to traverse the solution. (DEJ, FSL)

Brucellosis An illness caused by a bacterium of the Genus *Brucella*. *Brucella* are Gram-negative, nonmotile bacilli. They are at least partially host-adapted. The species that affect man include *B. abortus*, *B. melitensis*, *B. suis*, and *B. canis*. Those species have as hosts cattle, goats, swine, and dogs, respectively. The incubation period may range from 5 to more than 21 days. Most frequent symptoms include fever, chills, headache, muscle aches, malaise, weakness, loss of appetite, and loss of weight. Mortality is rare. The course of the disease may vary from a few days to years in duration. Contact with infectious material such as blood is an important mode of infection for livestock growers, veterinarians, and processing-plant workers. Intact skin provides effective protection against invasion of the organism, but cuts or abrasions may provide a route for infection. Human infection may arise from exposure via inhalation or ingestion in addition to direct contact. Foods involved in transmission of the bacterium through ingestion include improperly cooked meats from infected animals. *Brucella* organisms may be shed in milk from apparently healthy dairy cows. Shedding may occur through multiple successive lactation periods. Consumption of pasteurized milk is an important preventive measure. (HMB)

Bubble A system under which owners of existing emissions sources can propose alternative means to comply with a set of emissions limitations. Under the bubble concept, sources can control more effectively than is required at one emission point, where control costs are relatively low, in return for a comparable relaxation of controls at a second emission point, where costs are higher. (FSL)

Bubble policy *See* Emissions trading.

Buddy system A work practice commonly used in situations in which hazardous conditions, such as high concentrations of contaminants or flammable materials, and confined space entry, may exist. The purpose of the buddy system is to provide rapid assistance to employees in the event of an emergency. In such a system, it is designated that employees are organized into work groups, such that each employee is accompanied by another employee. The people in these pairs remain together for the duration of the task. If a backup system is required, it too, consists of pairs of persons. (SAN)

Buffer compounds, soil Soil fractions such as clay, organic matter, and carbonates that allow soil to resist appreciable change in pH. (SWT)

Buffer strips Strips of grass or other erosion-resisting vegetation between or below cultivated strips or fields. (FSL)

Buffet food service Any food service facility that provides a counter from which meals or refreshments are served. Buffet-style restaurants offer foods under different conditions than do other types of facilities. Specific criteria exist in food service and sanitation regulations that address these differences. For example, serving lines, salad bar protector devices, and display cases are designed and constructed to facilitate sanitation measures, and to protect against consumer contamination. Temperature control is critical. Sufficient hot or cold food storage facilities must be available to maintain adequate temperature control of foods on display. Specific conditions must be met in the same regard for dispensing devices for various potentially hazardous foods—e.g., ice, milk, and cream. (HMB)

Build-up factor The ratio of the intensity of X-radiation or gamma radiation (both primary and scattered) at a point in an absorbing medium to the intensity of the primary radiation alone. This factor has particular application for "broad-beam" attenuation. "Intensity" may refer to energy flux dose or energy absorption. (RMB)

Bulb A short underground stem surrounded by fleshy leaves or scales. Bulbs store nutrients and energy and protect the apical meristem from freezing or desiccation when the plant is dormant. Bulbs can also provide protection from fire and quick regrowth after it. Examples of plants with bulbs are members of the lily and amaryllis families. (SMC)

Bulk density, soil Dry soil mass per unit of bulk volume including air space. Bulk density is a weight measurement of soil. Soils that are loose, well-granulated, and high in organic matter tend to have a lower bulk density. (SWT)

Bulk sample As related to asbestos, a small portion of suspect building material that is collected and sent to a laboratory for analysis by polarized light microscopy coupled with dispersion staining, or by electron microscopy for verification. (MAG)

Bulk volume The volume of a soil mass that includes solids and pores. (SWT)

Bulldoze An explosive charge that is unconfined or covered with mud or other detritus and that is fired in contact with a rock surface without the use of a bore hole. (FSL)

Buoyancy The upward force exerted on a volume of fluid by the ambient fluid. Buoyancy depends upon various factors. The buoyancy of a gas, for example, depends upon its molecular weight and temperature. (FSL)

Burden The amount of soil and/or rock, usually measured in cubic yards or tons, to be blasted through use of a given hole. (FSL)

Burden of proof Requirement that a particular party in a dispute affirmatively prove a fact or facts at issue. For example, the state has the burden of proof in a criminal trial to prove the defendant committed the crime charged. (DEW)

Bureau of Alcohol, Tobacco and Firearms (BATF) An organization within the U.S. Treasury Department that has the primary responsibility for the enforcement of regulations related to the control of firearms and the movement of explosives in interstate commerce. The authority for this is provided in Part 181 of Title 26 of the Code of Federal Regulations. (FSL)

Bureau of Explosives A component of the Association of American Railroads that provides technical consultation to the U.S. Department of Transportation on matters relating to the classification of explosive materials to be transported in interstate commerce. (FSL)

Bureau of Labor Statistics (BLS) The federal agency responsible for administering and maintaining the OSHA record-keeping system, and for collecting, compiling, and analyzing work injury and illness statistics. *See also* OSHA Log of Occupational Injuries and Illnesses. (SAN)

Bureau of Mines The organization that, until 1972, was responsible for approving all forms of respiratory protection devices. At that time, the Mine Safety and Health Administration assumed that responsibility. *See also* Mine Safety and Health Administration. (DEJ)

Burette A glass column, fitted with a stopcock on the lower end, that is used for titration. (FSL)

Burial ground (radiation) A disposal site (graveyard) for radioactive waste materials that uses earth or water as a shield. (FSL)

Buried soil Soil that has been covered by a deposit, such as an alluvial or loessal deposit, usually to a depth greater than the thickness of the solum. (SWT)

Bursitis An inflammation of the joints of the body. *See also* Cumulative trauma disorder. (SAN)

Business establishment A distinct and separate economic activity that is performed at a single physical location. (FSL)

By-product **(1)** Material, other than the principal product, that is generated as a consequence of an industrial process. **(2)** (Radiation) Any radioactive material (except special nuclear material) yielded in, or made radioactive by exposure to, the radiation incident to the process of producing or utilizing special nuclear material. Also, the tailings or wastes produced by the extraction or concentration of uranium or thorium from any ore processed primarily for its source material content. 42 USC §2014 (e). (DEW, FSL)

Byssinosis A disease of the lungs, caused by chronic exposure to cotton dust. *See also* Bagasse. (FSL)

C

Cadmium (Cd) A soft, white element with an atomic weight of 112.40, associated with zinc ore. It is a soft blue-white or grayish-white powder. Cadmium is flammable in powder form and is toxic by the inhalation of dust or fumes. A known carcinogen, cadmium is used in alloys, storage batteries, pigments, and photoelectric cells. Ingestion of cadmium usually induces emesis, which minimizes the risk of fatal poisoning. This material, used in electroplating, was responsible for an outbreak of a disease called *itai-itai* in Japan. (FSL)

Caffeine A stimulant, derived from coffee and tea, that may serve as a mild diuretic. (FSL)

Caisson A wood, steel, concrete or reinforced concrete airtight and watertight chamber in which it is possible for individuals to work under pressure greater than that of the atmosphere for the purpose of excavating material below water level. *See also* Dysbarism, Sandhog. (SAN)

Caisson disease A condition caused by the rapid movement from high pressure to normal-pressure or low-pressure conditions. Under high-pressure conditions, inert gases will stay in solution in the blood and tissues within strictly defined pressure reductions. If limits are exceeded, however, the gases will come out of solution to form bubbles in tissue and blood, causing a series of pathophysiological events culminating in death or permanent paralysis. Local tissue distortion can cause pain, and gas emboli can cause congestive infarction of the spinal cord or gas pulmonary embolism, clotting, loss of capillary integrity, edema, and vasoconstriction. *See also* Caisson; decompression sickness; dysbarism. (DEJ)

Calcan A cutan composed of carbonates. (SWT)

Calcareous Refers to a material that contains calcium carbonate. (FSL)

Calcareous soil Soil containing enough calcium carbonate and/or magnesium carbonate to effervesce visibly when treated with 0.1-normal hydrochloric acid. (SWT)

Calcariuria The presence of calcium salts in the urine. (FSL)

Calcic horizon A horizon with secondary carbonate accumulation of greater than 6 inches in thickness, with a calcium carbonate equivalence of more than 15% and at least 5% more calcium carbonate than in an underlying horizon. (SWT)

Calcite The main carbonate mineral constituent of limestone and marble and an important element in many sedimentary rocks. (FSL)

Calcium (Ca) An element with an atomic weight of 40.08 that tarnishes quickly in air. It occurs widely as calcium carbonate and has many industrial uses. It is a necessary nutrient for plants and animals, and it is the most abundant mineral in the body. The presence of calcium is necessary for the normal clotting of blood, for the maintenance of normal heartbeat, and for neuromuscular and metabolic activities. (FSL)

Calcium hydroxide [$Ca(OH)_2$] So-called slaked lime, which is produced by reacting water with calcium oxide. Is used in the manufacture of plastics and mortars, and, as lime, is an important agricultural commodity. *See also* Calcium oxide. (FSL)

Calcium oxide (CaO) Known as quicklime, this chemical reacts rapidly with water to produce calcium hydroxide. It is widely used in the food industry, petroleum processing, and the manufacture of paper. *See also* Calcium hydroxide. (FSL)

Calibration Determination of variation from standard, or accuracy, of a measuring instrument, to ascertain necessary correction factors. (FSL)

Calorie The amount of heat required to raise the temperature of one gram of water one degree Celsius. Abbreviation: cal. *See also* Kilocalorie. (FSL)

Cambium The tissue in some vascular plants that gives rise to the secondary vascular tissues, phloem and xylem. (SMC)

Cambric horizon A mineral soil that has a texture of loamy, very fine sand, has a soil structure rather than rock structure, contains some weatherable minerals, and is characterized by the alteration of mineral material, as indicated by mottling or grayish colors, stronger chromas or redder hues than in underlying horizons, or the removal of carbonates. The cambric horizon lacks cementation or induration and has too few evidences of illuviation to meet the requirements of classification as argillic or spodic horizon. (SWT)

Campylobacteriosis An illness caused by the bacterium *Campylobacter fetus* subsp. *jejuni*. This organism is spiral-shaped, microaerophilic, non-sporeforming, motile, and Gram-negative. *C. jejuni* is considered now by some as the most common cause of bacterial diarrhea in the United States. The illness is commonly a foodborne disease, but it is also known to be waterborne. The incubation period may range from 2 to 10 days. Acute symptoms vary from mild to severe enteritis, with diarrhea, vomiting, abdominal pain, and fever. Bloody stools may occur in severe cases. Most outbreaks of campylobacteriosis have been associated with the consumption of raw milk and undercooked poultry. However, it is a common resident of domestic animals, and no single food animal source can be excluded as a potential vehicle for infection in humans. An estimated 50% to 70% of the human campylobacteriosis cases can be associated with poultry, usually undercooked. The organism does not grow on refrigerated meats but may be present because of fecal contamination of the product. One study of fresh chicken at the retail market showed 98% of all broilers to be positive for *C. jejuni*. Feral populations may carry the bacterium and spread it to domestic livestock. Wild bird populations, rabbits, rodents, and flies have been identified as sources. Dogs and especially cats may be implicated in the transmission of *C. jejuni* to humans. Little is known of the epidemiology of the organism in all food animals. Enough is known of the transmission of campylobacteriosis associated with poultry to begin studies of intervention methods to control transmission and substantially reduce the public health hazard. (HMB)

Canadian Electrical Code Sponsored by the Canadian Standards Association, this code outlines minimum standards for the installation and maintenance of electrical equipment. (FSL)

Cancellation Refers to Section 6(b) of the Federal Insecticide, Fungicide and Rodenticide Act (FIFRA), which authorizes cancellation of a pesticide registration if unreasonable adverse effects to the environment and public health develop when a product is used according to widespread and commonly recognized practice. *See also* Special review. (FSL)

Cancer Any malignant neoplasm. (FSL)

Candida A genus of fungi that are a normal part of the flora of the skin, mouth, intestinal tract, and vagina. These yeastlike organisms can cause infections, with the usual agent being *Candida albicans*. (FSL)

Candlepower A luminous intensity, expressed in candelas; a candle one inch in diameter produces one candela in a horizontal direction. (FSL)

Canine Linked to or characteristic of dogs. (FSL)

Cannabis The dried tops of the hemp plant (*Cannabis sativa*), which are classified as hallucinogens and marketed as hashish or marijuana. Its principal chemical constituent is tetrahydrocannabinol. (FSL)

Canopy The layer in a plant community that is on top, closest to the sun. In a forest, the canopy is composed of the branches and leaves of the tallest trees. The canopy intercepts most of the light, so subcanopy plants receive much less light. (SMC)

Canopy hood An exhaust hood designed to capture contaminants or heated air rising from an open tank some distance above the tank. (DEJ)

Canopy tree A tree that reaches and forms part of the canopy; a tree of a species whose mature members are tall enough to reach and form the canopy. (SMC)

Cap A layer of clay or other highly impermeable material installed over the top of a closed landfill to prevent entry of rainwater and to minimize production of leachate. (FSL)

Cap cloud The type of cloud formation that remains stationary on or over a mountain peak during stormy winds. (FSL)

Capillary A small, thin-walled blood vessel connecting an artery with a vein. (FSL)

Capillary porosity The pores in soil that hold water against a tension of greater than 60 cm of water. (SWT)

Cap rock The upper seal of impervious rock for an underground space containing either crude oil or natural gas. (FSL)

Captan ($C_9H_8Cl_3NO_2S$) A compound used to treat seed, as a fungicide in paints, plastics, leather, fabrics, and fruit preservation, and to prevent the growth of bacteria. Toxic by inhalation and an irritant of skin and mucous membranes; TLV 5 mg/m^3 of air. (FSL)

Capture cross-section The area of high probability that a nucleus will capture an incident particle. The unit of cross-section is commonly the barn (10^{-24} square cm). (RMB)

Capture efficiency The fraction of all organic vapors generated by a process that is directed to an abatement or recovery device. (FSL)

Capture, radiative The process by which a nucleus obtains an incident particle and loses its excitation energy immediately by the emission of gamma radiation. (RMB)

Capture velocity The air velocity at any point in front of a hood or at the hood opening necessary to overcome opposing or ambient air currents and to capture air contaminants at that point by causing them to flow into the hood. *See also* Laboratory hood. (DEJ)

Carbamate Any of a group of organic pesticides, based on one of the numerous derivatives and salts of carbamic acid (NH_2COOH), with a wide range of biological activity and used as herbicides, insecticides, and fungicides. These pesticides have a relatively low degree of toxicity to mammals and do not persist in the soil for long periods of time. Carbaryl (Sevin, $C_{10}H_7OOCNHCH_3$), an insecticide, is used to control fleas on dogs, turf insects, and caterpillars. The carbamates in general are harmful to bees and fish but have minimal impact on mammals or the environment. (FSL)

Carbolic acid The common name for phenol, which is a caustic material obtained by the distillation of coal tar. Phenol is typically used as an antiseptic disinfectant. (FSL)

Carbon adsorber An add-on device that uses activated carbon to adsorb volatile organic compounds from a gas stream. These compounds are later recovered from the carbon. (FSL)

Carbon cycle Global circulation and utilization of carbon atoms. (SMC)

Carbon dioxide A minor component of air, representing about 0.4% of the atmosphere, that is released by respiration and removed from the atmosphere by photosynthesis. (FSL)

Carbon fixation The conversion of carbon dioxide (CO_2) into organic compounds during photosynthesis. (SMC)

Carbon monoxide A colorless, odorless, tasteless gas formed as a product of the incomplete combustion of organic materials. This gas has an affinity for red blood cells about 220 times that of oxygen. Carbon monoxide figures prominently as a suicidal agent, with decreased pulmonary or cardiac function being the main physiologic effect. Long-term effects associated with short-term, low-level exposures are not known. *See also* Carboxyhemoglobin. (FSL)

Carbon-nitrogen ratio Ratio of the weight of organic carbon to the weight of total nitrogen in soil or in organic material. A low ratio is important for maintenance of soil organic matter. Plants with a low carbon-nitrogen ratio are more easily digestible and more nutritionally desirable to herbivores and decomposers than are plants with a high carbon-nitrogen ratio. (SMC) (SWT)

Carbon tetrachloride (CCl_4) A chemical used in fire extinguishers as early as 1908, and shown in later years to produce phosgene and hydrogen chloride in the presence of heat. This chemical was used as an industrial solvent and as a dry cleaning agent, as well as in fire extinguishers, where it was known as pyrene. As a result of the toxic properties of the material, most industrial uses have been eliminated. Studies that showed an association between prolonged exposures and liver damage contributed to decreased use of this chemical. (FSL)

Carboxyhemoglobin Hemoglobin in which the iron is associated with carbon monoxide (CO). The affinity of hemoglobin for CO is about 220 times greater than for oxygen. *See also* Carbon monoxide. (FSL)

Carboy Large glass bottles (5–10 gallons) surrounded by a protective framework of wood or steel used to hold chemicals. Increased regulation on the transportation of hazardous chemicals has practically eliminated the use of this type of container in favor of 5-gallon metal cans or thick-walled plastic bottles. This term also refers to 25-gallon plastic drums, which are frequently used to transport and store reactive chemicals. (FSL)

Carcinogen A substance known to cause cancer in humans and animals. Examples are vinyl chloride and aromatic hydrocarbons such as benzene. It is generally accepted that the term "carcinogen" refers to all chemicals in which there is some evidence of carcinogenicity in experimental animals, even if the test data are inconclusive. Carcinogens represent a broad range of organic and inorganic chemicals, hormones, immunosuppressants, and solid-state materials. A substance is considered carcinogenic if properly designed studies indicate carcinogenesis in humans, in two different animal species, in one animal species in at least two separate studies, or in one animal species in addition to multiple in-vitro mutagenicity tests. Two stages of carcinogenesis have been identified. They are *initiation*, which occurs after exposure of cells or tissue to a dose of a carcinogenic substance, and *promotion*, the process whereby the altered cells proliferate. *See also* Teratogen; Mutagen. (FSL)

Carcinogen Assessment Group A unit within the Environmental Protection Agency's Office of Health and Environmental Assessment that prepares lists of chemical substances for which substantial or strong evidence exists that exposure to these chemicals causes cancer in humans or can cause cancer in animal species. (FSL)

Carcinogenic Describes agents known to induce cancers. *See also* Carcinogen. (FSL)

Carcinoma Malignant neoplasm composed of epithelial cells, regardless of their derivation. (FSL)

Cardiac Relating to the heart. (FSL)

Cardinal data Data on a scale in which the distance between possible data values is significant. (VWP)

Cardiovascular Pertaining to the heart and the blood vessels or the circulation. (FSL)

Carnivore **(1)** An organism that feeds on animal tissue; **(2)** taxonomically, a member of the Order Carnivora, Class Mammalia. (SMC)

Carpal tunnel syndrome A disorder most often associated with work or sports involving flexing or extending the wrists or repeated force on the base of the palm and wrist. This condition is frequently seen among persons whose job involves repetitive assembly line work or who have jobs such as tree logging, butchering, and typing. It also affects those whose work requires the use of a hammer and is common among tennis players and people who use sign language.

The carpal tunnel is an opening in the wrist under the carpal ligament on the palmar side of the carpal bones. The median nerve, the finger flexor tendons, and blood vessels pass through this tunnel. Exterior swelling exerts pressure on this tunnel, pinching the median nerve and associated blood vessels.

The accompanying symptoms are often tingling, pain, or numbness in the thumb and first three fingers; this affects manipulative skills. Treatment generally involves immobilization of the affected wrist and hand, possible surgery to remove the pressure on the tunnel sheath, and possible redesign of the work procedure. (MAG) (SAN)

Carpenter's elbow A type of cumulative trauma disorder associated with repeatedly pushing the palm downward in such a way that a deviation of the ulnar nerve occurs. Symptoms include pain in the elbow, forearm, and hand. The condition was initially identified in carpenters and was shown to be associated with the consistent use of hammers. *See also* Cumulative trauma disorder. (SAN)

Carrier **(1)** A person or animal that harbors a specific infectious agent (but manifests no discernible clinical disease) and is a potential source of infection for people or animals. The carrier state can occur in an individual with an infection inapparent throughout its course. If the infection leads to disease, the carrier state may occur during the incubation, convalescent, or postconvalescent periods. While in these states, a person or animal is commonly referred to as an incubatory carrier or convalescent carrier. The carrier state can be of short or long duration (for temporary carriers and chronic carriers, respectively). **(2)** (Radiation) (a) A quantity of nonradioactive isotopes of an element that may be mixed with radioactive isotopes of that element, giving a ponderable quantity to facilitate chemical operations. (b) A substance in ponderable amount that when associated with a trace of another substance, will carry the trace with it through a chemical or physical process, especially a precipitation process. If the added substance is an element different from the trace, the carrier is called a nonisotopic carrier. (FSL) (RMB)

Carrier free An adjective applied to one or more radioactive isotopes of an element in minute quantity, essentially undiluted with stable isotope carriers. (RMB)

Carrier gases Gases such as nitrogen, helium, argon, and hydrogen that are used in gas chromatography or other laboratory procedures to sweep another gas or vapor into or through a system. (FSL)

Carrying capacity **(1)** In recreation management, the amount of use a recreation area can sustain without deterioration of its quality. **(2)** In wildlife management, the maximum number of animals an area can support during a given period of the year. (FSL)

Carson, Rachel (1907–1964) Author and naturalist who wrote *The Sea Around Us*, *The Edge of the Sea*, and *Silent Spring*. *Silent Spring* focused on the misuse of pesticides and brought to the attention of the general public the potential problems associated with the biomagnification and bioaccumulation of pesticide residues in food chains. *See also* Bioaccumulative; biological magnification. (FSL)

Cartilage Human or animal connective tissue, also known as gristle, characterized by a firm consistency and non-blood-carrying properties. (FSL)

Cascade impactor A device used to measure the size range of airborne particles, based on the principle that when a high-velocity airstream strikes a flat surface at a 90° angle, the sudden change in direction and momentum will cause the dust in the air to be deposited on a plate and to be separated from the air. A series of plates are used to capture different-sized particles, which can then be analyzed for total weight, particle count, and chemical composition. (DEJ)

Case An infected or diseased person or animal having specific clinical, laboratory, and epidemiologic characteristics. A *confirmed case* is a person or animal from which a disease-producing agent has been isolated and identified or from which has been obtained other laboratory evidence of the presence of an etiologic agent—e.g., a fourfold or greater rise in antibody titers between acute and convalescent serum specimens, whether or not that person or animal has a clinical syndrome indicative of the disease caused by the agent. A *presumptive case* is a person or animal with a clinical syndrome compatible with a disease but without laboratory confirmation of the etiologic agent. *See also* Epidemiology. (FSL)

Case fatality rate *See* Rate.

Case hardening The technique used to produce a hard surface layer on a nonhardening type of steel by heating the steel in contact with certain carbon compounds. (FSL)

Casein A milk protein used in the production of glues and foodstuffs, and a major component of cheese. It is insoluble in water and soluble in dilute alkaline and salt solutions. (FSL)

Caseinogen Term used, primarily in the United Kingdom, to describe casein. (FSL)

Cask A thick-walled container (usually lead) used to transport radioactive material. Also called a coffin. (FSL)

Catalase An iron-containing enzyme that breaks down hydrogen peroxide to oxygen and water, and is used in definitive tests in certain bacteriological procedures. (FSL)

Catalyst A substance that increases the rate of a chemical reaction but is not permanently changed by the reaction. (FSL)

Catalytic converter An air pollution abatement device that removes pollutants from motor vehicle exhaust, either by oxidizing them into carbon dioxide and water or by reducing them to nitrogen and oxygen. (FSL)

Catalytic incinerator A control device that oxidizes volatile organic compounds by using a catalyst to promote the combustion process. Catalytic incinerators require lower temperatures than do conventional thermal incinerators, with resultant fuel and cost savings. (FSL)

Catanadramous Refers to fish that swim downstream to spawn. (FSL)

Cataract A clouding of the crystalline lens of the eye. The clouding obstructs the passage of light. (FSL)

Cataractogenic Cataract-producing. (FSL)

Cat clay Wet soils that contain ferrous sulfide and that become highly acidic when drained. (SWT)

Categorical exclusions For purposes of the National Environmental Policy Act, a category of actions that do not individually or cumulatively have a significant effect on the human environment, that have been found to have no such effect in procedures adopted by a federal agency, and for which no environmental assessment nor environmental impact statement is required. 40 CFR Part 1508.4. *See also* National Environmental Policy Act; environmental impact statement. (DEW)

Categorical pretreatment standard A technology-based effluent limitation for an industrial facility that discharges into a municipal sewer system.

Category, soil Any of the ranks of the system of soil classification in which soils are grouped according to their characteristics. (SWT)

Catena A sequence of soils derived from the same parent material that are of about the same age and occur under similar climatic conditions but have different characteristics, due to variations in relief and in drainage. (SWT)

Cathode A negative electrode; an electrode to which positive ions are attracted. (FSL)

Cation A positively charged ion. (FSL)

Cation exchange The exchange between a cation on the surface of any surface-active agent, such as clay or organic matter, and a cation in solution. Exchange reactions are stoichiometric. Cation exchange is important in plant nutrient uptake and soil fertility. *See also* Adsorption; anion exchange. (SWT)

Cation exchange capacity The total of exchangeable cations that a soil can adsorb. Sometimes called total exchange capacity or cation adsorption capacity. Expressed in milliequvalents per 100 grams (meq/100 g) of soil. *See also* Cation exchange. (SWT)

Cationic wetting agent A compound that has the properties of a wetting agent and is also considered to be a synthetic detergent. It is an agent with a positively charged ionic group. The most commonly used cationic agents are quaternary ammonium compounds (quats), such as alkyl dimethyl benzyl ammonium chloride. They possess strong bacteriocidal properties, but are seldom used for cleaning. *See also* Surfactant. (HMB)

Cause-specific death rate *See* Rate. (VMP)

Caustic soda *See* Sodium Hydroxide. (FSL)

CBOD5 The amount of dissolved oxygen consumed in 5 days from the carbonaceous portion of biological processes in an effluent. (FSL)

Ceiling exposure limit The concentration of a chemical to which workers should not be exposed, even instantaneously, during any part of the working day. The ceiling exposure limit is often used for irritant gases, and is set by both the Occupational Safety and Health Administration (for permissible exposure limits) and by the American Conference of Governmental Industrial Hygienists (for threshold limit values). (DEJ)

Cell **(1)** The fundamental unit of structure and function in organisms. **(2)** In solid-waste disposal, holes in which waste is dumped, compacted, and covered with layers of dirt on a daily basis. (FSL)

Cellophane A strong and flexible transparent film made from cellulose nitrate or acetate and used as a wrapping material. (FSL)

Cellulitis An inflammation of tissues that produces pain, edema, swelling, and functional difficulties. It may be caused by streptococcal, staphylococcal, or other organisms. *See also* Erysipelas. (FSL)

Cement *See* Portland cement. (FSL)

Cemented soil Soil that has a hard, brittle consistency because the particles are held together by cementing substances such as humus, calcium carbonate, or oxides of silicon, aluminum, or iron. (SWT)

Centers for Disease Control and Prevention (CDC) A federal agency under the Department of Health and Human Services, with headquarters in Atlanta, Georgia. The agency is responsible for coordinating and conducting public health campaigns aimed at combatting communicable, environmental, and occupational diseases, and injuries. The agency also compiles statistics on various reportable diseases, and provides guidance to physicians, industrial hygienists, and other public health practitioners. (DEJ)

Central limit theorem If all possible random samples of size n are taken from a population with mean μ, and standard deviation σ, and the mean and standard deviation of the sampling distribution of means are denoted by $\mu_{\bar{x}}$ and $\sigma_{\bar{x}}$, respectively, then:

$$\textbf{(1)}\quad \mu_{\bar{x}} = \mu$$

$$\textbf{(2)}\quad \sigma_{\bar{x}} = \frac{\sigma}{\sqrt{n}}$$

The sampling distribution of sample means will be normally distributed when the population is normally distributed or will be approximately normally distributed for large samples ($n \geq 30$) when the population is not normally distributed. (VWP)

Centrifugal collector A mechanical system using centrifugal force to remove aerosols from a gas stream or to de-water sludge. (FSL)

Centrifugal pump A pump often used in small water supply systems that contains a rotating impeller mounted on a shaft turned by the power source. The rotating impeller increases the velocity of the water and discharges it into a surrounding casing shaped to slow down the flow of the water and convert the velocity to pressure. As the flow decreases, the pressure increases. (FSL)

Centrifuge A laboratory device used to subject substances in solution to centrifugal forces 20,000–25,000 times gravity. (FSL)

Centroid In a set of data points, the point $(\bar{x}, \bar{y})$, where $\bar{x}$ is the mean of x values and $\bar{y}$ is the mean of y values. (VWP)

Certified industrial hygienist (CIH) An individual certified by the American Board of Industrial Hygienists to practice industrial hygiene. *See also* American Board of Industrial Hygienists. (DEJ)

Certified occupational health nurse (COHN) *See* American Board for Occupational Health Nurses. (DEJ)

Certified safety professional (CSP) A certification credential that a practicing safety professional can receive after meeting certain educational and experience qualifications and successfully passing two examinations. There are several areas of emphasis, including management, engineering, and comprehensive aspects of safety. The certification process is administered by the Board of Certified Safety Professionals. *See also* Safety engineer. (SAN)

Certiorari A writ issued by a higher court to a lower court to review the proceedings of the lower court. (DEW)

Cesium (Cs) A silver-white element, with an atomic weight of 132.905, that is highly reactive. It is used in photoelectric cells and as a catalyst. (FSL)

Cesspool A covered pit with open-jointed lining into which raw sewage is discharged. The liquid seeps into the soil and the solids or sludge are retained in the pit. This method of disposal has been banned in most states because of the high potential for ground-water contamination. *See also* Septic tank; soil absorption system. (FSL)

Cestodes A class of flatworms or tapeworms that infect land and aquatic vertebrates. The larvae stage of these parasites is maintained in one or more intermediate hosts, such as sheep or fleas. (FSL)

Cetalkonium chloride [$C_6H_5CH_2N(CH_3)_2(C_{16}H_{33})Cl$] A quaternary ammonium compound used as a disinfectant and topical antiseptic. (FSL)

Cetology The study of cetaceans (Order Cetacea, Class Mammalia). Order Cetacea includes marine mammals such as whales, dolphins, and porpoises). (SMC)

Chain of custody In any activity that may be used to support litigation (e.g., collection of samples in an investigation of a foodborne-illness outbreak), the regulatory agency must be able to provide the chain of possession and custody of any sample offered for evidence or that forms the basis of analytical test results introduced into evidence. Procedures to ensure this involve creation of an accurate written record that can be used to trace the possession and handling of the sample from the moment of its collection through analysis and its introduction as evidence. Samples in someone's ''custody'' include: those in one's actual physical possession; those in one's view after being in one's physical possession; those in one's physical possession, then locked up to prevent tampering; and those kept in a secured area restricted to authorized personnel. Without a continuous record of chain of custody, the validity of any sample or results obtained from the analysis of that sample may be questioned. (HMB)

Chain of infection A series of related factors or events that must occur before an infection will occur. These factors can be identified as host, agent, source, and transmission factors. (HMB)

Chain reaction Any chemical or nuclear process in which some of the products of the process have an effect on additional particles of the reactants. *See also* Neutron chain reaction. (RMB)

Chain saw vibration syndrome *See* Raynaud's disease; vibration syndrome. (DEJ)

Chalcosis The deposition of copper in tissue. (FSL)

Chalicosis A type of pneumoconiosis due to the inhalation of stone particles. (FSL)

Characteristic (discrete) radiation The monochromatic radiation emitted by an atom when an orbital electron is removed and then returns, or that follows excitation of the atom. Each element may emit a number of characteristic radiations, each of a constant wavelength and different from the characteristic radiations of all other elements. (RMB)

Charcoal **(1)** Carbon obtained by heating substances such as wood in the presence of little or no oxygen. This material is highly effective as a deodorant and absorbant. **(2)** A substance with a large capacity for absorbing and adsorbing a large number of organic chemical contaminants. Activated charcoal is frequently used to collect air samples and to clean air in heating–ventilation systems. It is also used in certain chemical cartridges for air-purifying respirators. (DEJ) (FSL)

Charge The fuel (active material or fertile material, such as ^{238}uranium, ^{232}thorium) placed in a reactor to produce a chain reaction. (RMB)

Chasmogamous A normal open flower. (SMC)

Châtelier mixture rule *See* Le Châtelier mixture rule. (DEJ)

Chebyshev's theorem The part of any distribution of data that lies within K standard deviations of the mean is $1 - (1/K^2)$, where K is any positive number greater than one. (VWP)

Checklist A series of questions specific for the work process of interest. Checklists are often used during audits for process safety and personnel safety and they incorporate the major requirements needed to evaluate equipment, procedures, materials, and work practices. A checklist may be built around certain OSHA standards that apply to an operation. Checklists are best used in situations in which most of the hazards have been identified, eliminated, or reduced and the purpose of the audit is to maintain the integrity of a system. *See also* Health and safety audit; process safety. (SAN)

Chelate A chemical compound in which a metallic ion is combined with a molecule with multiple chemical bonds. (SWT)

Chelating agent A type of organic sequestering agent that reduces water hardness and inactivates certain metal ions in water. One such agent is ethylene diamine tetraacetic acid (EDTA). These agents may be used in detergent formulations to reduce the ill effects of metals in water. *See also* Sequestrant. (HMB)

Chelation A treatment that removes harmful substances from the body. The chelating agent bonds to the contaminant, which, due to the resulting poor absorption, is excreted from the body. Chelating agents have been used with varying degrees of success for heavy metals, notably lead, and radioactive substances. (DEJ)

Chemical asphyxiant A substance that chemically interferes with the uptake or transport of oxygen. There may be sufficient atmospheric oxygen present, but the body is unable to utilize it because the physiological mechanism for use and transport has been blocked, as is the case with exposure to carbon monoxide. The heme molecule responsible for uptake and transfer of oxygen in the blood has a greater affinity for carbon monoxide and thus becomes saturated over a period of time, causing an oxygen deficit, which, in turn, affects the body's metabolic processes. (MAG)

Chemical cartridge respirator An air-purifying respirator capable of filtering out chemical contaminants from air that is to be breathed. It usually acts by the use of chemical sorbant beds housed in cartridges that are attached to the respirator facepiece. *See also* Air-purifying respirator. (DEJ)

Chemical compound A substance composed of two or more elements combined in a fixed and definite proportion by weight. (FSL)

Chemical dosimeter A self-indicating device for determining total (or accumulated) radiation exposure dose based on color changes accompanying chemical reactions induced by the radiation. (RMB)

Chemical emergency An occurrence, such as a transportation accident, equipment failure, rupture of a container, or failure of control equipment, that results in an uncontrolled release of a hazardous chemical into the environment or workplace. (FSL)

Chemical (isotopic) exchange A process in which atoms (isotopes) of the same element in two different molecules exchange places. (RMB)

Chemical foam A stable aggregation of minute bubbles that flow freely over a burning liquid surface and form a coherent blanket that seals combustible vapors, thus suppressing the fire. *See also* Aqueous film-forming foam. (SAN)

Chemical hood A device, commonly known as a fume hood and usually located in a laboratory, enclosed on five sides with a movable sash for access or a fixed panel, and constructed and maintained to siphon air from the laboratory and to prevent or minimize the escape of contaminants into the laboratory. Hoods allow chemical manipulations to be conducted in the enclosure without insertion of any portion of the employee's body other than hands and arms. Airflow rates from the room through the hoods should range between 75 and 125 linear feet per minute. Walk-in hoods with adjustable sashes meet the definition, provided that the sashes are adjusted during use so that the airflow and the exhaust of air contaminants are not compromised and employees do not enter the enclosure during the release of volatile, airborne hazardous chemicals. *See also* Biological safety cabinet. (FSL)

Chemical hygiene officer An employee who is so designated by the employer. This employee must be qualified by training or experience to provide technical guidance in the development and implementation of the provisions of the Chemical Hygiene Plan, as required by the Occupational Safety and Health Administration regulation "Occupational Exposures to Hazardous Chemicals in Laboratories" (CFR 1910.1400). *See also* Chemical hygiene plan. (FSL)

Chemical Hygiene Plan A requirement under the Occupational Safety and Health Administration regulation ''Occupational Exposure to Hazardous Chemicals in Laboratories'' (CFR 1910.1400), which became effective May 1, 1990. The plan must be written and implemented by employers. It must outline procedures, equipment, personal protective equipment, and work practices that are capable of protecting employees from the health hazards presented by chemicals in the workplace. *See also* Chemical hygiene officer. (FSL)

Chemical oxygen demand (COD) A measure of the oxygen required to oxidize all compounds in water, both organic and inorganic. (FSL)

Chemical protective clothing Clothing made from various materials that exhibit chemical-resistant properties to an offending agent. The clothing may be composed of one or multiple layers of varying materials that resist permeation, degradation, and penetration by an agent. The selection of the clothing will depend on the substance(s) present and the task to be accomplished. Examples of protective clothing include gloves, aprons, splash suits, face shields and hardhats. *See also* Saranex®. (MAG)

Chemical sanitizer Chemical sanitizers include chlorine compounds, iodine compounds, bromine compounds, quaternary ammonium compounds, glutaraldehyde, and acid sanitizers. Several chemical sanitizers are commercially available for use in food processing or in food service operations. These differ in chemical composition and activity; their respective characteristics are known and are selected for use dependent on the particular application. Their efficacy is affected by factors such as concentration, exposure time, temperature, pH, surface cleanliness, and water hardness. (HMB)

Chemical substances For purposes of the Toxic Substances Control Act, any organic or inorganic substances of particular molecular identity, including: any combination of such substances occurring in whole or in part as a result of a chemical reaction or occurring in nature, and any element or uncombined radical. Does not include: any mixture, any pesticide, tobacco, any course material, special nuclear material, or by-product material, food, food additive, drug, cosmetic, or device under the Federal Food, Drug, and Cosmetic Act, and other specified articles. *See also* By-product materials; Federal Food, Drug, and Cosmetic Act [15 USC §2602 (2)]; pesticide; source material; special nuclear material; Toxic Substances Control Act. (DEW)

Chemical Transportation Emergency Center (CHEMTREC) A service that operates 24 hours a day, seven days a week, and provides emergency information on chemicals that may have been spilled as the result of a transportation accident. (FSL)

Chemical treatment Any of a variety of technologies that use chemicals or chemical processes to treat water for drinking purposes. (FSL)

Chemiluminescence Emission of light as a result of a chemical reaction. Chemiluminescence has been employed in instrumentation to monitor ozone, oxides of nitrogen, and other contaminants. (DEJ)

Chemistry The area of science that deals with the elements and atomic structure of matter and the compounds of the elements. (FSL)

Chemoprophylaxis The administration of a chemical, including antibiotics, to prevent the development of an infection or the progression of an infection to clinical disease. (FSL)

Chemosterilant A process or material used to sterilize all or part of an insect population, especially the males. Gamma radiation, for example, was successfully employed to eradicate populations of screwworm flies (Genus *Callitroga*) from the southeastern United States. This insect pest was responsible for health and economic damage to people and domestic animals. *See also* Pesticide. (FSL)

Chemotroph A microorganism that has the ability to obtain energy from the oxidation of organic materials. (FSL)

Cheney syndrome *See* Acroosteolysis. (FSL)

Chernobyl Location in Ukraine (formerly part of the Soviet Union) that was the site of a nuclear reactor accident on April 26, 1986. On that date, and for a period of about 10 days, the Chernobyl-4 reactor released a wide spectrum of fission and activation products, including ^{131}I, ^{137}Cs, and ^{134}Cs. Portions of the radioactivity released by the initial explosion and the subsequent graphite fire lofted into the middle troposphere and were transported over the northern hemisphere. Prevailing winds carried part of the discharge to Norway, causing contamination of sheep, reindeer, game animals, cattle, and the milk from goats and cattle. Freshwater fish were also shown to be contaminated. The health effects on human populations in close proximity to the reactor site are being evaluated, but it is predicted that numerous deaths from radiation sickness will occur, as well as excess cases of radiation-induced disorders. (FSL)

Cherry picker An elevating powered industrial work platform in the form of a basket or cage. The basket can move vertically or horizontally. Workers riding in such a basket should wear a safety belt. The platform should lock in place at the desired height and remain there in the event of a power failure. *See* Powered industrial platforms. (SAN)

Chert Structureless form of silica that breaks into angular fragments. Closely related to flint. (SWT)

Chicken ladder A board used for work on roofs that has cleats spaced and secured at regular intervals to allow worker movement up and down the board. Such ladders are not designed to carry materials. When used, they must be secured to the roof's surface by a hook. OSHA requires that a lifeline be strung beside the board as a safety precaution. (SAN)

Chick's law Named for the investigator who hypothesized that bacteria in an unfavorable environment die at a constant rate—that is, that a given percentage of the residual population dies during each successive time unit. (FSL)

Chigger The so-called red-bug or larva of the mite family *Trombiculidae* whose bite produces a wheal with itching and dermatitis. (FSL)

Chigoe The sand flea (*Tunga penetrans*), which burrows into the skin of the feet, legs, and other portions of the body, causing ulceration. This insect is prevalent in tropical and subtropical areas. (FSL)

Chilling effect The lowering of the earth's temperature because of increased particles in the air blocking the sun's rays. *See also* Greenhouse effect. (FSL)

Chimney effect The tendency of air or gas in a duct or other vertical passage to rise when heated, due to the fact that its density is lower than that of the surrounding air or gas. Often employed to exhaust combustion sources (burners, hot-water heaters, furnaces) in buildings. If another powered source of exhaust ventilation is added (e.g., a laboratory exhaust hood), the passive chimney effect can be overcome, resulting in a reversal of airflow in the chimney and entry of combustion products into the building, instead of out through the chimney. (DEJ)

Chinese restaurant disease Often called the Chinese restaurant illness or syndrome, this condition is due to the ingestion of large amounts of food containing monosodium glutamate, a flavoring additive. Ill individuals typically suffer from headaches, tightness in the face, and lightheadedness. (FSL)

Chiseling Breakage or shattering of compact soil or subsoil layers by use of a tillage implement called a chisel. (SWT)

Chi-square (χ^2) distribution If a random sample of a size n is taken from a normally distributed population with variance σ^2 and sample variance S^2, then the random variable

$$\chi^2 = \frac{n - 1}{\sigma^2} \cdot S^2$$

has the chi-square distribution with degrees of freedom $(df) = n - 1$. The probabilities for χ^2 can be determined using areas under the chi-square curve with $df = n - 1$. (VWP)

Chlamydia Gram-negative nonmotile bacteria that are among the most prevalent organisms causing sexually transmitted disease. (FSL)

Chloracne A disfiguring skin condition noted among workers who have had significant contact with certain chemicals, such as chlorinated diphenyls, chlorinated dioxins, and chlornaphthalenes. The condition resembles the acne of adolescence and is characterized by the appearance of blackheads, cysts, folliculitis, and scars. (DEJ) (FSL)

Chloramine-T An organic chloride sanitizer. This type of sanitizer is not as stable or soluble as an inorganic chlorine compound such as sodium hypochlorite, but it is much less corrosive and not irritating to the skin. Organic chlorides release chlorine into solution at a much slower rate than do inorganic chlorides. Therefore, they are used only if longer exposure times are practical. (HMB)

Chlorates Various strong oxidizing agents, such as sodium chlorate ($NaClO_3$), that can form from explosive mixtures with organic materials. (FSL)

Chlordane ($C_{10}H_6Cl_8$) A chlorinated hydrocarbon insecticide widely used for the control of soil insects such as ants and termites. This chemical has been recovered in soils 15 to 20 years after application. Because of the tendency of this chemical to remain in the ecosystem and bioaccumulate in food-chain insects and mammals, most of its uses have been eliminated. *See also* Chlorinated hydrocarbon insecticides. (FSL)

Chlorinated hydrocarbon insecticides Organic compounds containing chlorine that have pesticidal properties. These compounds, such as DDT, aldrin, dieldrin, and endrin, are noted for their persistence in soil and low solubility in water. Many were used throughout the world for the control of insects, but as resistance to many pest species became known and environmental considerations were given more importance, the use of these pesticides has virtually ceased. *See also* Pesticide. (FSL)

Chlorination The application of chlorine to drinking water, sewage, or industrial waste to disinfect or oxidize undesirable compounds. *See also* Chlorine; Disinfection. (FSL)

Chlorinator A device that adds chlorine, in gas or liquid form, to water or sewage to kill infectious bacteria. (FSL)

Chlorine (Cl) A dense, greenish-yellow gas that is noncombustible but supports combustion and has a pungent, irritating odor. This chemical, with an atomic weight of 35.453, is the ninth-highest-volume chemical produced in the United States. Chlorine is extremely toxic to humans and dangerous when it comes in contact with hydrogen, ether, or ammonia. Chlorine is used in the manufacture of hydrogen chloride, ethylene dichloride, and carbon tetrachloride, and in food processing and water purification. *See also* Chlorination; chlorine disinfectants; disinfection, calcium hypochlorite; sodium hypochlorite. (FSL)

Chlorine, combined Chlorine is a highly reactive chemical, and in aqueous solution will react quickly with a wide variety of substances that are present as impurities. Ammonium ion in equilibrium with ammonia and hydrogen ion serves as a ready substrate for chlorine action. Any combined form of chlorine, with ammonia or with any other substrate upon which it acts (especially amines), is commonly referred to as combined chlorine or combined chlorine residual. *See also* Chloramines; chloramine-T; chlorine compounds; residual chlorine. (HMB)

Chlorine contact basin A detention basin provided to diffuse chlorine through treated wastewater to provide adequate contact time for disinfection. Its use occurs typically at the last step of a conventional wastewater treatment process prior to discharge. *See also* Conventional treatment. (HMB)

Chlorine disinfectants Any of a number of forms of chlorine that can function as sanitizers. These include hypochlorites, organic chloramines, chlorine dioxide, and liquid chlorine. Most chlorine compounds are highly reactive but vary in their antimicrobial activity.

The chemical behavior of chlorine in water results in hydrolysis to form hypochlorous acid, which dissociates to form hydrogen ion and hypochlorite ion. The most bactericidal form is thought to be the undissociated hypochlorous acid, which, upon entering the cell, and among other actions, oxidizes sulfhydryl groups of certain enzymes. However, the whole mode of action of chlorine as a bactericidal agent is not fully understood.

These chemicals are widely used for disinfection purposes in water treatment systems, swimming pools, and hospitals. One of the commonly used forms is calcium hypochlorite, which is commercially available in powder and tablet form. It is classed as high-test hypochlorite and contains 65% to 75% available chlorine by weight. Sodium hypochlorite is available at grocery stores and through chemical or swimming-pool suppliers. The most common form is household bleach, which has a concentration of approximately 5% available chlorine by weight. Other mixtures vary in strength from 3% to 15% available chlorine by weight. When hypochlorite powders are used, fresh solutions need to be prepared at frequent intervals because the concentration of the chlorine in solution deteriorates gradually. Hypochlorite solutions are used either full strength or diluted to solution strength, as appropriate for the dispenser involved and the rate of water flow.

The efficiency of the process is improved by several factors: **(1)** high concentration, **(2)** long contact time, **(3)** high temperature, the higher the temperature, the more effective is the chlorine, **(4)** a larger quantity of free chlorine rather than combined chlorine, and **(5)** low pH. *See also* Disinfectants; water treatment. (FSL)

Chlorofluorocarbons (CFCs) Compounds composed of carbon, fluorine, chlorine, and hydrogen, such as trichlorofluoromethane. Regulations against the use of CFCs have been in effect since 1979 because of their deleterious effects on the stratospheric ozone layer. (FSL)

Chlorophyll The green pigment contained in plant cells and in some protists and prokaryotes. Chlorophyll absorbs light energy in wavelengths necessary for photosynthesis. (SMC)

Chlorosis Discoloration of normally green plant parts. The condition can be caused by disease, lack of nutrients, or pollutants. (FSL)

Chloruria The presence of a high level of chlorides in the urine. (FSL)

Chocking Blocks placed under the wheels of railroad cars and trucks to prevent movement while parked at loading docks, loading ramps, and stations. (SAN)

Cholinesterase An enzyme that hydrolyzes acetylcholine within the central nervous system and at peripheral neuroeffector junctions. This enzyme can be depressed following exposure to organophosphate pesticide compounds. Workers using organophosphate pesticides should be routinely monitored for cholinesterase levels. (FSL)

Chroma (soil) Relative purity, strength, or saturation of a color; directly related to the dominance of the determining wavelength of the light and inversely related to grayness; one of the three variables of color (the others being hue and value on the Munsell color system). (SWT)

Chromate A salt of chromic acid—e.g., potassium chromate—used in paint pigments and often combined with lead. Many chromates are designated as carcinogens. (FSL)

Chromatography A practical analytical methodology involving the separation of complex mixtures and the detection of each component of the mixture. Most forms of chromatography involve a stationary phase and a mobile phase. The mobile phase carries the analyte through the stationary phase, with each component having a different rate of travel, resulting in separation. The main forms of chromatography employed in the environmental sciences are gas chromatography, gas-liquid chromatography, ion-exchange chromatography, liquid-liquid partition chromatography, thin-layer chromatography, gas chromatography/mass spectroscopy (GC/MS), and high-performance liquid chromatography (HPLC). Various detectors, including flame ionization detectors, are used to analyze each component as it is eluted. Portable gas chromatographs are now routinely employed in the field. When combined with mass spectroscopy, they provide a powerful analytical technique for analysis of unknown substances. (DEJ)

Chromic acid (CrO_3) A brilliant-orange caustic chemical that reacts with organic matter. This chemical was widely used, in the past, for cleaning laboratory glassware. However, because of carcinogenic properties and environmental prohibitions against discharging the material into sanitary sewers, substitute chemicals are now being used. It is still used in chrome-plating operations and as a degreaser. (FSL)

Chromium (Cr) A hard, white metal, with an atomic weight of 51.996, that occurs as chromite ore and is used in the manufacture of stainless steel. (FSL)

Chromosome One of the threadlike bodies (normally 46 in humans) of chromatin that are found in the nucleus and that are the bearers of genes. (FSL)

Chronic carrier A person who continues to harbor an infectious agent without showing symptoms of the disease. Chronic carriers are possible in many illnesses. Salmonellosis is an example. *See also* Carrier. (HMB)

Chronic exposure **(1)** (Chemical) Continual exposure to low levels of a chemical over a long period of time, which can produce symptoms and disease. **(2)** (Radiation) Radiation exposure of long duration, by fractionation or protraction. Generally any dosage absorbed over a period of 24 hours or longer. *See also* Acute exposure. (FSL) (RMB)

Chronic toxicity The capacity of a substance to cause long-term poisonous human health effects. *See* Acute toxicity. (FSL)

Chronosequence (soil) A sequence of related soils that differ in properties, one from the other, primarily as a result of differences in the amount of time since they were formed. (SWT)

Cilia Short, hairlike processes on the surface of protozoans or certain metazoan cells that by their motion accomplish locomotion or produce a water current. (FSL)

Cimex Bedbugs of the Order Hemiptera that are bloodsucking insects that feed on animals and people. (FSL)

Cinnabar The principal ore of mercury (mercuric sulfide) found in shallow hydrothermal deposits. The mineral is used in scientific instruments such as thermometers and in electric watches. (FSL)

Circadian rhythm A biological activity that recurs in periods of about 24 hours under natural environmental conditions. Sleep patterns in mammals and leaf movements in some plants are examples of circadian rhythms. *See also* Biological clock. (SMC)

Circle graph A pictorial representation of data using a circle divided into proportional parts. The data are often reported in percentages. (VWP)

Circumaural protector A form of hearing protector commonly known as an earmuff, consisting of two cup-shaped devices that fit over the entire external ear, and are sealed against the side of the head. *See also* Earmuff. (DEJ)

Cistern A covered tank used for the collection and storage of water from a controlled catchment area. A controlled catchment is a defined surface area from which rainfall runoff is collected. A properly located and controlled catchment and cistern in combination with a filtration unit and disinfection device will provide water of safe quality. (FSL)

Citation A reference to a regulation or statute. Also, a description of a violation of a regulation or statute. For example, an OSHA compliance officer finding a worker exposed above the permissible exposure limit issues a citation to the employer for failure to comply with the applicable OSHA standard. The citation is in writing and describes with particularity the nature of the violation, including a reference to the provision of the chapter, rule, regulation, or order alleged to have been violated. Citations may or may not carry financial or criminal penalties. (DEJ)

Citizen suit Provision found in many environmental statutes that allows citizens to bring suit to enforce provisions of the laws and regulations. Suits generally may be brought against the government for failure to perform nondiscretionary duties, or against permit holders and polluters to enforce appropriate requirements. (DEW)

Citrine A type of quartz that is yellow because of the presence of impurities. (FSL)

Citronella Grass that is the source of a fragrant volatile oil used as an insect repellent. (FSL)

Civil action Generally, an adversary proceeding for the declaration, enforcement, or protection of a right, or the redress or prevention of a wrong. Distinguished from a criminal action. May be brought in either state or federal court systems, depending on the action. Many government actions enforcing environmental laws are civil actions. (DEW)

Clarification The removal of ''settleable'' and floating solids from sewage. It may involve any process or combination of processes whose purpose is to reduce the concentration of suspended matter in a liquid. It is a sedimentation of the more dense settleable solids from wastewater held under quiescent conditions. The solids are removed for separate processing. Floatable materials—e.g., oil, grease, and paper—are removed by ''skimmers'' for separate processing. Primarily, suspended and dissolved solids remain in the clarified wastewater. *See also* Primary wastewater treatment. (HMB)

Clarifier Any large circular or rectangular sedimentation basin used to remove settleable solids in wastewater. *See also* Clarification. (HMB)

Class **(1)** A category of grouping data. **(2)** The taxon in the classification hierarchy between phylum and order. **(3)** A specific type of chemical—for example, carcinogens. (SMC) (VWP)

Class action Generally, an action brought on behalf of other persons similarly situated. In federal court, a class action may be brought only if: **(1)** the class is so numerous that joinder of all members is impracticable, **(2)** there are questions of law or fact common to the class, **(3)** the claims or defenses of the representative parties are typical of the claims or defenses of the class, and **(4)** the representative parties will fairly and adequately protect the interests of the class.

In addition, one of the following conditions must also be met. **(1)** The prosecution of separate actions by or against individual members of the class would create a risk of: (a) inconsistent or varying adjudications with respect to individual members of the class that would establish incompatible standards of conduct for the party opposing the class, or (b) adjudications with respect to individual members of the class that would, as a practical matter, be dispositive of the interests of the other members not parties to the adjudication or would substantially impair or impede their ability to protect their interests. **(2)** The party opposing the class has acted or refused to act on grounds generally applicable to the class, thereby making appropriate final injunctive relief or corresponding declaratory relief with respect to the class as a whole. **(3)** The court finds that the questions of law or fact common to the members of the class predominate over any questions affecting only individual members, and that a class action is superior to other available methods for the fair and efficient adjudication of the controversy. (DEW)

Classification **(1)** (Biology) The process of dividing organisms into groups based on their similarities. Modern classification systems are hierarchical and reflect the evolutionary relations of organisms. Below is the general classification hierarchy listing the major taxa, with two examples, showing the classification of human beings and of Kudzu. Kingdom is the largest, most inclusive group and species is the smallest, most exclusive group. There are sometimes super- or subgroups associated with the major taxa—e.g., superphylum, subphylum.

Taxon	Example 1	Example 2
Kingdom	Animalia	Plantae
Phylum	Chordata	Tracheophyta

Taxon	Example 1	Example 2
Class	Mammalia	Angiospermae
Order	Primata	Dicotyledoneae
Family	Hominidae	Fabaceae
Genus	*Homo*	*Pueraria*
Species	*Homo sapiens*	*Pueraria lobata*

(2) (Soil) The systematic arrangement of soils into groups or categories based on their characteristics. Broad groupings are made on the basis of general characteristics, and subdivisions are made on the basis of more detailed differences in specific properties. (SMC) (SWT)

Class, soil Group of soils having a definite range in a certain property, such as texture, slope, acidity, structure, land use capability, degree of erosion, or drainage. (SWT)

Class width In mathematical computations, the difference between the lower class limit of a given class and the lower limit of the next-higher class. (VWP)

Claustrophobia Fear of closed or confined places. (FSL)

Clay As a soil separate, clay is a mineral soil particle less than 0.002 millimeter in diameter. As a soil textural class, clay soil is a soil material that is 40% or more clay, less than 45% sand, and less than 40% silt. (SWT)

Clayey soil Soil that contains large amounts of clay or that has properties similar to those of clay. (SWT)

Clay films Clay coatings on the surface of soil beds and mineral grains, and in soil pores. (SWT)

Clay mineral Naturally occurring inorganic material that usually has a crystalline structure. The particles are of clay size (less than 0.002-millimeter in diameter). (SWT)

Claypan A subsoil layer that has a much higher clay content than the overlying material, from which it is separated by a sharply defined boundary. Claypan layers are compact and slowly permeable, and are usually hard when dry and plastic and sticky when wet. (SWT)

Clean Free from soil, dirt, stain, or impurities. Soil must be effectively removed from surfaces because it provides nutrients for microorganisms to survive and grow. Soil can be removed using cleaning compounds and mechanical force, so that, ideally, a microbiologically clean environment results. Soil may deposit on uneven surfaces, in cracks and crevices, making it difficult to remove. Such deposits are complex in nature, making the choice of method of cleaning difficult. Not all cleaning and sanitizing can be done with a simple cleaner or sanitizer. Therefore, compounds are formulated for specific cleaning applications. An effective sanitation program cleans using a cleaning compound and sanitizer appropriate for the specific application, resulting in a surface free of soil and microorganisms. *See also* Cleaning compound; sanitizer. (HMB)

Cleanability Ease of cleaning and/or sanitizing. Equipment and utensils intended for use in food processing or food service are designed and fabricated of materials that facilitate the process of cleaning and sanitizing. That is, items should be designed so that all surfaces can be readily accessed with cleaning and sanitizing agents. Surfaces should be smooth and free of breaks, open seams, cracks, chips, pits, and difficult-to-clean internal corners and crevices. Construction should be chosen for durability under conditions of normal use and should be resistant to the action of cleaners or sanitizers, denting, buckling, pitting, chipping, and crazing. Equipment and utensils with such properties have good cleanability. (HMB)

Clean Air Act (CAA) Federal law, administered by the EPA, that regulates the emission of pollutants into the air in order to protect the public health and natural resources. It generally utilizes a three-pronged approach requiring adoption of: a control strategy for emissions from existing stationary sources of pollutants; technology-based performance standards limiting emissions from new stationary sources of pollutants; and tailpipe emission standards for mobile sources. Programs are implemented by the Environmental Protection Agency or approved state programs. 42 USC§7401 *et seq*; 40 CFR Parts 50–80. *See also* Ambient air; hazardous air pollutants; national ambient air quality standards; national primary ambient air quality standards; national secondary ambient air quality standards; national emission standard for hazardous air pollutants; prevention of significant deterioration; state implementation plan. (DEW)

Cleaned-in-place (CIP) systems Cleaning procedures can be employed either in a manual or in an automated system. A CIP system includes a fully controlled, consistent cleaning schedule that may be partially or fully automated, the purpose of which is to provide thorough cleaning and sanitizing of equipment surfaces without having to disassemble and reassemble pipes, tanks, or other pieces of equipment. Properly designed systems can clean and sanitize as effectively as cleaning by hand after dismantling. CIP systems have been used in dairy processing operations and in breweries for a long time. They may be used less frequently in other food processing plants because of cost limitations or difficulties encountered in their use with certain pieces of equipment. CIP systems are usually custom-designed and vary in complexity, ranging from simple timing devices to fully automated, computer-controlled systems.

There are two basic CIP designs: single-use and reuse systems. Single-use systems are generally small systems that use cleaning solution only once. Reuse systems recover and reuse cleaning compounds and solutions. A third alternative is to combine the elements of the two basic designs into a so-called multiuse system. In either CIP system a typical cleaning cycle would include: (a) preliminary rinse, (b) detergent wash, (c) rinse, (d) sanitization, and (e) final rinse (optional dependent upon the sanitizer used). (HMB)

Cleaned-out-of-place (COP) system Systems usually involving cleaning of small pieces of equipment and utensils that must be cleaned by dismantling and removed from their normal location in a processing plant. The advantage of COP equipment is that it can do an effective job of cleaning parts that are disassembled, as well as small equipment and utensils. A COP unit may be designed as a single-compartment vat that can be set up as a recirculating-parts washer, or used to sanitize utensils by extended soaking in sanitizer solutions. More frequently, the unit is set up with a double-compartment sink equipped with motor-driven brushes. The first compartment is used for cleaning with the brushes. Cleaned parts or utensils are rinsed with a spray nozzle in the second compartment, and are air dried in the unit or on a separate drying rack or drainboard. (HMB)

Cleaning The physical removal of both infectious and noninfectious agents and organic matter from surfaces and other locations suitable for the survival and multiplication of the agents. (FSL)

Cleaning compound A compound that cleans but does not disinfect and usually consists of several cleaning agents because no single component has all the functional properties necessary for effective cleaning. Cleaning compounds function in four basic steps: wetting and penetrating to bring the detergent in contact with the soil; displacing soil from the surface being cleaned; dispersion of the soil in the cleaning solution through emulsification, dispersion, or deflocculation; and prevention of redeposition of the soil onto the cleaned surface. Cleaning compounds are based on either alkaline or acid cleaning agents with added sequestrants and wetting agents. Sodium hydroxide (caustic soda) removes fats and proteins, and is effective in CIP systems, but is too harsh for hand cleaning. Sodium carbonate (soda ash) is an alkali often used for hand cleaning. Chlorinated alkalis are most commonly used for CIP system cleaning. Acid cleaners are most effective in removing mineral deposits. Sequestrants used in cleaning compounds are the polyphosphates, sodium hexametaphosphate, and sodium tetrametaphosphate. Wetting agents or surfactants function by lowering the surface tension of the cleaning solution in order to better penetrate the soil. Alkyl aryl sulfonates and sulfated alcohols are commonly used surfactants. *See also* Acid cleaner; alkali cleaner; sequestrant; surfactant; wetting agent. (HMB)

Cleanser A cleaning compound that is usually available in powder form and that contains an abrasive, a bleach, and a surfactant. The bleach aids in stain removal and may disinfect, the surfactant acts to facilitate suspension of soil, and the abrasive provides scouring and polishing properties. The particular cleanser selected must be one that provides the type of cleaning action needed. Many formulations are commercially available. (HMB)

Cleanup An action taken to deal with a release or threat of release of a hazardous substance that could affect humans and/or the environment. The term *cleanup* is sometimes used interchangeably with the terms *remedial action*, *removal action*, *response action*, or *corrective action*. (FSL)

Clean Water Act (CWA) Federal law (originally the Federal Water Pollution Control Act, FWPCA), administered by the EPA, that is intended to restore and maintain the chemical, physical, and biological integrity of the nation's waters, by regulation of the discharge of pollutants into water. Regulation is generally carried out through the imposition of technology-based effluent control standards and water-quality-based standards. Establishes the National Pollutant Discharge Elimination System (NPDES), which requires dischargers to obtain permits issued by either the federal government or an authorized state program that set out the technology and water quality standards applicable to the discharge. Establishes standards of performance that apply to new sources of pollutant discharge (NSPD), and effluent standards that apply to specified toxic pollutants. Also sets out pretreatment standards for discharges to publicly owned treatment works (POTW) for pollutants that are not susceptible to treatment or would interfere with the operation of the treatment works.

For the discharge of dredged and fill material, the act requires permits that are issued by the Army Corps of Engineers. This program, found at section 404 of the CWA, is the basis for the protection of wetlands.

The Act further establishes a mechanism for responding to the release of oil or hazardous substances into specified waters, and that is implemented by the National Oil and Hazardous Substances Pollution Contingency Plan (40 CFR Part 300). Specifies reportable quantities for discharge of dredged material, and limits hazardous-substances releases (40 CFR Parts 104–110, 400–470, 33 USC §1251–1387). *See also* Discharge of a pollutant; dredged material; effluent limitations; water quality standards; pollutant; toxic pollutants; discharge of filled material; new source; national pollutant discharge elimination system; point source; publicly owned treatment works; fill material; wetlands; navigable waters; national contingency plan; standards of performance. (DEW)

Clear cut A forest management technique that involves harvesting all the trees in one area at one time. (FSL)

Cleistogamous A self-pollinated flower that never opens. Basal flowers in species of violets are cleistogamous. (SMC)

Climax The final successional stage or sere for a certain habitat. A climax community remains unless there is a major perturbation, such as fire or a hurricane. The concept of a linear successional progression culminating in a climax community is common but there is considerable controversy about whether such climax communities are real and predictable. (SMC)

Cline A gradual change in population characteristics within a species over a geographic area, correlated with some gradual change in climate or other environmental conditions. (SMC)

Clinical Pertaining to the direct treatment and observation of patients as distinguished from experimental or laboratory study. (FSL)

Clinical laboratory A workplace where diagnostic or other screening procedures are performed on blood or other potentially infectious materials. (FSL)

Clod (Soil) A compact, coherent mass of soil varying in size, usually produced by the human actions (plowing, digging, etc.), especially when soils are either too wet or too dry. Clods are usually formed by compression or breaking off from a larger unit, as opposed to a building-up action, as in aggregation. (SWT)

Clonal Describes a group of plants with a relationship in which all plants come from one rootstock, rhizome, stolon, or root system. (SMC)

Clone A population of individuals descended from a single ancestor by asexual reproduction (mitotic division or vegetative propagation). (SMC)

Closed-loop recycling Reclaiming or reusing wastewater for not-potable purposes in an enclosed process. (FSL)

***Clostridium perfringens* food poisoning** An illness caused by the ingestion of food contaminated with the bacterium *Clostridium perfringens*. Clostridia are anaerobic, Gram-positive, spore-forming rods that are ubiquitous in nature and can frequently be recovered from meats and meat products. *C. perfringens* produces five distinct toxins, designated A–E; types A and C cause gastroenteritis. The enterotoxin causing gastroenteritis is thought to be spore-specific. The incubation period is approximately 12 hours, with a range of 8–24 hours. Common symptoms include abdominal cramps, diarrhea, and nausea. Typically, vomiting and fever are absent. The course of the disease is short, 12–24 hours, with mortality rare, except in the elderly, where deaths have occurred.

The foods implicated are meats or meat dishes prepared in large quantities, which allows anaerobic conditions and appropriate temperature conditions that permit the germination of spores that have survived cooking. Subsequent improper cooking time and temperature result in a *C. perfringens* foodborne outbreak. Many enterotoxin positive strains are heat resistant. Spores are highly resistant to conventional methods of cleaning and sanitation. (HMB)

Cloud chamber A device for observing the paths of ionizing particles, based on the principle that supersaturated vapor condenses more readily on ions than on neutral molecules. (FSL)

Coagulation The process of changing from a fluid into a thickened mass; a curdling or congealing, as with the formation of a clot of blood. *For water treatment application, see* Coagulation-flocculation. (FSL)

Coagulation-flocculation In water treatment, coagulation is the process of forming flocculent particles in a liquid by the addition of chemicals. Coagulation is often achieved by the use of hydrated aluminum sulfate (alum), whereby the chemical is mixed with turbid water and allowed to remain still. The suspended particles will combine physically and form a ''floc'' or mat, which will eventually settle to the bottom of the basin or which can be removed by filtration, as in swimming pools. (FSL)

Coal tar A black viscous liquid with a naphthalene-like odor that is obtained by the destructive distillation of bituminous coal, and that is used as a raw material for dyes, solvents, and many other products. Because coal tar contains many carcinogens, some uses of this material—e.g., as a disinfectant and in skin treatment—have been eliminated. (FSL)

Coastal zone Lands and waters adjacent to the coast that exert an influence on the uses of the sea and its ecology, or, inversely, whose uses and ecology are affected by the sea. (FSL)

Cobalt (Co) A hard, silvery-white element, with an atomic weight of 58.9332, that occurs in combination with sulfur and arsenic. It is used in alloys, and to produce shades of blue in glass and ceramics, and in cobalt-chromium high-speed-tool steels. The dust is flammable and toxic by inhalation. (FSL)

Cocarcinogen A substance that works symbiotically with a carcinogen in the development of cancer. (FSL)

Coccidioidomycosis A fungal disease caused by the organism *Coccidioides immitis*, which grows in soil and artificial media. The disease is characterized by lung lesions and abscesses throughout the body. The disease affects people, dogs, horses, and cattle. The organism can be found in acid and semiarid areas of the world, particularly where the proper temperature, moisture, and soil conditions exist. (FSL)

Coccus Spherical bacterial cells. The cocci (pl.) may appear singly, in pairs called diplococci, in chains, as is characteristic of the genus *Streptococcus*, or in grape-like clusters, as is typical of the genus *Staphylococcus*. (HMB)

Code of Federal Regulations (CFR) A compilation of the regulations of the various federal agencies. Volumes are divided into 50 titles according to subject matter. Titles are divided into chapters, which are, in turn, divided into parts and sections. Regulations promulgated by federal agencies are initially published in the *Federal Register*. *See also Federal Register*. (DEW)

Coefficient of haze (COH) A measurement of visibility interference in the atmosphere. (FSL)

Cohort Any defined group of persons selected for a special purpose or study. (FSL)

Cohort study *See* Prospective study.

Coincidence (Radiation) The occurrence of ionizing events in one or more detectors simultaneously or within an assignable time interval, as a result of the passage of a single particle or of several generically related particles. (RMB)

Coliform index A rating of the purity of water based on a count of fecal bacteria. (FSL)

Coliforms Any of a group of bacteria that have the following characteristics: aerobic or facultatively anaerobic, Gram-negative, non-spore-forming rods that ferment lactose, forming acid and gas within 48 hours at 35°C. Representatives of 20 or more species may conform to the definition for the coliform group. The organisms comprising this group are used routinely in laboratories to monitor foods and water supplies, environmental waters, or treated wastewater. They are used to measure treatment effectiveness, or, in the case of foods or water supplies, the possible presence of fecal contamination. The group includes organisms of both fecal and nonfecal origin. It is necessary to consider separately those of fecal and nonfecal origin in order to assess the public health significance of their presence. Food or water-safety regulations may be based in part on the maximum number of coliforms permitted. *See also* Fecal coliforms; fecal streptococci; indicator organisms. (HMB)

Collagen A filarous protein that constitutes one of the main skeletal substances in many animals. It is a component of cartilage, bone, and connective tissues in vertebrates. (FSL)

Collateral estoppel Legal doctrine that indicates that facts determined in a litigation action are conclusive with respect to the same parties in a subsequent suit. *See also* Res judicata. (DEW)

Collection efficiency The fraction of the contaminant removed from the environmental medium (e.g., air, water, soil) passing through a sampling device. Ideally, the collection efficiency of environmental sampling instrumentation should be constant and close to 100%, although this is only rarely achieved. Collection efficiency is known to vary for certain air contaminants by volume of air sampled, relative humidity, atmospheric pressure, airflow rate, temperature, and other factors that, when combined, describe the sampling and analytical error associated with each air-sampling technique. (DEJ)

Collector The person or persons who, under verbal or written agreements, with or without compensation, does the work of collecting and/or transporting solid wastes from industries, offices, retail outlets, businesses, institutions, residential dwellings, and similar locations. *See also* Person. (FSL)

Collier's disease *See* Anthracosis; black lung disease, Shaver's disease. (FSL)

Collimation The confining of a beam of particles or rays to a defined cross-section. (RMB)

Collision, elastic A billiard-ball-type collision in which energy and momentum are conserved and in which there is no change in the internal energy of the colliding objects. (RMB)

Collision, inelastic Collision in which at least one system gains internal excitation energy at the expense of the total kinetic energy of the center-of-gravity motion. (RMB)

Colloid, soil Inorganic and organic matter with very small particle size and a correspondingly large surface area per unit mass. (SWT)

Colluvium A deposit of rock fragments and soil material that accumulates at the base of a steep slope as a result of gravity. (SWT)

Colonization Propagation of a microorganism on or within a host without causing cellular injury. A colonized host can serve as a source of infection. (FSL)

Colonizing species A species whose members colonize new areas, regions, or habitats. Typical colonizers are easily dispersed into new areas and can germinate or grow under a wide range of conditions to establish new populations. Pioneer species are often colonizing species. *See also* Pioneer species. (SMC)

Colony **(1)** A stand, group, or population of plants of one species; **(2)** a population of organisms—e.g. bacteria of a single species—that have been introduced into a new area or habitat. (SMC)

Colustrum The milk that is produced the first few days after a mammal has given birth and that is rich in proteins and antibodies. (FSL)

Combinations The number of combinations of n things taken K at a time is:

$$nC_K = \binom{n}{K} = \frac{n!}{K!(n-K)!}$$

It represents the number of ways of selecting K items from n items wherein the order of selection does not matter. *See also* Factorial. (VWP)

Combined sewers A sewer system that carries both sewage and storm-water runoff. Normally, its entire flow goes to a waste treatment plant, but during a heavy storm the storm-water volume may be so great as to cause overflows. When this happens, untreated mixtures of storm water and sewage may flow into receiving waters. (FSL)

Combustible Refers to any gas, vapor, or liquid that has a flash point at or above 100°F (37.8°C) but less than 200°F (93.3°C). (FSL)

Combustible dust An airborne concentration of finely dispersed particles having a large ratio of surface area to mass, and often generated through mechanical means, such as grinding and pulverizing. When exposed to a sufficient energy source, such materials will ignite. The ignition of the dust cloud will be propagated throughout its entirety and result in an explosion only if the dust concentration in air is between certain lower and upper explosive limits. The explosive limit for many dusts is approximately 0.1 ounce per cubic foot of air, roughly equivalent to 100,000 milligrams per cubic meter. This concentration is 10,000 times the threshold limit value of 10 milligrams per cubic meter recommended by the American Conference of Governmental Industrial Hygienists (ACGIH) for nuisance dust. (MAG)

Combustible-gas meter Also known as a combustible-gas instrument, this is a device that measures the concentration of combustible gases in air, usually expressed as a percentage of the lower explosive limit (LEL). Typically, the instrument consists of a heating element, a catalytic agent that permits combustion at a low activation energy, a flashback arrestor, electronic circuitry (Wheatstone bridge circuit) to display and amplify the combustion energy produced, and an air pump. It is used to determine the presence or absence of combustible vapors or gases in confined spaces prior to entry. Such devices are not useful for measuring concentrations of combustible or explosive dusts, or for measuring concentrations of combustible or explosive dusts, or for measuring levels of toxic gases or vapors. Because the lower explosive limit is always greater than the permissible exposure limit, such devices are also dependant on the presence of adequate oxygen to support combustion. (DEJ)

Combustion An exothermic, self-sustaining reaction involving a liquid, solid, and/or gas fuel; the reaction, which takes place at a high temperature, of a substance with oxygen, and that produces heat as a by-product. *See also* Endothermic; exothermic. (FSL)

Combustion product Substance produced during the burning or oxidation of a material. (FSL)

Command post Facility that is located at a safe distance upwind from an accident site and at which the on-scene coordinator, responders, and technical representatives can make response decisions. (FSL)

Comment period Time provided for the public to review and comment on a proposed action or rule after it is published in the *Federal Register*. (FSL)

Comminuter A shredder installed in a wastewater treatment process as part of preliminary treatment. The device is installed directly in a flow channel, usually immediately after bar screens. The comminuter/shredder cuts solids in the wastewater to about 0.25 inch. The shredded solids are removed from the waste stream and landfilled or returned to the wastewater. (HMB)

Comminution Mechanical shredding or pulverizing of waste. Used in solid waste management and wastewater treatment. (FSL)

Commission of European Communities (CEC) An organization that established air pollution criteria and levels of action similar in many respects to those set by the World Health Organization since 1976. (DEJ)

Common law As distinguished from laws created by legislation, the body of principles and rules of action that are derived from usage and customs or from the judgments and decrees of the courts recognizing, affirming, and enforcing such usages and customs. (DEW)

Communicability (period of) The interval during which a person or animal with an infectious disease is a potential source of infection. (FSL)

Communicable disease An illness that is caused by a specific infectious agent or its toxic products, and that arises through transmission of that agent or its products from a reservoir to a susceptible host. Such transmission can be either direct, as from an infected person or animal, or indirect, through the agency of an intermediate plant or animal host or vector, or the inanimate environment. (FSL)

Community **(1)** All the organisms inhabiting a common environment and interacting with one another; **(2)** an association of interacting populations defined by the nature of their interaction or the place in which they live. (SMC)

Community Awareness and Emergency Response Program (CAER) A program developed by the Chemical Manufacturer's Association to assist chemical plant manufacturers in cooperating with local communities to develop integrated plans for responding to hazardous-materials release. (FSL)

Community ecology **(1)** The ecology of sympatric organisms or populations within an identifiable habitat or community; **(2)** the study of the interactions between groups of organisms or populations within a particular common environment or habitat. Community ecology usually focuses on interactions with the biotic environment rather than interactions with the abiotic environment. *See also* Synecology. (SMC)

Community Health and Environmental Surveillance System (CHESS) A series of studies conducted by the U.S. Environmental Protection Agency to evaluate the health effects of ambient air pollution in a quantitative fashion among large populations. (DEJ)

Community sewerage system A system for the collection and disposal of sewage, serving two or more properties and including the various devices for the treatment of such sewage. (FSL)

Compaction Reduction of the bulk of solid waste by rolling and tamping. (FSL)

Compatibility As applied to chemicals, the capability of substances to be mixed without undergoing destructive chemical changes or exhibiting mutual antagonism. As applied to chemicals, the ability of materials to be stored in close contact without a chemical reaction occurring. (FSL)

Competent person According to OSHA, an individual who is capable of identifying existing and predictable hazards in surroundings or working conditions that are unsanitary, hazardous, or dangerous to employees, and who has authorization to take prompt corrective measures to eliminate them. (SAN)

Complaint The initial pleading filed by a plaintiff in a civil action. In federal practice, the complaint must set out in a short and plain statement: **(1)** grounds for the court's jurisdiction, **(2)** a claim showing plaintiff is entitled to relief, and **(3)** a demand for judgment by the court for that relief. FRCP, Rule 8(a). (DEW)

Complement of an event The set of all outcomes that do not belong to an event. The complement of event A is denoted $\bar{A}$. (VWP)

Compliance coating A coating whose volatile organic compound content does not exceed that allowed by regulation. (FSL)

Compost A mixture of garbage and degradable trash with soil in which certain bacteria in the soil break down the garbage and trash into organic fertilizer. Organic residues, or a mixture of organic residues and soil, that have been mixed, piled, and moistened, with or without the addition of fertilizer materials and lime, and generally allowed to undergo biological decomposition. The final product can be used as a soil amendment. *See also* Garbage. (FSL, SWT)

Compound Chemical combination of two or more elements combined in a fixed and definite proportion by weight. (FSL)

Compound event An event formed by combining two or more simple events. (VWP)

Comprehensive emergency response plan Under the Emergency Planning and Community Right-to-Know Act, a plan prepared by the local emergency planning committee to establish procedures for local response to hazardous-substance emergencies. The plan must include identification of covered facilities and likely routes of transport of specified hazardous substances, response procedures, designation of coordinators, notification procedures, methods for determining occurrences and likely affected areas, description of available equipment and facilities, evacuation plans, training programs, and methods and schedules for exercising the emergency plan. 42 USC §11003. *See also* Emergency Planning and Community Right-to-Know Act; Emergency Response Commission; local emergency planning committee. (DEW)

Comprehensive Environmental Response, Compensation, and Liability Act (CERCLA) Federal law enacted in 1980 and substantially amended in 1986 by the Superfund Amendments and Reauthorization Act (SARA), to provide a means to clean up and otherwise respond to releases of hazardous substances into the environment, primarily from past disposal practices.

Responses are generally referred to as removal actions or remedial actions, and are carried out by the government and funded by the Hazardous Substance Superfund, or are carried out and funded by the responsible parties. Response activities must be in accordance with the National Contingency Plan (40 CFR §300 *et seq.*). Money expended from the Superfund is recoverable from responsible parties. Liability applies under a standard of strict liability, and multiple parties are joint and severally liable with a statutory right of contribution. Responsible parties may include: the current owner or operator of the site, any past owner or operator at the time hazardous substances were disposed, any transporter of hazardous substances for disposal or treatment, or any person who arranged for the transport, disposal, or treatment of hazardous substances in his or her possession. Defenses to liability are limited to an act of God, an act of war, and/or an act of a third party not under contract to the defendant, if defendant took reasonable precautions against these acts and exercised due care.

EPA may also order private parties to undertake response actions when there may exist an ''imminent and substantial endangerment to the public health or welfare of the environment because of an actual or threatened release of a hazardous substance from a facility.''

Other provisions require the creation of a national priorities list (NPL) of known releases or threatened releases, establishes reportable quantities (RQ), which are regulations setting the quantity of a specific hazardous substance the release of which shall be reported to the National Response Center. This act also established the Agency for Toxic Substances and Disease Registry within the Public Health Service of the Department of Health and Human Services, to implement health-related authorities of ths statute 42 USC §9601 *et seq*. *See also* Agency for Toxic Substances and Disease Registry; applicable or relevant and appropriate requirements; CERCLIS; feasibility study, National Contingency Plan; pollutant or contaminant; preliminary assessment; record of decisions; remedial action; remedial investigation; remedial actions reportable quantity; site inspections; technical assistance grant. (DEW)

Comprehensive Environmental Responsibility, Compensation and Liability Information System (CERCLIS) EPA's comprehensive data base and management system that inventories and tracks releases addressed or needing to be addressed by the Superfund program. Contains the official inventory of CERCLA sites and supports EPA's site planning and tracking functions. Inclusion of a specific site or area in the CERCLIS data base does not represent a determination of any party's liability, nor does it represent a finding that any response action is necessary. 40 CFR §300.5. (DEW)

Comprehensive general liability Type of insurance policy that provides indemnification for damages owed to third parties for injury or property damage caused by an occurrence. Such policies usually contain a pollution exclusion clause that excludes from coverage damage caused by the discharge or release of pollutants, unless the discharge or release is sudden and accidental. (DEW)

Compressed gas A gas or mixture of gases having, in a container, an absolute pressure exceeding 40 psi at 70°F (21.1°C), or a gas or mixture of gases having, in a container, an absolute pressure exceeding 104 psi at 30°F (54.4°C), regardless of the pressure at 70°F (21.1°C), or a liquid having a vapor pressure exceeding 40 psi at 100°F (37.8°C), as determined by the American Society for Testing Materials document D-323-72. (FSL)

Compressibility, soil The property of a soil pertaining to its susceptibility to decrease in volume when subjected to load. (SWT)

Compton effect The interaction of a photon with matter wherein part of the photon energy is transferred to an orbital electron of an atom, the photon proceeding with altered direction and diminished energy. The electron is excited and ejected from the atom. The Compton effect is important in the attenuation of X-rays or gamma rays, being especially effective for low- and intermediate-energy photons and for light-shielding materials. (RMB)

Compton electron An electron of increased energy ejected from an atom as the result of a Compton interaction with a photon. *See also* Compton effect. (RMB)

Computer packages A collection of computer programs that can be used to perform tedious calculations *See also* Statistical Packages. (VWP)

Concealment Generally, the withholding of something known that one has a duty to reveal. For purposes of insurance law, concealment pertains to the intentional withholding of a fact that is material to the risk and that the insured should reveal in good faith to the insurer. (DEW)

Concrete A mixture of cement, water, and aggregate materials bound into a hardened mass. The active material in this mixture is cement, which was invented in the 1820s by Joseph Aspden. *See also* Portland cement. (FSL)

Concretion (soil) A solidified mass of a chemical compound, such as iron oxide or calcium carbonate, that is in the form of a nodule or grain. Concretions vary in size, color, shape, and hardness. (SWT)

Condensed phase detonation Detonation of condensed-phase materials such as high explosives and propellants as a result of decomposition. Examples of condensed-phase materials include acetylene, hydrogen, certain metallic azides, and ammonium nitrate fertilizer. *See also* Explosion. (SAN)

Condenser R-meter An instrument consisting of an "air wall" ionization chamber together with auxiliary equipment for charging and measuring its voltage; it is used as an integrating instrument for measuring the quantity of X-radiation or gamma radiation, in roentgens. *See also* Ionization chamber. (RMB)

Conditionally exempt generator One of three types of hazardous-waste generators, under the Resource Conservation Recovery Act. In summary, a generator is considered conditionally exempt if, in a calendar month, it does not accumulate more than: **(1)** 1 kilogram of acutely hazardous waste, **(2)** 100 kilograms of a spill residue generated from the cleanup of an acutely hazardous waste, or **(3)** 1000 kilograms of hazardous waste. (FSL)

Conditional probability If A and B are events, the probability that the event A occurs, given that the event B occurs. The probability is denoted $P(A/B)$. (VWP)

Conditional registration The Federal Insecticide, Fungicide, and Rodenticide Act (FIFRA) permits registration of pesticide products that is conditional upon the submission of additional data. These special circumstances include a finding by the EPA administrator that a new product or use of an existing pesticide will not significantly increase the risk of unreasonable adverse effects. (FSL)

Conduction The transfer of heat by direct contact from one body to another. (FSL)

Conductive hearing impairment A loss of hearing due to interference with the transmission of sound to the cochlea (inner ear). Often due to wax in the external auditory canal, large perforation of the eardrum, blockage of the Eustachian tube, interruption of the ossicular chair due to trauma or disease, fluid in the middle ear, or fixation of the stapedial footplate (otosclerosis). (DEJ)

Conductive hearing loss A physical defect or condition of the outer or middle ear that interferes with the passage of sound. This can be the end result of physical obstruction within the outer or middle ear, a birth defect, the aging process, or disease, all of which can affect the conversion of sound energy into mechanical energy. This type of hearing loss involves a reduction in the perception of loudness and not in clarity. (MAG)

Conductive heat load The transfer of heat from one point to another within a body, or from one body to another when both bodies are in physical contact with each other, in the absence of motion in the medium. (DEJ)

Conductivity The ability of a mixture or substance to transfer heat or electricity. This property is the reverse of resistivity. (FSL)

Cone index (soil) The force per unit basal area that is required to push a cone penetrometer through a specified increment of soil. (SWT)

Cone of depression A conical depression in a water table surrounding a well. (FSL)

Cone penetrometer A cylindrical instrument that has a cone-shaped tip designed for penetrating soil to measure the end-bearing component of penetration resistance. The resistance to penetration developed by the cone equals the vertical force applied to the cone divided by its horizontally projected area. (SWT)

Confidence coefficient The probability that a selected value lies within a specified neighborhood of the parameter being estimated. *See also* Neighborhood. (VWP)

Confidence interval An interval estimate with a specified level of confidence. (VWP)

Confined aquifer *See* Aquifer. (FSL)

Confined space Any space that has a limited or restricted means of entry or exit, and that is subject to the accumulation of toxic or flammable contaminants or has an oxygen-deficient atmosphere. Confined or enclosed spaces include, but are not limited to, storage tanks, process vessels, bins, boilers, ventilation or exhaust ducts, sewers, silos, hoppers, mixing tanks, underground utility vaults, tunnels, pipelines, and excavations more than 4 feet in depth, such as pits, tubs, vaults, and vessels. Confined spaces are not designed for continuous occupancy. Confined spaces must be large enough and so configured that an employee can enter and perform assigned work. Where confined spaces are categorized, there are two levels of classification: permit-required confined space and low-hazard permit space. Also called an enclosed space. *See also* Excavation; low-hazard permit space; permit-required confined space. (SAN)

Conflict of laws A set of principles by which the court determines which laws are applicable to a controversy when more than one state or jurisdiction may be involved. Also referred to as choice of law. (DEW)

Confluence The juncture or flowing together of two or more surface-water streams, rivers, etc. (FSL)

Confounding variable A variable that may be associated with success or failure of an event. Such a variable must usually be controlled before consideration of the probability of an event. (VWP)

Congenital Refers to certain mental or physical traits, abnormalities, malformations, or diseases that may be either inherited or due to an influence that occurred between conception and birth. (FSL)

Conjunctivitis The inflammation of conjunctiva—i.e., the mucous membrane lining the inner surface of the eyelids and covering the front part of the eyeball. (FSL)

Connector As related to fall protection, a device used to couple parts of a system together. It may be an independent component of the system (such as a carabiner), or an integral component or part of the system (such as a buckle or D-ring sewn into a body belt or body harness, or a snap-hook spliced or sewn to a lanyard or self-retracting lanyard). *See also* D-ring; fall protection; self-retracting lanyard; safety belt; safety harness; snap-hook. (SAN)

Consent decree In settlement of a dispute, a negotiated agreement entered as an order of the court and enforceable by the plaintiff in that court. (DEW)

Conservation The protection of human and natural resources or habitats. Generally, conservation implies some use or management of resources. (SMC)

Consistence (soil) Combination of soil properties that determine resistance to crushing and the ability of soil to be molded or change shape. Terms that describe consistence include *loose*, *friable*, *firm*, *soft*, *plastic*, and *sticky*. (SWT)

Consistency (soil) The degree of cohesion or adhesion of the soil mass. Terms used for describing consistency at various soil moisture contents are: **(1)** (for wet soil) nonsticky, slightly sticky, sticky, very sticky, nonplastic, slightly plastic, plastic, and very plastic; **(2)** (for moist soil) loose, very friable, friable, firm, very firm, and extremely firm; **(3)** (for dry soil) loose, soft, slightly hard, hard, very hard, and extremely hard; **(4)** (for cementation) weakly cemented, strongly cemented, and indurated. (SWT)

Constructional Surface A land surface retaining the form or forms developed by the process of deposition. (SWT)

Construction/demolition waste Waste building materials and rubble resulting from construction, remodeling, repair, and demolition operations on pavement and on houses, commercial buildings, and other structures. Such wastes include, but are not limited to, wood, bricks, metal, concrete, wallboard, paper, cardboard, inert waste landfill material, and other nonputrescible wastes that have a low potential for groundwater contamination. *See also* Refuse; rubbish. (FSL)

Consumer *See* Heterotroph. (SMC)

Consumer Product Safety Act of 1972 An Act of Congress that created the Consumer Product Safety

Commission and regulates consumer products that may pose unreasonable risk of injury to the public. It mandates proper labelling and use limitations, and gives the authority to ban products. The act also requires manufacturers to notify the public of the risk and to repair, replace, or refund the purchase price of the product affected. Codified at 15 USC 2051. (DEJ)

Contact A person or animal that has been in such association with an infected individual or a contaminated environment as to have had an opportunity to acquire the etiologic agent. (HMB)

Contact dermatitis A delayed type of induced sensitivity of the skin resulting from cutaneous contact with a specific allergen. (FSL)

Contact herbicide *See* Herbicide.

Contact insecticide An insecticide that kills the target pest by penetrating the cuticle or blocking the insect's spiracles or breathing ports. *See also* Contact pesticide; pesticide. (FSL)

Contact irritants Chemicals that produce visible signs of skin or eye irritation, including erythema, edema, vesicles, and blanching (whitening). Often produce pain, soreness, tingling, itching, and the sensation of heat or cold. Rubber, plastics, resins, glues, cement, oil, and organic solvents such as trichloroethylene, acetone, and petroleum distillates are all contact irritants. (DEJ)

Contact pesticide A chemical that kills pests when it touches them, rather than by being eaten (as is the case with stomach poisons). Also, soil that contains the minute skeletons of certain algae that scratch and dehydrate waxy-coated insects. *See also* Contact insecticide; pesticide. (FSL)

Contact plate *See* Agar contact plate. (HMB)

Contaminant **(1)** Foreign material not usually found in a substance. **(2)** Any physical, chemical, biological, or radiological substance or matter that has an adverse effect on air, soil, or water. (FSL)

Contaminated sharps Any contaminated objects that can penetrate the skin, including, but not limited to, needles, scalpels, broken glass, broken capillary tubes, and exposed ends of dental wires. (FSL)

Contamination **(1)** (Radiation) The deposition of radioactive material on the surfaces of structures, areas, objects, or personnel. This material can consist of fallout in which fission products and other weapon debris have become incorporated with particles of dirt, etc. Contamination can also arise from the radioactivity induced in certain substances by the action of neutrons. **(2)** The presence of an etiologic agent on a body surface, or on or in clothing, bedding, toys, surgical instruments, dressings, and other inanimate articles or substances, including water, milk, and food. **(3)** Action by which something is made impure. The process of introducing impurities into a substance or an environment. In microbiology, it refers to the presence of living microorganisms in or on a material. However, the term may also be used with regard to the presence of unusually unwanted chemicals or radioactive substances. Sanitation measures attempt to reduce levels of unwanted microorganisms, chemicals, or radioactive materials to an acceptable level to protect the public health. *See also* Decontamination; fallout; induced radioactivity; weapons debris. (HMB) (RMB)

Contingency plan A document setting out an organized, planned, and coordinated course of action to be followed in case of a fire, explosion, or other accident that releases toxic chemicals, hazardous wastes, or radioactive materials. (FSL)

Contingency table An arrangement of data into a two-way classification. The data are sorted into cells and the number of data in each cell is reported. The contingency table involves two variables. The common use of such tables is to determine whether the data indicate dependence or interdependence of the variables. (VWP)

Continuous monitoring Usually refers to air sampling or radiation monitoring conducted at locations where leaks may occur, or where hazardous materials are handled in high quantities. The monitoring device may be interfaced with an alarm system to warn workers or nearby residents to evacuate. (DEJ)

Continuous tape sampler An air-sampling device that collects a contaminant on moving paper tape that may or may not be chemically impregnated, depending on the type of analysis to be performed. Analysis may be performed by attenuation or reflection of light, or may be colorimetric. (DEJ)

Contour An imaginary line connecting points of equal elevation on the surface of the soil. A contour terrace is laid out on sloping soil at right angles to the direction of the slope and is nearly level throughout its course. *See also* Contour plowing. (SWT)

Contour plowing Farming methods that break ground following the shape of the land in a way that discourages erosion. *See also* Contour. (FSL)

Contract labs Laboratories under contract to the Environmental Protection Agency that analyze samples taken from wastes, soil, air, and water, or carry out research projects. (FSL)

Contractor owned/contractor operated facility (COCO or [Equipment] COCOE) A non-government-owned, privately operated facility that provides goods and/or services to a federal agency under contract, and that may be furnished government equipment to manufacture a product or provide a service. (DEW)

Contrails Long, narrow clouds caused when high-flying jet aircraft disturb the atmosphere. (FSL)

Contribution The right of one liable party to force other liable parties to reimburse him or her for the share of the compensation he or she has paid. (DEW)

Contributory negligence Conduct by a plaintiff who has initiated a tort action for negligence and who was also negligent and contributed in causing the harm at issue. *See also* Assumption of Risk. (DEW)

Controlled area (radiation) A defined area in which the occupational exposure of personnel to radiation or radioactive material is under the supervision of an individual responsible for radiation protection. (RMB)

Controlled catchment *See* Cistern. (FSL)

Control rod A rod used to control the nuclear power of a nuclear reactor. The reactor functions through the fission of nuclear fuel by neutrons. The control rod absorbs neutrons that would normally produce fission in the atoms of the fuel. Pushing the rod in reduces the release of nuclear power. Pulling the rod out increases it. Another type of control rod operates with a rotary motion. (RMB)

Convalescent carrier An individual who harbors an infectious agent for various periods of time after recovering from apparent symptoms of the illness. For example, stool specimens of such individuals may be positive for salmonellae for up to 10 weeks in some cases. *See also* Carrier. (HMB)

Convection The transfer of heat from one place to another by moving fluid (a gas or liquid). Natural convection results from differences in temperature. (DEJ)

Convective heat load The amount of heat energy transferred between the skin and the air. Human skin is normally 95°F (35°C). Air in excess of that temperature will warm the body, whereas air below that temperature will cool the body. (DEJ)

Conventional pollutants Statutorily listed pollutants that are well-understood by scientists. These pollutants may be in the form of organic waste, sediment, acid, bacteria and viruses, nutrients, oil and grease, or heat. (FSL)

Conventional wastewater treatment Well-established wastewater treatment process that excludes advanced or tertiary treatment methods. Conventional processes usually include primary and secondary treatment. *See also* Primary wastewater treatment; secondary wastewater treatment. (HMB)

Converter A nuclear reactor that converts fertile atoms into fuel by neutron capture. Two different ways of using the words "converter" and "breeder" are in use at present. **(1)** A converter is a reactor that uses one kind of fuel and produces another—e.g., it may consume ^{235}Uranium and produce plutonium. **(2)** A breeder is a converter that produces more fissionable atoms than it consumes. (RMB)

Coolant A liquid or gas used to reduce the heat generated by power production in nuclear reactors, electric generators, various industrial and mechanical processes, and automobile engines. (FSL)

Cool flame The visual phenomenon associated with the low-temperature oxidation of an organic material. It is usually accompanied by temperature and pressure changes that are small compared to those accompanying normal or hot-flame ignition and is often referred to as a partial or intermediate oxidation reaction. (FSL)

Cooling table A table designed with its own refrigeration unit intended for the display and service of food items that must be maintained at low temperatures. They are designed to maintain temperatures of 45°F or lower, as specified by food service regulations. These tables are usually intended for use in buffet-style or other similar customer self-service food service operations. (HMB)

Cooling tower A structure that helps remove heat from water used as a coolant—e.g., in electric-power-generating plants. (FSL)

Cooperative agreement A financial assistance mechanism used by the federal government in lieu of a grant when substantial federal programmatic involvement during performance of the activities is anticipated. *See also* Grant. (DEW)

Copper An element with an atomic weight of 63.54 that is red, malleable, and a good conductor of electricity. Copper occurs as the element, as cuprite (Cu_2O), and as copper sulfate (Cu_2S). It is used in electrical wires and in many alloys. (FSL)

Coppice mound A small mound (microrelief feature) of stabilized soil around desert shrubs. (SWT)

Coprolite Fossilized scat or feces. (SMC)

Coprophage An organism that feeds on excrement or feces. (SMC)

Core The heart of a nuclear reactor, where the nuclei of the fuel undergo fission (split) and release energy. The core is usually surrounded by a reflecting material that bounces stray neutrons back to the fuel. (RMB)

Coriolis force The force created by the rotation of the earth in relation to objects at or above the surface. The Coriolis force, plus the heating and cooling of water and air, drives the prevailing wind patterns and ocean currents. Prevailing winds between 30°N and 30°S latitude are easterlies (from the southeast in the southern hemisphere and from the northeast in the northern hemisphere). Between 30° and 60° in each hemisphere the prevailing winds are from the west. Ocean currents generally move clockwise in the northern hemisphere and counterclockwise in the southern hemisphere. (SMC)

Corm A bulblike structure in which the fleshy portion is predominantly stem tissue covered by membranous scales. Corms function much like bulbs and are found, for example, in plants in the orchid family. (SMC)

Corpuscle A blood cell. *See also* Erythrocycte; leukocycte. (FSL)

Corrective action Under the Resource Conservation and Recovery Act, the Environmental Protection Agency may take action at permitted and interim-status facilities to respond to releases of hazardous wastes. 42 USC §6928. *See also* Cleanup. (DEW)

Correlation analysis The investigation of the strength of a linear relationship between two variables. (VWP)

Corrosion **(1)** The dissolving and wearing away of metal, caused by a chemical reaction between water and pipes that the water contacts, chemicals touching a metal surface, or contact between two metals. Ordinary corrosion is the oxidation-reduction process by which metals are oxidized by oxygen, O_2, in the presence of moisture. (Oxidation-reduction reactions, or redox reactions, are those in which atoms undergo changes in oxidation number.) **(2)** An electrochemical reaction in which metal deteriorates or is destroyed when in contact with environmental factors such as air, water, or soil. When reaction occurs, there is a flow of electric current from the corroding portion of the metal toward the electrolyte or conductor of electricity, such as water or soil. Factors in water that can affect its corrosiveness include temperature, carbon dioxide, oxygen, conductivity, and acidity. **(3)** The action of air and water on iron and other metals to form oxides and carbonates. If allowed to continue, it leads to the ultimate destruction of the metal. (FSL)

Corrosive Refers to a chemical agent that reacts with the surface of a material, causing it to deteriorate or wear. (FSL)(HMB)

Corrosivity **(1)** Term used to define a property of a waste that is either aqueous or has a pH less than 2 or greater than or equal to 12.5, as determined by a pH meter using either an EPA test procedure or an equivalent test method approved by the Administrator, EPA, under the procedures outlined in Sections 260.22 and 260.211 of the Resource Conservation and Recovery Act. **(2)** An ability of a substance to produce corrosion. Corrosivity is a property of some sanitizers—e.g., chlorine compounds and acids—that is undesirable when these compounds are improperly used to sanitize equipment surfaces. *See also* Chlorine compounds; sanitizer. (FSL) (HMB)

Cortisone A steroid compound isolated from the adrenal glands that acts upon carbohydrate metabolism and influences the nutrition and growth of connective tissue. It is used in the treatment of rheumatoid arthritis and certain inflammatory conditions. (FSL)

Coruscative Term that describes pairs of chemicals that react with each other without the formation of gas. (FSL)

Cosmic radiation Penetrating ionizing radiation, both particulate and electromagnetic, originating in outer space. Secondary cosmic rays, formed by interactions in the earth's atmosphere, add to the general background radiation and contribute perhaps 1 rad or more to the gonads of each individual of the general population in 30 years. (RMB)

Cost-benefit analysis The technique for estimating the effect of a course of action that involves delineating in monetary terms both the advantages and disadvantages of a proposal and making economic-outcome predictions of the proposed action. (FSL)

Cost-effect alternative An alternative control or corrective method identified after analysis as being the best available in terms of reliability, permanence, and economic considerations. (FSL)

Cost recovery A legal process by which potentially responsible parties who contributed to contamination at a Superfund site can be required to reimburse the Trust Fund for money spent during any cleanup actions by the federal government. (FSL)

Coulomb Unit of electrical charge in the practical system of units. A quantity of electricity equal to 3×10^9 electrostatic units of charge. (FSL)

Coumarin ($C_9H_6O_2$) A fragrant, neutral component obtained from the tonka bean or made synthetically from salicylic aldehyde. It has an odor that resembles vanilla and is used as an odor-masking compound. (FSL)

Council on Environmental Quality (CEQ) A presidential council created by the National Environmental Policy Act (NEPA) in 1969. The CEQ provides oversight on the environmental implications of all federal projects and develops guidelines for the preparation of environmental impact statements, which are required in all major federal actions. *See also* National Environmental Policy Act. (FSL) (DEW)

Count (radiation measurements) The external indication of a device designed to enumerate ionizing events. It may refer to a single detected event or to the total registered in a given period of time. The term often is erroneously used to designate a disintegration, ionizing event, or voltage pulse. (RMB)

Counter A device for counting nuclear disintegrations, used to measure radioactivity. The signal that announces a disintegration is called a count. (RMB)

Counterclaim Cause of action, brought by a defendant in a civil suit against the plaintiff, that opposes or diminishes the original claim. (DEW)

Count rate meter A device that gives a continuous indication of the average number of ionizing events detected. Most survey instruments have a count rate meter incorporated with a suitable detector. (RMB)

Counting error Specifies the reliability of a measurement. (RMB)

Coupling agent A chemical, such as N,N′-dicyclohexylcarbodiimide (DCC), that when mixed with an amine and an acid, links the two together to form an amide. Substances such as DCC are therefore used as peptide coupling agents. (FSL)

Course texture (soil) The texture of sands, loamy sands, and sandy loams that are not very fine. (SWT)

Court of appeals See federal courts. (DEW)

Covariance For a set of data points (x,y), the sum of the products of the distances of all values of x and y from the associated centroid values, divided by one less than the sample size $(n - 1)$.

$$Cov\ (x,y) = \frac{\Sigma(x - \bar{x})(y - \bar{y})}{n - 1}$$

(VWP)

Cracking Opening a valve slowly to allow a small amount of material to flow. (FSL)

Cradle knoll A small knoll (microrelief feature) formed by earth that is left raised by an uprooted tree. (SWT)

Creep (soil) A slow mass movement of soil down a slope, primarily under the influence of gravity, but facilitated by alternate freezing and thawing, and saturation with water. (SWT)

Creosote A mixture of phenols from coal or wood tar that are used as wood preservatives or disinfectants. (FSL)

Cristobalite A form of crystalline-free silica capable of causing pneumoconiosis and silicosis. (DEJ)

Criteria Specific limits or characteristics of a physical, chemical, or biological nature that are defined as a part of a hazard-analysis critical-control-point program (HACCP). These are tolerances that must be met to ensure that a critical control point effectively controls an identified hazard. *See also* HACCP. (HMB)

Criteria pollutants The 1970 amendments to the Clean Air Act required the Environmental Protection Agency (EPA) to set national ambient air quality standards for certain pollutants known to be hazardous to human health. EPA has identified and set standards to protect human health and welfare for six pollutants: ozone, carbon monoxide, total suspended particulates, sulfur dioxide, lead, and nitrogen oxide. (FSL)

Critical control point (HACCP) A practice or a step of an operation (process, procedure, or location) at or by which a preventive measure can be exercised that will prevent, eliminate, or minimize a hazard. Examples of critical control points in a food-service operation might be the cooking, cooling, and storage processes. For example, in cooking, the time and temperature of cooking a roast can assure killing of microorganisms. Rapid cooling of a cooked meat product is essential so that the temperature of the food moves quickly through the danger zone within which a spoilage or pathogenic organism can multiply. Once cooled to the appropriate refrigeration temperature, the meat should be stored at that temperature to minimize microbial growth. It is essential in a hazard-analysis critical-control-point approach to monitoring food-handling procedures that all critical control points be identified and monitored, and that control measures be taken immediately to correct problems whenever monitoring results indicate that criteria for conditions or operations at a control point are not correct. *See* Hazard analysis critical control point (HACCP). (HMB)

Critical habitat Under the Endangered Species Act, the critical habitat for a threatened or endangered species includes specific areas within the geographical area that are occupied by the species at the time of its listing, and on which are found special physical or biological features essential to the conservation of the species. These features may require special management consideration or protection. Other areas outside the geographical area that are also determined to be essential are included in the critical habitat. 42 USC §1532(5). *See also* Endangered Species Act; endangered species; threatened species. (DEW)

Critical mass The minimum mass of a fissionable material that will just maintain a fission chain reaction under precisely specified conditions. Conditions may include the nature of the material and its purity, the nature and thickness of the tamper (or neutron reflector), the density (compression), and the physical shape (geometry). For an explosion to occur, the system must be supercritical—i.e., the mass of material must exceed the critical mass under the existing conditions. *See also* Supercriticality. (RMB)

Critical organ (radiation) The body organ receiving the radionuclide that results in the greatest overall damage to the body. Usually, but not necessarily, it is the organ receiving the greatest concentration. (RMB)

Critical range As applied to aquatic environments and particularly fish populations, those environmental factors such as the availability of food and the concentration of dissolved oxygen, carbon dioxide, ammonia, hydrogen sulfide, nitrous oxide, and hydrogen that are crucial for the survival of fish species.

Certain chemicals, particularly insecticides, are extremely toxic to fish. The critical range for the survival of some species involves acute toxicity values from 5 to 100 μg/liter. (FSL)

Criticality A term used in weapon and reactor physics to describe the state of a given fission system when the specified conditions are such that the mass of active material present in the system is precisely a critical mass. Thus, the fission neutron production rate is a constant and is exactly balanced by the total of neutron loss and utilization rate, so the neutron population remains constant. The word "criticality" alone is often used improperly to describe a system's *degree of criticality*, which is a relative term describing a variable physical property of the fissionable assembly. The degree of criticality is the ratio of the mass of active material actually present in the system to the critical mass under the identical conditions, and is usually expressed as a decimal. *See also* Supercriticality. (RMB)

Critical size For a fissionable material, the minimum amount of material that will support a chain reaction. (RMB)

Crocidolite A form of asbestos, sometimes called blue asbestos, that is the fibrous form of riebecktite, a hydrated silicate. (DEJ)

Crop rotation A method of shifting from one type of crop growing in a given area to another crop in order to maintain fertility of the soil and to disrupt the life cycles of various plant pests. Crop rotation increases soil fertility, decreases erosion, and aids in control of insects and disease. (FSL)

Cross-connection A physical connection of pipes whereby water may flow between two separate water supply systems, one of which contains potable water and the other, water of unknown or questionable safety. *See also* Backflow. (FSL)

Cross-contamination The transfer of harmful microorganisms from one food to another by means of a nonfood surface such as utensils, equipment, or human hands. (HMB)

Cross-section (nuclear) A measure of the probability that a certain reaction between a nucleus and an incident particle or photon will occur. It is expressed as the effective area that the nucleus presents for the reaction. *See also* Barn. (RMB)

Cross-sectional study An investigation of data collected at one point in time. (VWP)

Crude death rate *See* Rate. (VMP)

Crude oil A descriptive term for the wide range of liquid petroleum derivatives. Oil containing less than 0.57% sulfur is considered more desirable than higher-sulfur-content materials. (FSL)

Crust (soil) A surface layer on soil, with a thickness of between a few millimeters and an inch, that is much more compact, hard, and brittle than the material directly beneath it. (SWT)

Cryogenic gas A liquefied gas that exists in its containers at temperatures far below normal atmospheric temperatures. Fluorine, helium, and hydrogen under these conditions are examples. (FSL)

Cryogenic liquid A refrigerated liquid gas with a boiling point below −130°F (−90°C). (FSL)

Cryometer A thermometer used to measure very low temperatures. (FSL)

Cryptococciosis An infection by the fungal organism *Cryptococcus neoformans* that usually involves the skin, lungs, brain, and meninges. (FSL)

Cryptosporidiosis Infection with the protozoan *Cryptosporidium*, which causes diarrhea in humans. The organisms are parasitic in the intestinal tracts of many vertebrates, including birds, mammals, and reptiles. (FSL)

Crystal A homogeneous inorganic substance of definite chemical composition bounded by plane surfaces that form definite angles with each other and that give the substance a regular geometric form. (SWT)

CS gas Chemical agent (ortho-chlorobenzylidene malonitrile) used as a powder for riot control because of its propensity to induce eye, nasal, respiratory-system irritation, coughing, and nausea. (FSL)

Cubic feet per minute (CFM) A measure of the volume of a substance flowing through air within a fixed period of time. (FSL)

Culicfuge A chemical agent, such as diethyltoulamide (DEET), that is used to repel mosquitoes. (FSL)

Culm The scrap material resulting from the processing of anthracite (hard coal). Of environmental significance because of air pollution problems created when the material ignites and because of other impact connected with the formation of sulfuric acid as the result of water contact with the burning material. (FSL)

Cultural eutrophication The death of water bodies by pollution from human activities. *See also* Eutrophication. (FSL)

Culture The deliberate propagation of microorganisms on or in substances (media), such as blood, broth, and agar, especially prepared for this purpose. (FSL)

Cumulation frequency distribution A display of each value in a data set or class boundary and the accumulated frequency associated with each value or class boundary. (VWP)

Cumulative dose (radiation) The total dose resulting from repeated exposures to radiation of the same region or of the whole body. (RMB)

Cumulative infiltration (soil) Total volume of water infiltrated per unit area during a specified period of time. (SWT)

Cumulative trauma disorder A collective term for syndromes characterized by discomfort, impairment, disability, or persistent pain in joints, muscles, tendons, and other soft tissues, with or without physical manifestations. It is often caused, precipitated, or aggravated by repetitive or forced motions, which may occur in many different occupational activities, such as in assembly, manufacturing, meat processing, packing, sewing, and typing. Others—for example, "tennis elbow"—have been associated with various sports. Many of these disorders are named after the occupations in which they were first identified. *See also* Carpal tunnel syndrome; bursitis; carpenter's elbow; deQuervain's syndrome; trigger finger, Guyon tunnel syndrome. (SAN)

Curare A wide range of toxic chemicals derived from botanical sources such as *Strychnos*, a tropical tree that is the source of strychnine. *See also* Strychnine. (FSL)

Curie Named for French chemists and physicists Marie (1867–1934) and Pierre (1759–1906) Curie, this was formerly defined as a quantity of any radioactive nuclide producing 3.7×10^{10} disintegrations per second. Now the curie (Ci) is officially a unit of activity rather than quantity—i.e., a unit of radioactivity, which is a measure of the rate at which a radioactive material emits particles. The new definition is $1\ \text{Ci} = 3.7 \times 10^{10}$ disintegrations $\times\ s^{-1}$. The higher the rate of disintegrations, the greater is the hazard. (MAG) (RMB)

Cutan A modification of the texture, structure, or fabric at natural surfaces in soil materials, due to concentration of particular soil constituents or in-situ modification of the plasma. Often refers to the clay "skins" that form in illuvial horizons. (SWT)

Cutaneous Relating to the skin; dermal. (FSL)

Cutting oils Petroleum products that contain heat-resistant chemicals that enable them to resist high temperatures and lubricate various types of mineral cutting machines. (FSL)

CWA (Clean Water Act, 1972). Federal law regulating the discharge of pollutants into surface waters. (FSL)

Cyanide, hydrogen (HCN) Hydrocyanic acid, also known as prussic acid, a water-soluble liquid that has a faint odor of bitter almonds. The material is obtained by reacting ammonia and air with methane or natural gas, by the recovery from coke oven gases, or from bituminous (soft) coal and ammonia. This chemical is flammable, and toxic by either inhalation, ingestion, or skin absorption. It is used in the manufacture of acylates, dyes, rodenticides, and chelates. In recent years, the use of cyanides as rodent burrow gassing agents has been restricted because of the poisonous nature of these materials and their environmental consequences. (FSL)

Cyclodiene insecticides So-called "hard" or persistent pesticides, such as chlordane, heptachlor, aldrin, or dieldrin. The use of these materials (with the exception of heptachlor) for insect control has been suspended because of environmental concerns over such effects as biomagnification. Heptachlor is still used in many areas as a termiticide, owing to the capability of the material to remain in the environment at levels toxic to termite larvae. (FSL)

Cyclone A storm with strong winds that rotate clockwise in the southern hemisphere, counterclockwise in the northern hemisphere, about a moving center of low atmospheric pressure. (FSL)

Cyclotron A device for accelerating charged particles to high energies by means of an alternating electric field between electrodes placed in a constant magnetic field. (RMB)

Cysticercosis An intestinal infection with the larval stage of the large tapeworm *Taenia solium*. (FSL)

Cytology Study of the function and structure of living cells. (FLS)

Cytolosis Disruption of cells, resulting in the destruction and breakdown of the cell membrane. (FSL)

Cytoplasm All the protoplasm of a cell except for the nucleus. (FSL)

Cytotoxic An agent that brings about destructive action on certain cells. (FSL)

Cyturia The presence of cells in the urine. (FSL)

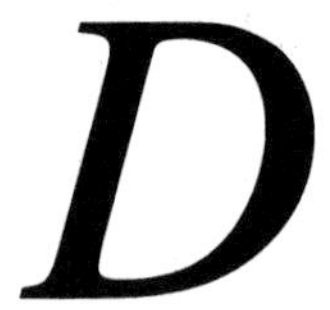

Damages Monetary compensation or indemnity that may be recovered in the courts by any person who has suffered loss, detriment, or injury to his or her person, property, or rights through the unlawful act, omission, or negligence of another. Compensatory damages are damages to compensate the injured party for the wrong sustained, and nothing more. Exemplary or punitive damages are damages in addition to compensatory damages and are awarded when the wrong done is aggravated by circumstances of violence, oppression, malice, fraud, or wanton or wicked conduct on the part of the defendant. Damages are intended to solace the plaintiff for mental anguish, laceration of feelings, shame, degradation, or other aggravations resulting from the original wrong, or to punish the defendant for evil behavior or make an example of him or her. (DEW)

Damper A device used to regulate airflow in ducts, often used to balance airflow in branch ducts. Proper sizing of branch ducts often eliminates the need for dampers in heating, ventilation, or air conditioning systems. *See also* Blast gate. (DEJ)

Dander Scales, dust, and dirt from the fur or feathers of animals, that may cause allergic reactions in susceptible persons. (FSL)

Data (Continuous) Information collected and analyzed through measurement on some scale. (Discrete) Information collected and analyzed through enumeration or counting. (VWP)

Data collection The gathering of information including: **(1)** definition of the objectives of the experiment; **(2)** definition of the variable and the population to be investigated; **(3)** determination of suitable descriptive or inferential data-analysis methods. (VWP)

Daughter products Isotopes formed by the radioactive decay of some other isotope. In the case of radium-226, for example, there are 10 successive daughter products, ending in the stable isotope lead-206. (RMB)

Davit A device, used singly or in pairs, for suspending a powered industrial platform from work, storage, and rigging locations on the building being serviced. Unlike outriggers, a davit transfers its operating load into a single roof socket or carriage attachment. A ground-rigged davit cannot be used to raise a suspended working platform above the building face being serviced. Roof-rigged davits are used to raise the suspended working platform above the building face being serviced. This type of davit can also be used to raise a suspended working platform that has been ground rigged. *See also* Powered industrial platforms. (SAN)

dBA Refers to decibels measured on the A scale, which is a frequency weighting network that approximates the response of the human ear. B and C scales are also used in noise control engineering, and are written as *dBB* and *dBC*, respectively. The A scale is most widely used in measuring the risk of noise-induced hearing loss. (DEJ)

Dead-man control On a power tool, a switch that, when released, turns off the power. In order to operate the tool, the operator must depress the switch manually for the entire time. For some tools, there is a secondary stop that prevents them from spinning after the switch is released. Drills, saws, mowers, hedge trimmers, grinders, and other power tools have this type of control. If the tool falls from the operator's grasp, the chance of injury is minimized with this type of control. (SAN)

Dead time (Radiation) The time during which a Geiger-Mueller detector is insensitive to incoming radiation. (RMB)

Death phase The phase that follows a stationary phase in the growth curve of a bacterial culture, and in which there is a progressive decline in the number of viable cells in the culture. At this stage, the dying cells far outnumber those still viable in the population. The rate of decline of the population is characteristic of the particular organism and the cultural conditions. (HMB)

Debility A weakness or lack of strength. (FSL)

Decay constant The fraction of the number of radioisotope atoms that decay in a unit time. The decay constant is $0.693/T$, where T is the half-life. (RMB)

Decay curve A curve showing the relative amount of radioactive substance remaining after any time interval. (RMB)

Decay product A nuclide resulting from the radioactive disintegration of a radionuclide or series of radionuclides. A decay product may be either stable or radioactive. (RMB)

Decay, radioactive The decrease in activity of any radioactive material with the passage of time, due to the spontaneous emission from the atomic nuclei of either alpha or beta particles, sometimes accompanied by gamma radiation. *See also* Half-life; radioactive. (RMB)

Decay rate The rate of decrease of emission of radioactive material with time. The decay rate of fission products is such that the dose rate in 7 hours will decrease to $1/10$ of its H + 1-hour dose rate and to $1/100$ of its H + 1-hour dose rate at H + 14 hours. This is a convenient rule of thumb (H equals initial time). (RMB)

Deceleration device Any mechanism, such as a rope grab, ripstitch lanyard, woven lanyard, tearing or deforming lanyard, or automatic self-retracting lifeline/lanyard, that serves to dissipate a substantial amount of energy during a fall, or that otherwise limits the energy imposed on a worker during fall arrest. *See also* Fall protection; lanyard; lifeline; safety belt; safety harness. (SAN)

Deceleration distance In fall protection, the additional vertical distance a falling worker may travel, excluding lifeline elongation and free-fall distance, before stopping, from the point at which the deceleration device begins to operate. It is measured as the distance between the location of a worker's body belt or body harness attachment point at the moment of activation of the deceleration device during a fall (that is, at the onset of fall-arrest forces), and the location of that attachment point when the employee comes to a full stop. In transportation safety, deceleration distance is the minimum distance between two vehicles that is required for safe stopping. *See also* Deceleration device; fall protection. (SAN)

Decibel (dB) A means for expressing the logarithmic level of sound intensity, sound power, or sound pressure above an arbitrary reference value (20 micropascals in air).

$$\text{Sound pressure level (in dB)} = 20 \log p/p_0 \text{ dB}$$

where p is the measured rms sound pressure, p_0 is the reference sound pressure, and the logarithm is to the base 10. By convention, the reference sound pressure level is 20 micropascals (0.0002 dynes/square centimeter, 0.00002 newtons per square meter, or 4 ten-millionths pound per square foot), and is defined as the lower threshold of hearing. *See also* dBA. (DEJ)

Deciduous Refers to plants that shed all their leaves at a certain period, usually the cold or the dry season. (SMC)

Deciles Numbers that divide a set of ranked data into ten equal parts. (VWP)

Declaration on the Human Environment A document signed at the 1972 United Nations Conference on the Environment, at which there was agreement to address environmental issues on a world-wide scale. (FSL)

Declaratory judgment Judgment by the court that serves to set out the rights of the parties involved or clarifies the court's opinion as to a matter of law. (DEW)

Decolorizer Refers to an agent used to remove color, or bleaches such as sodium hypochlorite. *See also* Chlorine. (FSL)

Decomposer An organism that feeds on dead organic matter and breaks it down into simpler parts. *Decomposer* refers especially to bacteria and fungi that feed on dead organic matter. *See also* Detritivore. (SMC)

Decomposition The breakdown of dead organic material into smaller or simpler parts that are then recirculated. Bacteria, fungi, heterotrophic protists, and saprophagous insects are important in the process of decomposition. (SMC)

Decompression sickness (DCS) Illness or injury associated with exposure to high-pressure atmospheres, followed by rapid exposure to normal atmospheric pressure. This is to be distinguished from dysbarism, which is related to exposure to low-pressure atmospheres. Type I DCS is characterized by pain, numbness, paresthesia, edema, fatigue, rashes, fever, and sweating. Type II is characterized by substernal distress, coughing, dyspnea, asphyxia, circulatory obstruction, visual disturbances, deafness, slurred speech, paralysis, unconsciousness, and death. Chronic effects include avascular bone necrosis, paralysis, and focal lesions of the brain. *See also* Bends; caisson disease. (DEJ)

Decontamination **(1)** The removal of an etiologic agent (in this case, a contaminant) from an animate or an inanimate surface. **(2)** (Radiation) The reduction or removal of contaminating radioactive material from a structure, area, object, or person. Decontamination of radioactive material may be accomplished by: (a) treating the surface to remove or decrease the contamination; (b) allowing the material to stand so that the radioactivity is decreased as a result of natural decay; or (c) covering the contaminated area to attenuate the radiation emitted. Any radioactive material removed under process *a* must be disposed of in accordance with federal and state laws and regulations. **(3)** (Sanitation) The treatment of materials in order to eliminate or reduce the number of unwanted microorganisms, unwanted chemicals, or radioactive substances. The term "decontamination" may be used instead of "disinfection'' when referring to the reduction or removal of microorganisms. *See also* Contamination; disinfection. (HMB, RMB)

Deenergize As applied to electricity, the process of removing a source of electromotive force or current to a circuit or system of conductors. (FSL)

Deep ecology An ideology based on the belief that all species and natural systems have an intrinsic right to existence. (SMC)

Default judgment Judgment rendered in favor of a plaintiff because of a defendant's failure to make a timely response to initiation of the suit. In federal court, default may be entered pursuant to Federal Rule of Civil Procedure 55. At its discretion, the court may set aside a default judgment for good cause. (DEW)

Defendant The party against whom a civil action is brought, or the accused in a criminal action. (DEW)

Deflagration A reaction (usually violent) that occurs as the result of the transfer of energy ahead of the flame front by conduction, convection, and/or radiation. (FSL)

Deflagration point The temperature at which a small sample of an explosive mixture detonates violently when heated or decomposed. (FSL)

Deflation (soil) The process whereby fine soil particles are removed by wind. (SWT)

Deflocculate To separate individual component particles by chemical and/or physical means. To suspend the particles of the disperse phase of a colloidal system dispersion medium. (SWT)

Defoliant A chemical (herbicide) designed to remove the leaves from trees and shrubs. Herbicides such as 2,4-D produce lethal metabolic difficulties in plants. *See also* Dichlorophenoxyacetic acid. (FSL)

Degradation The process by which a chemical is reduced to a less complex form. (FSL)

Degreaser **(1)** A chemical, such as trichloroethylene, used to remove grease and oil from machinery. The use of trichloroethylene, a suspected carcinogen, has been reduced in recent years and a substitute chemical, perchloroethane, is being used in its place. Prolonged exposure to trichloroethylene, followed by consumption of alcohol, leads to a skin inflammation known as degreaser's flush. **(2)** The removal of surface grime, oil, and grease from metal, using either vapor-phase degreasing or soak cleaning. Vapor-phase degreasing is performed in a tank containing a heated solvent. The vapors rise to a point determined by the location of a condenser, which causes the vapor to return to the liquid state. The pieces to be degreased are lowered into the vapor, cleaned by the condensing vapor, and then dried when thermal equilibrium is achieved. Soak cleaning usually is performed by applying solvents by brushing or wiping. Solvents typically used include petroleum distillates, ketones, esters, detergents, and alkaline cleaners. (DEJ, FSL)

Degrees of freedom (df) An index used for determining the distribution to be applied in a statistical analysis. The value usually used is the sample size minus one. (VWP)

Dehiscence The opening of a fruit or anther, whereby seeds or pollen is shed and dispersed. (SMC)

Dehumidifier A device for lowering the moisture content in the atmosphere. (FSL)

Delamination The physical separation or deterioration of a bonding material that causes the pulling apart of plastic, wood, and other materials generally used in buildings and other structures. Sprayed-on asbestos products, for example, can separate from the surfaces onto which they were originally applied, thus creating a respiratory hazard. *See also* Lamination. (FSL)

Delaney clause *See* Food, Drug, and Cosmetic Act. (DEW)

Delead To remove lead from the tissues by the use of chelating agents such as EDTA (ethylenediaminetetraacetic acid). It is then excreted in the urine. The same term is applied to the removal of lead-based paint in dwelling units, whereby the contaminated paint is physically removed from doors and trim and other surfaces in order to eliminate and/or minimize lead exposure. *See also* Plumbism. (FSL)

Deleterious Refers to an agent (physical, chemical, or microbial) that is injurious, or capable of causing harm. (FSL)

Delineation The outlining or marking of areas contaminated with biologic, radiologic, or chemical materials. (FSL)

Delineation (soil) The portion of a landscape shown as a closed boundary on a soil map that defines the area, shape, and location of one or more component soils plus inclusions and/or miscellaneous land area. (SWT)

Delist To use the petition process to attempt to have rescinded a facility's designation as toxic. (FSL)

Delisted wastes Site-specific wastes that are excluded from reporting under 40 CFR 260.20 and 260.22. A waste at a particular generating site may be excluded or delisted from the lists of hazardous waste in Subpart D of Part 261 by petitioning of the EPA administrator for regulatory amendments. (FSL)

Deluge shower Shower, often called an emergency deluge shower, used to dilute and remove spilled chemicals from the body. When properly designed and installed, it should provide a minimum of 30 gallons of water per minute at low velocity, in order to keep potential tissue damage to a minimum. (FSL)

"De minimis" violation In OSHA proceedings, noncompliance with a regulation or standard that presents no immediate or direct threat to an employee's health and safety. A state of noncompliance with a government provision or standard, commonly used in OSHA enforcement, in which no direct or immediate threat to the safety or health of employees is present, and in which no penalty is assessed or citation issued. Such a violation is issued by OSHA when an employer complies with the intent of a standard but deviates from its particular requirements in a way that has no direct or immediate relationship to health or safety, or when the employer complies with a proposed amendment to the standard and the amendment provides equal or greater protection than the standard itself. The violation may also be issued when an employer's workplace is technically advanced beyond the requirements of a standard. (DEJ)

Demyelination Deterioration, destruction, or loss of the myelin sheath of an individual neuron or group of neurons. (FSL)

Denature To change the normal composition of a substance—e.g., to add methanol (wood alcohol) to ethanol (grain alcohol) to render it unsuitable for drinking. (FSL)

Dendrochronology The study of the growth rings in trees. Examining the relative width of these rings can provide information about growth and past weather patterns. Measurements of elements in the wood of rings representing different years can provide information about substances in the environment. For example, measurements of lead content in tree rings can show changes in the levels of atmospheric lead over time. (SMC)

Dendron One of the branching processes of a nerve cell or neuron that conveys impulse. (FSL)

Dengue A viral disease carried by *Aedes* mosquitoes. Also known as breakbone fever because of the intense joint pain associated with it. The best methods of preventing the disease involve premises sanitation and control of the mosquito's habitat, including the removal of refuse and containers that may hold water and facilitate mosquito breeding. (FSL)

Densitometer An instrument utilizing the photoelectric principle to determine the degree of darkening of developed photographic film. (FSL)

Density **(1)** The ratio of the mass of a material to its volume, e.g., g/cm^3 or lb/ft^3. **(2)** (Photographic) Logarithm of opacity of exposed and processed film.

Opacity is the reciprocal of transmission. Transmission is the ratio of transmitted intensity to incident intensity. Density is used to denote the degree of darkening of photographic film. (FSL)

Denude To strip or remove vegetation from an area. *See also* Clear cut. (FSL)

Deodorant Chemical used to mask or conceal offensive odors. (FSL)

Deoxidation The process whereby oxygen is removed from a chemical compound. (FSL)

Deoxyribonucleic acid (DNA) The type of nucleic acid that contains deoxyribose as a sugar component and that is found mainly in the chromosomes of animal and vegetable cells. DNA is considered to be the repository of hereditary characteristics and the autoreproducing constitutent of chromosomes and many viruses. (FSL)

Department of Agriculture (USDA) In conjunction with the Food and Drug Administration (FDA), the USDA administers aspects of food processing dealing with wholesomeness, safety, and sanitary practices involving meats, poultry, and egg products. The laws covering these three product areas are the Poultry Products Inspection Act (1957), the Egg Products Inspection Act (1970), and the Meat Inspection Act (1967). Foods containing over 2% poultry or 3% meat are regulated by the USDA. Federal jurisdiction usually applies to those products in interstate commerce, but the acts dealing with poultry, eggs, and meats extend the USDA responsibility to the intrastate level if states are not able to provide adequate enforcement as required by law. Mandatory inspection of animals for slaughter, of slaughtering conditions, and of meat processing facilities is required by the Wholesome Meat Act (1967) and the Wholesome Poultry Act (1968). These acts provide that all red meat and red-meat products, and all poultry products, whether in interstate commerce, in intrastate commerce, or imported, must be processed under federal standards. (HMB)

Department of Defense (DOD) Federal agency responsible for the administration of military programs designed to protect the nation from hostile forces. This agency manages numerous arsenals and other facilities that may contain hazardous materials and wastes. (FSL)

Department of Energy (DOE) Federal agency responsible for research and development of energy technology, sale of power from federal generating facilities, administration of the nuclear weapons program, and regulation of energy. (FSL)

Department of Transportation (DOT) Federal agency that enforces regulations relating to the transportation of hazardous and nonhazardous materials. (FSL)

Depauperate **(1)** smaller than usual natural size or short of usual development; **(2)** deficient, lacking in some factor; **(3)** describes a habitat that has fewer individuals, fewer species, or fewer trophic levels than it could be expected to support. (SMC)

Dependent events Two events for which the occurrence or nonoccurrence of one affects the occurrence or nonoccurrence of the other. (VWP)

Depilatory Chemical having the ability to remove or destroy hair. *See also* Alopecia. (FSL)

Depolymerization The breakdown of an organic compound into two or more less complex molecules. (FSL)

Deposit Material that has been transported by agents such as water, wind, ice, gravity, or human activity. (SWT)

Deposition Taking of testimony from a witness outside of court for purposes of discovery or to perpetuate testimony that may be unavailable at trial. (DEW)

Depth of the median The number of positions of the median from either end of the data when the data are arranged in ranked order. (VWP)

DeQuervain's disease Named after the Swiss surgeon Fritz DeQuervain (1868–1940), this is a type of tenosynovitis of the exterior tendons of the wrist or the abductors of the thumb. The inflammation of the synovial lining is often pronounced under conditions of highly repetitive hand usage in new employees, or is due to poor design of the workplace or of tools. This condition often results from repetitive work that involves moving the wrist and gripping with the thumb. It is also associated with tasks such as wringing clothes, buffing, grinding, and polishing. *See also* Cumulative trauma disorder. (DEJ) (SAN)

Dermal Of or pertaining to the skin. (FSL)

Dermatitis Inflammation of the skin as a result of exposure to various microbial, chemical, radiologic, or physical agents. Skin abnormalities related to occupation outnumber all other work-related illnesses. Also known as industrial dermatitis, occupational contact dermatitis, professional eczema, cement dermatitis, chrome ulcers, oil acne, rubber itch, tar warts, and other cutaneous maladies. The most efficient way to prevent dermatitis is to keep the skin free of contact with irritants and allergenic materials, by using gloves and practicing good personal hygiene. Agents causing dermatitis can be classified as primary irritants, cutaneous sensitizers, and mechanical, physical, and biological irritants, carcinogens, and ulcerations. Primary irritants include organic and inorganic acids, alkalies, alcohols, ketones, chlorinated organics, petroleum, coal tar, turpentine, and metal salts (such as those of chromium and arsenic). Cutaneous sensitizers are those substances that, through an immunological mechanism, cause a skin reaction upon subsequent exposure. They include epoxy resins, phenol-formaldehyde plastics, potassium dichromate, cement, and poison ivy. Dermatitis can also result from photosensitivity effects, such as sunburn. Mechanical irritants cause blisters or calluses, and include fiberglass. They can also be produced during operation of power tools. Physical irritants include heat, cold, electricity, and ultraviolet light. Biological agents include bacteria, fungi, viruses, and parasites. Discoloration of the skin can occur from either inhalation or contact with various agents. *See also* Chloracne; dermatosis. (DEJ)

Dermatobia Refers to botflies, such as *Dermatobia hominis*, whose larvae infect mammals and birds. (FSL)

Dermatophagoides pteronyssimus The house dust mite, which causes asthmatic difficulties in susceptible individuals. (FSL)

Dermatophytoses A group of diseases caused by fungi, and often found among farmers, animal handlers, pet and hide handlers, wool sorters, cattle ranchers, athletes, lifeguards, gymnasium employees, and animal laboratory workers. Ringworm of the hands and feet is the most common form, and is usually prevented by recognition of the disease in animals, sterilization and proper laundering of towels, general cleanliness of showering facilities, and proper personal hygiene. (DEJ)

Dermatosis Generic term for skin disorders, particularly those not involving inflammation. *See also* Dermatitis. (FSL)

Derris The active ingredient in the insecticide rotenone extracted from the root of the *Derris elliptica* plant. An effective insecticide against aphids and mites, rotenone is not harmful to mammals or birds. It is extremely toxic to fish and for that reason is frequently used to remove ''trash fish'' from lakes and ponds. In its uses as an insecticide, it has largely been supplanted by the synthetic organic insecticides. (FSL)

Desalination The extraction of salt from sea water in order to create a palatable source of drinking water. (FSL)

Descriptive statistics Methods used to collect, organize, analyze, and summarize information. These descriptive methods may include the construction of graphs, charts, and tables, and the computation of various kinds of averages and indices. (VWP)

Desert crust A hard layer exposed at the surface of desert regions that is composed of calcium carbonate or other binding material. (SWT)

Desertification The process of habitats becoming desert, usually caused by a combination of drought and overgrazing or deforestation. The removal of vegetation through overgrazing or deforestation reduces the soil's ability to trap water, increases erosion, and increases the reflectance (albedo) of the surface, which raises the temperature and aggravates the drought. Drought results in a reduction of vegetation, and can convert desirable habitat into desert. (SMC)

Desert pavement A layer of gravel or stones that is left behind after wind has removed fine soil material, and that usually occurs in arid regions. (SWT)

Desiccant Chemicals, such as silica gel, calcium chloride, zinc chloride, and activated alumina, that absorb moisture and are used to promote dryness. Some desiccants are capable of drying out and killing plants or insects. Others are used to maintain dry conditions in containers used for packaging food and chemical reagents. (FSL)

Desiccator A laboratory device with a tight-fitting lid and chamber in which substances can be placed for the gradual extraction of water. Silica gel is frequently used as the desiccant in these vessels. (FSL)

Designated Agency Safety and Health Official (DASHO) The executive official of a federal department or agency who is responsible for safety and occupational health matters and is so designated or appointed by the head of the agency. (FSL)

Designated pollutant An air pollutant that is not a hazardous pollutant, as described in the Clean Air Act, but for which new source performance standards exist. The Clean Air Act does require states to control such pollutants, which include acid mist, total reduced sulfur (TRS), and fluorides. (FSL)

Designated uses Those water uses identified in state water quality standards that must be achieved and maintained as required under the Clean Water Act. Uses can include cold-water fisheries, public water supply, and agriculture. (FSL)

Designer bugs Popular term for microbes developed through biotechnology that can degrade specific toxic chemicals at their source in toxic waste dumps or in ground water. (FSL)

Destroyed regulated medical waste Regulated medical waste that is no longer generally recognizable as such because the waste has been torn apart, or mutilated (although not by compaction) through processes such as: **(1)** thermal treatment or melting, during which treatment and destruction could occur; or **(2)** shredding, grinding, tearing, or breaking, during which only destruction would take place. *See also* Medical waste. (FSL)

Destruction or removal efficiency (DRE) A measure of a technology's effectiveness in the removal of contaminants from hazardous chemical wastes. (FSL)

Desulfurization Removal of sulfur from fossil fuels to reduce pollution. (FSL)

Detailed radiation survey A survey used: **(1)** to find the radiation level of specific objects or locations within the contaminated area of interest; **(2)** to locate regions of higher-than-average intensity, or hot spots; or **(3)** to establish with greater accuracy the position of the outer line of contamination and the danger line. (RMB)

Detector tube An air sampling device used to measure the concentration of various air contaminants that consists of a glass tube filled with a solid chemical that changes color when it reacts with the air contaminant under question, and a hand-operated piston or bellows-type pump. A measured volume of air is admitted into the tube; the longer the length of the resulting stain, the more concentrated is the contaminant in the air sampled. The glass tube has graduated markings that delineate the concentration of the air contaminant. Short-term detection tubes can be used and read in approximately 2 minutes, whereas long-term detector tubes are attached to a battery-operated air pump and read after several hours. Some detector tubes rely on diffusion. Detector tubes have a relatively high sampling and analytical error, in the range of 25–35% at concentrations in the range of the threshold limit value, and are also subject to a number of interferences. However, they do provide a quick, economical method of air sampling. Also known as length-of-stain detector tubes, short-term detector tubes, diffusion tubes, and long-term detector tubes. (DEJ)

Detergent **(1)** Synthetic washing agent that helps to remove dirt and oil. Some contain compounds that kill useful bacteria and encourage algae growth when they are in wastewater that reaches receiving waters. **(2)** Washing and cleaning agents with a composition other than soap that clean by a mechanism similar to soap. These products are synthesized chemically from a variety of raw materials derived from petroleum, fatty acids, and other sources. They may also contain other ingredients, such as fragrances, foaming control agents, corrosion inhibitors, wetting agents, and saponifiers. These ingredients produce different types of products—e.g., heavy-duty detergents, light-duty detergents, and cleansers. (HMB)

Detergent auxiliaries Additives included in cleaning compounds to protect sensitive surfaces or improve the cleaning properties of the cleaner. For example, acid compounds may be added to a cleaner to protect alkaline-sensitive surfaces such as varnishes, or for protection during light metal cleaning. Oxalic acid removes iron oxide rust without attacking the metal. Phosphoric acid is used to clean metals before painting. Sequestrants, also called chelating agents, combine with magnesium and calcium ions, which reduces water hardness. This action improves the cleaning properties of the detergent. Another example is the addition of surface-active agents that facilitate the transport of cleaning and sanitizing compounds over the surface to be cleaned. Surfactants are classified as synthetic detergents, but may be used as auxiliaries. (HMB)

Detergent sanitizer A blend of sanitizer and cleaning agent capable of the functions of both simultaneously. Commercially available products are usually more expensive and many times more limited in their application. For best results, sanitizing should be carried out separately from cleaning because the effectiveness of the sanitizer can be affected by the soil removed by the cleaner. It is best for sanitizing to be completed after a thorough cleaning of surfaces. *See also* Detergent; sanitizer. (HMB)

Determinate Of limited, definite growth or size; refers to growth that stops at a certain time or after a certain size has been reached. (SMC)

Detonation The outcome resulting from the propagation at supersonic rate (from 1000 to 9000 meters per second) of the reaction zone through a chemical substance. Detonations involve shock waves that produce gaseous reaction products of high temperature and pressure and reverberate through surrounding media. (FSL)

Detoxification Diminution or removal of the toxic qualities of any substance; reduction of the virulence of any pathogenic organism; metabolic conversion of pharmacologically active components to less active components. (FSL)

Detritivore An organism that feeds on detritus or dead organic matter and breaks it down into smaller parts. The term *detritivore* usually refers only to organisms other than bacteria and fungi. *See also* Decomposer. (SMC)

Detritus **(1)** Dead or partly decomposed organic matter. Litter on the forest floor or leaves and dead animals at the bottom of a pond are examples of detritus. **(2)** Collective term for trash or other debris in a specific environment—e.g., in a pond, lake, or stream. (FSL, SMC)

Deuterium A heavy, stable isotope of hydrogen having one proton and one neutron in the nucleus. It can be used in thermonuclear fusion reactions for the release of energy. Abbreviation: D or ${}^{2}_{1}H$. *See also* Fusion; thermonuclear reaction. (RMB)

Deuteron The nucleus of a deuterium atom. Mass: about 2 amu (atomic mass unit). Abbreviation: d. (RMB)

Developer A person, government unit, or company that proposes to build a hazardous-waste treatment, storage, or disposal facility. (FSL)

Dew point The temperature at which air at a constant pressure and constant water-vapor content will be saturated; the temperature at which the saturation vapor pressure of the parcel of air is equal to the actual vapor pressure of the contained water vapor. Cooling below the dew point usually results in frost or dew. When this temperature is below 0°C, it is sometimes called the frost point. (DEJ)

Diagnose To isolate or recognize a disease. (FSL)

Diagnostic soil horizons Specific soil characteristics that indicate soil classes. Horizons that occur at the surface of the soil are called epipedons and those below the surface are called diagnostic subsurface horizons. (SWT)

Diagnostic specimen As defined in the regulations for the interstate transport of etiologic agents, any human or animal material including, but not limited to, excreta, secreta, blood and its components, tissue, and tissue fluids being shipped for purposes of diagnosis. 42 CFR §72. *See also* Etiologic agents. (DEW)

Dialysis The passage of solute molecules through a semipermeable membrane from a higher concentration to a solution of a lower concentration. (FSL)

Diarrhea Fecal discharge of an abnormal frequency and liquidity. A condition often associated with foodborne illnesses and one of the most common symptoms of gastroenteritis or the gastrointestinal syndrome. *See also* Gastroenteritis; salmonellosis; staphylococcal intoxication. (HMB)

Diatomaceous earth A geologic deposit of fine, friable siliceous material that is composed chiefly or entirely of the skeletal remains of diatoms. It may occur as a powder or as a porous, rigid material. It is one of the types of filtration media used in water treatment and swimming-pool operations. (FSL)

Diatoms Algae with siliceous cell walls that persist as a skeleton after death. They occur in fresh and salt waters and their remains are widely distributed in soils. (SWT)

Diazinon **($[C(CH_3)_2CHC_4N_2H(CH_3)O]PS(OC_2H_5)_2$)** An insecticide. In 1986, the Environmental Protection Agency banned its use in open areas such as sod farms and golf courses because it posed a danger to migratory birds, which gathered on them in large numbers. The ban did not apply to its use in agriculture or on lawns of homes and commercial establishments. (FSL)

Dibromochloropropane (DBCP) *See* Nematocide. (FSL)

Dichlorodiphenyltrichloroethane [(DDT), $(ClC_6H_4)_2CHCCH_3$] A chlorinated hydrocarbon insecticide synthesized in the 1930s by Paul Mueller (1899–1965). This chemical was first used in the 1940s and was shown to be highly effective against human-body lice (*Pediculus humanus corpis*) in Naples, Italy, where it stemmed an epidemic of typhus. Use was continued until the 1970s, when its propensity for biomagnification in the environment became a concern. Studies showed that the material bioaccumulated in many components of the ecosystem, and this factor has led to a virtual ban on the use of DDT. The material was not known to be acutely toxic, as demonstrated by human feeding studies and epidemiologic investigations of persons who ingested large amounts of the chemical. DDT is known, however, to be harmful to fish, and to accumulate in the fatty tissues of animals. (FSL)

Dichlorophenoxyacetic acid (2,4-D) ($Cl_2C_6H_3OCH_2COOH$) A herbicide used for the control of broadleaf weeds and noxious plants. The material is highly selective for broadleaf plants and thus is excellent for use on lower pastures and other areas where mixed stands of plants occur. 2,4-D is highly water soluble and, with the exception of oral ingestion, has a low order of toxicity for humans and other mammals. (FSL)

Dichlorvos [DDVP, $(CH_3O)_2P(O)CH{:}CCl_2$] An organophosphate compound (2,2-dichlorovinyl dimethyl phosphate) that is sold under the trade name Vapona. This chemical provides excellent control of aphids, spider mites, and many flying insects. It is also used as an anthelmintic in veterinary and human medicine. The chemical does suppress cholinesterase in persons chronically exposed, but is considered only moderately toxic to humans and other mammals. It does not persist in the environment. *See also* Cholinesterase. (FSL)

Dichotomy A division or branching into two parts. (FSL)

Dicofol A pesticide used mainly on citrus fruits. (FSL)

Dicumarol (3,3′-methylenebis[4-hydroxy-2H-1-benzopyran-2-one]) The active ingredient in the rodenticide warfarin that inhibits the synthesis of vitamin K and thus prevents blood clotting in the target mammals. *See also* Warfarin. (FSL)

Dieldrin ($C_{12}H_{10}OCl_6$) A so-called "hard" or persistent chlorinated hydrocarbon insecticide that was used widely in the 1950s and 1960s but is now banned because of its persistence in the environment. Its fat solubility causes it to bioaccumulate in the food chain. (FSL)

Dielectric A material that permits the passage of the lines of force of an electrostatic field but does not conduct current. Such materials include rubber, glass, and certain gases. A vacuum is also able to sustain an electrical field without conducting electricity. (FSL)

Diethylstilbestrol (DES, 3,4-bis(p-hydroxyphenyl)-3-hexane) A synthetic, crystalline, nonsteroid compound that has estrogenic activity when given orally or by hypodermic injection. It has been implicated as causing ovarian cancer in the daughters of women given the compound to prevent miscarriage. The compound is also used as a growth stimulant in food animals. Residues are thought to be carcinogenic. (FSL)

Differential absorption ratio Ratio of concentration of an isotope in a given organ or tissue to the concentration that would be obtained if the same administered quantity of this isotope were uniformly distributed throughout the body. (RMB)

Differentiation (Biology) The process by which single cells grow into particular forms of specialized tissue—e.g., leaf stems, roots. (Embryology) As pertaining to tissues, cells, or portions of cytoplasm, possession of a different character or function from that of the original type. (FSL)

Diffraction The bending or breaking of a ray of light into its individual parts. (FSL)

Diffused air A type of aeration that forces oxygen into sewage by pumping air through perforated pipes inside a holding tank and bubbling it through the sewage. (FSL)

Diffuse double layer A heterogeneous soil system consisting of a solid surface having a net electrical charge, together with an ionic swarm under the influence of the solid and in a solution phase that is in direct contact with the surface. (SWT)

Diffusion A spontaneous process of intermixing of different substances attributable to molecular motion and tending to produce uniformity of concentration. The rate of diffusion is dependent upon the size of the molecules of the material involved. Diffusion can be defined as:

$$m = DA\frac{(d_2 - d_1)t}{h}$$

where

m = the mass of the substance that diffuses through the cross-section of A in time t
d_1 = the concentration (mass of solid per unit volume of solution) at one surface of a layer of liquid
d_2 = the concentration (mass of solid per unit volume of solution) at the other surface of a layer of liquid
A = the area under consideration
D = the diffusion coefficient
t = time

(DEJ)

Digester In wastewater treatment, a closed tank; in solid waste conversion, a unit in which bacterial action is induced and accelerated in order to break down organic matter and establish the proper carbon-to-nitrogen ratio. (FSL)

Dike A barrier constructed to control or confine hazardous substances and prevent discharge into sewers, ditches, streams, or other flowing water. (FSL)

Dilution A reduction in the concentration of an ingredient by the addition of a neutral agent. (FSL)

Dilution ventilation The mixing of contaminated air with uncontaminated air for the purpose of controlling potential airborne health hazards, fire and explosion conditions, odors, and nuisance-type contaminants; also includes the control of airborne contaminants generated inside buildings. The formula for calculating dilution ventilation rates under normal temperatures and pressures is given by:

$$Q = \frac{403 \times 10^6 \times SG \times ER \times K}{MW \times C}$$

where

Q = volumetric airflow (in cubic feet per minute)
403 = the volume, in cubic feet, that one pint of liquid, when vaporized, will occupy at standard temperature and pressure
SG = specific gravity of the volatile liquid
K = safety factor to account for incomplete mixing
ER = evaporation rate of the volatile liquid (in pints/minute)
MW = molecular weight
C = the acceptable concentration of the airborne contaminant (e.g., the TLV or LEL), in parts per million

Dilution ventilation is to be distinguished from local exhaust ventilation. (DEJ)

Dimethoate [$(CH_3O)_2PSSCH_2CONHCH_3$] A systemic organophosphate insecticide used mainly for the control of aphids and other sucking insects. This compound is toxic to humans and animals. (RMB)

Dimethyl sulfoxide [DMSO, $(CH_3)_2SO$] A solvent, widely used as a degreasing compound, that has the ability to penetrate the intact skin. Although DMSO is used primarily for industrial purposes, it is known to be an effective anti-inflammatory agent and is often sold illegally for that purpose. Persons having contact with DMSO typically have a garliclike odor on their breath. (FSL)

Dinocap [$CH_3CH{:}CHCOOC_6H_2(NO_2)_2CH(CH_3)C_6H_{13}$] A fungicide and insecticide used to control spider mites and powdery mildew on agricultural crops. This compound is relatively toxic to humans, but breaks down rapidly in the environment after application. In 1986, the Environmental Protection Agency proposed restrictions on its use when laboratory tests found it caused birth defects in rabbits. (FSL)

Dioecious "Two houses"; unisexual; having staminate (male) and pistillate (female) flowers on separate plants. (SMC)

Dioxin Any of a family of compounds known chemically as dibenzo-p-dioxins. Concern about them arises from their potential toxicity and their contamination of commercial products. Tests on laboratory animals indicate that dioxins are among the more toxic synthetic chemicals known. Dioxins have an oral LD_{50} of 0.022 mg/kg in male rats and 0.045 mg/kg in female rats. *See also* Chloracne; chlorinated hydrocarbons; Times Beach; Seveso. (FSL)

Diphtheria An acute, contagious disease of children that normally affects the membranes of the throat. The causative agent *Corynebacterium diphtheriae* is spread from the nose, mouth, and throat of infected persons. (FSL)

Diphyllobothrium A genus of large fish tapeworms found in the intestines of humans, cats, dogs, and other fish-eating mammals. (FSL)

Diplopia An eyesight defect in which an object appears double. (DEJ)

Diptera The order of insects that includes flies and mosquitoes. (FSL)

Diquat [$(C_5H_4NCH_2)_2Br_2$] A quaternary ammonium-based contact herbicide used for the control of broadleaf weeds in ponds, bogs, and other aquatic environments. This pesticide is toxic to humans and other mammals and must be applied with care. (FSL)

Direct discharger A municipal or industrial facility that introduces pollution through a defined conveyance or system; a point source. (FSL)

Direct radiation Obsolete term for radiation, other than the useful beam, coming from within an X-ray tube and tube housing. Now designated leakage radiation. (RMB)

Direct-reading instruments Apparatus providing a direct readout of the contaminant level without further off-site laboratory analysis. Generally, direct-reading instruments have higher detection limits and lower accuracy and precision than do laboratory analytical techniques, but provide results much more rapidly. (DEJ)

Direct shear test (Soil) A test in which soil under an applied normal load is stressed to the failure point by movement of one section of the sample or sample container relative to the other section. (SWT)

Dirt Material deemed to be in the wrong place. It is a term relating to the cleaning and sanitation process and is not interchangeable with the term *soil. See also* Soil. (HMB)

Disability An impairment or defect of one or more organs or members. The following categories are defined by worker's compensation to classify the extent of damage an individual has received as a result of an injury or illness: **(1)** *Permanent partial:* A permanent physical impairment—e.g., the loss of an eye—that restricts the ability of a worker to perform certain jobs, but does not prevent him or her from working. The individual's benefits are dependent on what part of the body is injured and the degree of permanent injury to the part. This type of disability is required by OSHA to be recorded. **(2)** *Permanent total:* A disability that so limits a worker that competition in the job market is not possible. For OSHA purposes, this type of disability is recordable. **(3)** *Temporary partial:* A disability that leaves the worker capable of some work, such as light duties or part-time work, and is capable of improvement to the degree that the worker will return to former duties. For OSHA record-keeping this type of disability may result in restricted work days, but not necessarily lost time. **(4)** *Temporary total:* A disability that renders the worker incapable of working, but from which he or she is expected to recover fully. An example of this sort of disability is a broken leg. For OSHA record-keeping, this type of disability is recordable. (SAN)

Discharge The volume of water flowing through a specific point in a stream channel in a given period of time. (FSL)

Discharge of dredged material For purposes of the Clean Water Act, any addition of dredged material to the waters of the United States, including, without limitation, the addition of dredged material to a specified discharge site located in waters of the United States and the runoff of overflow from a contained land or water disposal area. Does not include: discharges resulting from the onshore subsequent processing of dredged material that is extracted for any commercial use, or for plowing, cultivating, seeding, and harvesting for the production of food, fiber, and forest products; or, *de minimis*, incidental soil movement occurring during normal dredging operations. 33 CFR §323.2(d). *See also* Clean Water Act; dredged material; wetlands. (DEW)

Discharge of fill material For purposes of the Clean Water Act, the addition of fill material into waters of the United States. Generally includes, without limitation, placement of fill for the following activities or structures: the construction of any structure in a water of the United States; the building of any structure of impoundment requiring rock, sand, dirt, or other material for its construction; site-development for recreational, industrial, commercial, residential, and other uses; causeways or roads; dams and dikes; artificial islands; property protection and/or reclamation devices such as riprap, groins, seawalls, breakwaters, and revetments; beach nourishment; levees; structures such as sewage treatment facilities, intake and outfall pipes associated with power plants and subaqueous utility lines; and artificial reefs. Does not include plowing cultivation, or seeding and harvesting for the production of food, fiber, and forest products. 33 CFR §323.2(f). *See also* Clean Water Act; fill material; wetlands. (DEW)

Discharge of pollutants For purposes of the Clean Water Act, the addition of any pollutant to navigable waters from any point source, or to the waters of the contiguous zone or the ocean from any point source other than a vessel or other floating craft. 33 USC §1362(12). *See also* Clean Water Act; pollutant; point source; navigable waters of the United States. (DEW)

Discovery Generally, the process by which parties to litigation obtain information relevant to the pending matter from each other or third parties. Methods of discovery include: depositions by oral or written examination, written interrogatories, production of documents or things, permission to enter upon land or other property for inspection or other purposes, physical and mental examinations, and requests for admissions. FRCP Rules 26–37. (DEW)

Discovery rule Legal doctrine that states that for some types of injury the statute of limitations does not begin to run until such time as the injured party discovers or reasonably should have discovered the injury. (DEW)

Disease A deviation from normal health status associated with a characteristic sequence of signs and symptoms and caused by a specific etiologic agent. (FSL)

Disease surveillance The continuing review of all aspects of occurrence and spread of a disease that are related to effective control. This may consist of the collection and review of data related to immunity levels in a population, morbidity, and mortality reports, and the laboratory isolation and identification of the agents. (FSL)

Disequilibrium, vibration-induced A loss of balance accompanied by swaying of the body and tremors, sometimes experienced by workers exposed to whole-body vibrations above 2 Hz. (DEJ)

Dishwashing machines Commercially available dishwashing machines suitable for use in food-service operations are of two basic designs: spray-type machines using hot water for washing, rinsing, and sanitizing; and spray-type machines using a hot-water wash and rinse and a final rinse containing a chemical sanitizing agent. Both of the two types of machines are commercially available with stationary dish racks or as a conveyor-type unit. They are also constructed as single- or multiple-tank systems.

A chemical sanitizing unit is a machine that consists of a tank for wash water that is circulated by an electric-motor-driven pump through spray pipes or nozzles above and below the dishes. The final rinsing is accomplished by water containing a sanitizing agent in an effective concentration applied through line pressure and/or pumped systems. The dishes are carried through the series of sprays by a timed conveyor. The single-tank units have water containing sanitizers circulated by the same electric-motor-driven pump through the same set of spray nozzles.

The hot-water unit is a machine that consists of one or more tanks for wash water, one or more tanks for pumped rinse water and a final rinse of fresh hot water. Electric-motor-driven pumps circulate the wash and pumped rinse water through spray pipes or nozzles above and below the dishes. The final rinse introduces fresh hot water through spray pipes or nozzles above and below the dishes. The dishes are carried through the series of sprays by a timed conveyor. In the single-tank units, final rinsing is done by fresh hot water from the line introduced from above and below through a separate set of spray pipes or nozzles. Any of the above units, either chemical or hot-water, may have a prewashing section ahead of the washing section.

Every dishwashing unit should be designed to provide adequate washing, rinsing, and sanitizing of all dishes being cleaned. The temperature and volume of water, time of spraying, spray pattern, nozzle size, and pressure are all important in terms of the effectiveness of any dishwashing machine operation. Machine design will vary as to the minimum time and temperature required for each cycle. For example, hot-water washers should provide washing for a minimum of 40 seconds, at temperatures that range from 150°F (65.6°C) to 165°F (74°C), and a rinse temperature of at least 180°F (82.2°C). Chemical sanitizer units, on the other hand, must provide a wash cycle for a minimum of 45 seconds, at a temperature of at least 120°F (48.9°C), and a rinse cycle at a temperature of 120°F (48.9°C), typically using a chlorine sanitizer of 50 ppm strength. These time–temperature relationships are based on standards developed by the National Sanitation Foundation. (HMB)

Disinfectant Chemicals employed to reduce or kill microorganisms present on inanimate objects. They are usually harsher chemicals than antiseptics and therefore are not used to decontaminate or disinfect plants or animals. *See also* Antiseptic; chlorination. (HMB)

Disinfection The destruction of infectious microorganisms on the surface of an object, usually by chemical means. Disinfectants are usually employed to reduce the possibility of infection from the handling of contaminated materials. Disinfection is the treatment of materials to eliminate or reduce the number of disease-causing organisms. *Disinfection* is not synonymous with *sterilization*; rather, it implies that some microorganisms will remain after treatment. *See also* Disinfectant; sterilization. (FSL, HMB)

Disinfestation A physical or chemical process intended to eliminate undesirable arthropods or rodents from an environment or to destroy arthropods present upon individuals or in their clothing. There are, for example, numerous chemical shampoos on the market that are used for the removal of head lice (*Pediculus capitis*). An important part of a general sanitation program in food processing or food-service operations. *See also* Insect infestation; rodent control. (FSL, HMB)

Disintegration (Radiation) The process of spontaneous breakdown of a nucleus of an atom, resulting in the emission of a particle and/or photon. The rate of disintegration of a quantity of any radioactive nuclide is a function of the number of atoms present and a disintegration or decay constant characteristic of the nuclide. When numbers of nuclei are involved, the process is characterized by a definite half-life. (RMB)

Disperse (soil) To distribute or suspend fine particles, such as clay or silt, in or throughout a dispersion medium. *See also* Dispersion medium. (SWT)

Dispersing agent A material that increases the stability of particles in a liquid. Examples are surfactants and carboxymethylcellulose. *See also* Soil suspending agent; surfactant. (HMB)

Dispersion medium The portion, such as water, of a colloidal system in which the disperse phase is distributed. (SWT)

Disposal Final placement or destruction of toxic, radioactive, or other wastes; surplus or banned pesticides or other chemicals; polluted soils; and drums containing hazardous materials from removal actions or accidental releases. Disposal may be accomplished through use of approved secure landfills, surface impoundments, land farming, deep well injection, ocean dumping, or incineration. (FSL)

Disposal facility Any facility or location where any treatment, utilization, processing, or deposition of waste occurs. (FSL)

Disposal operations Activities including administration, personnel, land, equipment, design, and other elements necessary or used in carrying out solid-waste disposal. (FSL)

Disposal site The location at which the final deposition of solid waste occurs. (FSL)

Dissection (soil) The partial destruction of a surface of land formed by erosion, such as gullies or canyons, leaving ridges or hills that are separated by drainage ways. *See also* Gully. (SWT)

Dissolved oxygen (DO) The oxygen freely available in water. Dissolved oxygen is vital to fish and other aquatic life and for the prevention of odors. Traditionally, the level of dissolved oxygen has been accepted as the single-most-important indicator of a water body's ability to support desirable aquatic life. Secondary and advanced waste treatment are generally designed to protect DO in waste-receiving waters. (FSL)

Dissolved solids Disintegrated organic and inorganic material contained in water. Excessive amounts make water unfit to drink or for use in industrial processes. (FSL)

Distemper A viral disease of dogs that is usually characterized by high fever, loss of appetite, and nasal discharge. (FSL)

Distillation The process of heating a mixture to separate its various volatile components by condensation. The act of purifying liquids through boiling, so that the steam condenses to a pure liquid and the pollutants remain in a concentrated residue. (FSL)

Distress As pertaining to excavation, a situation in which the soil is in a condition that makes a cave-in imminent or likely to occur. Distress is evidenced by such phenomena as: the development of fissures in the face of, or adjacent to, an open excavation; the subsidence of the edge of an excavation; the slumping of material from the face of an excavation or the bulging or heaving of material from the bottom of an excavation; the spalling (chopping or splitting) of material from the face of an excavation; and ravelling (pebbles or small clumps of material suddenly separating from the face of an excavation and trickling or rolling down into the excavation). *See also* Excavation; soil classifications. (SAN)

Distribution The movement of a chemical substance from the entry site and throughout the body. (FSL)

Distribution (Biostatistics) *See specific types:* Bernoulli, bimodal, binomial, chi-square, cumulative frequency, frequency, Gaussian, normal, Poisson, probability, sampling, skewed (negatively and positively), standard normal, symmetric, trimodal, unimodal. (VWP)

District court *See* Federal court. (DEW)

Disulfiram (tetraethylthiuram disulfide) $\{[(C_2H_5)_2NCS]_2S_2\}$ Known by the trade name Antabuse, this compound, when taken in the presence of alcohol, produces severe nausea and vomiting. It inhibits the oxidation of acetaldehyde, resulting in the buildup of this chemical in the body. It is frequently used in programs for the rehabilitation of alcoholics. *See also* Antabuse. (FSL)

Dithekite An explosive mixture of nitric acid, nitrobenzene, and water. (FSL)

Diuresis An increase in the volume of urine produced by the kidneys. (FSL)

Diuretic A chemical that promotes the secretion of urine. (FSL)

Diurnal An activity that occurs during daylight hours or during a period of light. (FSL)

Diuron $[C_6H_3Cl_2NHCON(CH_3)_2]$ A soil herbicide of the substituted urea group that is used for total weed control on acreage not intended for crops. Residuals can last up to one year after application. Diuron is a skin irritant and is harmful to fish. (FSL)

Diversion dam A structure or barrier built to divert part or all of a watercourse to a different route. (SWT)

DNA *See* Deoxyribonucleic acid. (FSL)

DNA hybridization Use of a segment of DNA, called a DNA probe, to identify its complementary DNA; used to detect specific genes. This process takes advantage of the ability of a single strand of DNA to combine with a complementary strand. (FSL)

Dock plate A temporary bridge between a surface and a temporarily docked vehicle or ship. Dock plates should be fixed in place and labelled with their load capacity. Some dock plates are built into loading platforms and power lifts for adjustment. *See also* Gangplank. (SAN)

Domestic wastes Collective term that includes human sewage, liquid wastes from sinks and dishwashers, and the so-called gray water from laundries. (FSL)

Domino theory Theory of accident causation, promoted by W.F. Heinrich, that states that an accident is the result of a sequence of events. One way of illustrating this theory is to picture a row of dominos, the tumbling of each of which represents an event. When the first event occurs, it leads to another, and so on, until the entire row is knocked over. Heinrich assigned categories to events that influence whether an accident will occur as a result. The first is the social environment; the second is undesirable traits that have been inherited by an individual as a result of social environment; the third comes into play when these undesirable traits lead to the practice of unsafe acts and conditions; the fourth is the occurrence of the accident itself; and the final is injury. (SAN)

Donora smog An air pollution episode that occurred in the industrial town of Donora, Pennsylvania, in October 1948. As the result of a series of inversions, pollutants such as sulfur dioxide and particulates from local industries accumulated in the atmosphere, causing about 6000 cases of illness and 18 deaths among the town's residents. (FSL)

Dosage *See* dose. (FSL)

Dose **(1)** (Radiation) A quantity (total or accumulated) of ionizing (or nuclear) radiation. The term "dose" is often used in the sense of the exposure dose, expressed in roentgens, which is a measure of the total amount of ionization that the quantity of radiation could produce in air. This should be distinguished from the absorbed dose, expressed in reps or rads, which represents the energy absorbed from the radiation per gram of specified body tissue. The biological dose, in rems, is a measure of the biological effectiveness of the radiation exposure. **(2)** (Toxicology) The total amount of a toxicant, drug, or other chemical administered to the organism. *See also* Rad; RBE dose; rem; rep; roentgen. (FSL) (RMB)

Dose rate As a general rule, the amount of ionizing (or nuclear) radiation to which an individual would be exposed or that he or she would receive per unit of time. It is usually expressed in roentgens, rads, or rems per hour, or in multiples or submultiples of these units, such as milliroentgens per hour. The dose rate is commonly used to indicate the level of radioactivity in a contaminated area. (RMB)

Dosimeter **(1)** A portable instrument for measuring and registering the total accumulated exposure to ionizing radiation. **(2)** A device capable of measuring a person's exposure to hazardous substances or agents, usually over a period of time, extending from less than one hour to several months. Dosimeters are commonly employed to measure occupational exposures to noise, radiation, and chemicals. Specific types of dosimeters include the following. (a) *Noise dosimeter:* A device that measures an individual's exposure to noise over a period of time. The device consists of a microphone worn near the ear (typically clipped to the shoulder of the shirt), a cable, an amplifier, and a memory chip or other recording device, which is typically worn on the belt. Modern dosimeters are capable of calculating an individual's exposure in decibels, weighted on the A scale, using several different thresholds, percent of allowable dose, peak exposure intensity, duration, time of occurrence, and other features. (b) *Passive dosimeter:* An air-sampling device that relies on diffusion (rather than dynamic pumping and sampling of air) through a sorbent bed to determine time-weighted average exposures to a variety of chemical substances, usually gases and vapors. The device often consists of a wind screen to provide a uniform airflow rate and constant face velocity, and a sorbent bed, which can be analyzed either on-site immediately following the sampling period, or in a laboratory after the dosimeter has been sealed and shipped, depending on the specific compound sampled. Passive dosimeters are available for formaldehyde, ethylene oxide, and various organic vapors. *See also* Dosimetry. (RMB, DEJ)

Dosimetry The theory and application of principles and techniques involved in the measurement and recording of radiation doses. Its practical aspect is concerned with the use of various types of radiation instruments with which measurements are made. *See also* Chemical dosimeter; film badge, survey meter. (RMB)

DOT Department of Transportation; federal agency that enforces the regulations governing the transport of hazardous and nonhazardous materials. (FSL)

Double block and bleed The closure of a line, duct, or pipe by locking and tagging a drain or vent that is open to the atmosphere in the line between two closed valves that have been locked. *See also* Blinding. (SAN)

Dowsing The practice of "water-witching" or search for water or other underground substances by supposed detecting of the motion of a forked stick or other item held in the hands. (FSL)

Drain tile A pipe designed to conduct water from the soil. (SWT)

Drawdown The condition that occurs when the level of the water table in the vicinity of a well is lowered, causing the water table to take the shape of an inverted cone, called the cone of depression. *See also* Cone of depression. (FSL)

Dredged material For purposes of the Clean Water Act, material that is excavated or dredged from waters of the United States. 33 CFR §323.2(c). *See also* Clean Water Act; discharge of dredged materials; wetlands. (DEW)

Dredging Removal of mud from the bottom of water bodies, using a scooping machine. Dredging of contaminated muds can expose aquatic life to heavy metals and other toxins. Dredging activities may be subject to regulation under Section 404 of the Clean Water Act. (FSL)

Drilled well Well constructed by either percussion or rotary hydraulic drilling. The selection of the method depends primarily on the geology of the site and the availability of equipment. This is the most popular type of well, and, when properly cased, grouted, and sealed, it will yield water of high quality, provided the aquifer is not contaminated. (FSL)

D-ring A ring that is flat on one side and used in fall protection systems. The flat portion of the ring attaches to a safety harness. *See also* Coupling. (SAN)

Drip irrigation A method of irrigation whereby water is applied to the soil slowly by use of nozzles with small orifices. (SWT)

Driven well Well constructed by driving into the ground a well point, which is fitted to the end of a series of pipe sections. The well is driven with the aid of a maul or special drive weight. The water yield from this type of well is generally small to moderate. (FSL)

Droplet nuclei The residues that result from evaporation of fluid from droplets emitted by an infected animal or human host. These dried residues of droplets may contain infectious microorganisms. In contrast to droplets, droplet nuclei can remain suspended in the air for long periods. *See also* Droplets. (FSL)

Droplets Liquid particles expelled into air during the act of talking, spitting, singing, coughing, or sneezing. Droplets are formed through aerosolization of secretions present in the mouth, nasopharynx, and bronchi. They can contain infectious microorganisms. *See also* Droplet nuclei. (FSL)

Drought An extended period of time in which less than the expected amount of precipitation falls. (FSL)

Dry-bulb temperature The temperature derived from a thermal sensor or a thermometer that is shielded from a direct radiant energy source. The dry-bulb temperature is used for estimating comfort conditions. It is one of three ambient indices used for heat-stress analysis. (MAG)

Dry chemical/dry powder An extinguishing agent composed of small particles of chemicals such as sodium bicarbonate, potassium bicarbonate, urea-based potassium bicarbonate, potassium chloride, or monoammonium phosphate, supplemented by special treatment to provide resistance to packing and moisture absorption (caking), as well as to provide proper flow capabilities. A multipurpose dry chemical is a dry chemical approved for use on class A, class B, and class C fires. Dry powders, also called combustible metal agents—e.g., talc and soda—are used to control class D (metal) fires. (SAN)

Drying agent Substances, such as silica gel or calcium oxide, that extract water. *See also* Desiccant. (FSL)

Dryland farming The practice of growing crops using only rainfall as water (without supplemental water from irrigation). (SWT)

Dry weight percentage (soil) Ratio of the weight of any soil constituent to the oven-dry weight of the soil. (SWT)

Due diligence (law) The degree of attention or care expected of a person in a given situation. *See also* Innocent landowner defense. (FSL)

Due process of law Constitutional concept derived from the Fifth Amendment to the United States Constitution, which states that no person shall "be deprived of life, liberty, or property, without due process of law." In addition to the federal coverage provided by the Fifth Amendment, similar language in the Fourteenth Amendment is applicable to state action. The practical applications are often left to the courts. Generally, due process is divided into procedural and substantive types. Procedural due process requires that the process by which one is deprived of property or liberty must be fair and provide adequate notice and a right to a fair hearing. The particular procedural elements that comprise fairness depend on the situation. Substantive due process requires that legislation be reasonably related to the furtherance of a legitimate government purpose, and carry out that purpose in the method least burdensome to the protected rights of those affected. (DEW)

Dug well Type of well, now rarely constructed, named because it is typically dug with pick and shovel. These wells, which are dug to a depth of 15 to 20 feet, are usually subject to surface contamination and are not considered reliable sources of water. (FSL)

Dump A site used to dispose of solid wastes without consideration of environmental factors. *See also* Landfill. (FSL)

Dunk tank A laboratory device containing liquid germicide used for the passage of biological materials into or out of a gas-tight containment chamber. (JHR)

Duripan (soil) A mineral soil that is cemented by silica, aluminum silicate, calcium carbonate, or iron. (SWT)

Dust Large airborne or settled particles usually formed by abrasion. Visible and/or invisible particles can be included. These particles are of widely varying size; they may arise from soil, bedding, or surfaces such as floors or walls. (DEJ)

Dust collector A device used in filtering air to remove particulate matter. The four main types of dust collectors used in exhaust ventilation systems are electrostatic precipitators, fabric collectors, wet collectors, and dry centrifugal collectors. (DEJ)

Dust explosion A very rapid combustion of a cloud or suspension of dust, caused by the presence of an ignition source, oxygen, and a finely divided airborne combustible material. Ranges of explosive airborne concentrations vary according to the particular material and its size distribution. The explosive limit for many dusts is approximately 100,000 mg/cubic meter, or about 0.1 ounce per cubic foot of air. Dust explosions often occur in series, with the initial explosion dislodging additional dust from roofs and support beams. Dust explosions have occurred in grain elevators, storage bins, and mines. (DEJ)

Dustfall jar An open container used to collect large particles from the air for measurement and analysis. (FSL)

Dust mulch A loose, finely granular, or powdery condition on the surface of the soil, usually produced by cultivation. (SWT)

D value The time in minutes required to kill 90% of the cells in a culture. D values should indicate with subscripts the temperature in degrees Celsius. The values reflect the heat resistance of bacteria at a given temperature. They make it possible to develop the exact time and temperature necessary to assure the destruction of microorganisms in heat treatments. *See also* Heat as a sanitizer; Z value. (HMB)

Dynamic measurement An aspect of anthropometry involving the correct location of controls, tools, and other items requiring worker manipulation. *See also* Anthropomorphic. (DEJ)

Dynamic penetrometer An instrument that is used to measure the hardness of the surface and that is forced into the soil by the use of a hammer or weights. Also, an instrument that measures the penetrating power of X-rays. (SWT)

Dynamite A commercial explosive, developed by Alfred Nobel (1833–1896), which was made of nitroglycerin bonded with nitrocellulose. (FSL)

Dynamometer An instrument used to measure the draft of tillage implements and the resistance of soil to penetration by them. (SWT)

Dyne The unit of force that, when acting upon a mass of 1 gram, will produce an acceleration of 1 centimeter per second per second. (FSL)

Dysbarism A term representing a number of disorders that result from compression. Also called compression sickness and caisson disease. *See also* Caisson; caisson disease. (SAN)

Dysphagia Difficulty in swallowing. (FSL)

Dyspnea Shortness of breath, or the sensation of it, due to labored breathing. (MAG)

Dystrophic lakes Shallow bodies of water that contain humus and/or organic matter; they tend to contain many plants but few fish and are generally highly acidic. (FSL)

Dystrophy The collective term for disorders due to nutritional disturbances. (FSL)

E

E^0 Symbol for oxidation-reduction potential. *See also* Oxidation-reduction potential; redox potential. (HMB)

Earmuffs Devices usually worn by an individual to guard against hearing loss in high-noise environments, or to protect against cold environmental conditions. *See also* Circumaural protectors. (DEJ)

Earplugs *See* Aural Insert Protectors and Superaural Protectors. (DEJ)

Ear Protection A device usually worn by an individual to guard against hearing loss in high-noise environments. Forms of hearing protection including aural insert protectors, superaural protectors, and circumaural protectors. These types of devices are used to protect the ears in cold environmental conditions, and to prevent entry of water into the ear. (DEJ)

Eccrine A form of sweat consisting mostly of dilute salt water. Its function is to help the body dissipate internal metabolic heat gain by evaporation from the surface of the skin. Eccrine sweat glands exist on all areas of the skin, except the lips and a few other areas. *See also* Apocrine. (DEJ)

Ecological impact The effect that a human or natural activity has on organisms and their nonliving (abiotic) environment. (FSL)

Ecology The relationship of living things to one another and their environment or the study of such relationships. (FSL)

Ectomycorrhiza A situation in which the fungal hyphae grow on the surface of the roots. This root–fungus association is of mutual benefit to the plant and fungus, particularly in low-fertility soils, in which the fungus can enhance nutrient uptake. (SWT)

Economic poisons Chemicals used to control pests and to defoliate cash crops such as cotton. (FSL)

Ecosphere The "bio-bubble" that contains life on the earth, air. (FSL)

Ecosystem The interacting system of a biological community and its nonliving environmental surroundings. (FSL)

Ecotome The transition zone between two distinctly different habitats; an edge habitat. (SMC)

Eczema Generalized term for an inflammatory process involving the epidermis and marked by itching, weeping, and crusting. (FSL)

Edaphic Relating to soil or influenced by soil conditions. (SMC)

Edaphology The study of the influence of soils on living things, particularly plants, and including use of land for plant growth. (SWT)

Eddy A current in a fluid that moves in a direction contrary to that of the main stream. (FSL)

Edema A condition in which body tissues contain an excessive amount of fluid. The presence of abnormally large amounts of fluid in the intercellular tissue spaces of the body or part of the body. (FSL)

Effective atomic number A number calculated from the composition and atomic numbers of a compound or mixture. A compound or mixture of this effective atomic number would interact with photons in the same way as the element of that actual atomic number. (RMB)

Effective dose (ED) The amount of a toxicant (or drug) required to cause a given functional change in an intact organism, at a biochemical site, or in isolated tissue. Expressed as a proportion of the population or target affected—e.g., ED_{50}, ED_{75}. (FSL)

Effective half-life The time required for a radioactive element fixed in the tissues of an animal body to be diminished 50% as a result of the combined action of radioactive decay and biological elimination.

$$\text{Effective half-life} = \frac{\text{Biological half-life} \times \text{Radiological half-life}}{\text{Biological half-life} + \text{Radioactive half-life}}$$

(RMB)

Effective precipitation The portion of total rainfall that becomes available for plant growth. (SWT)

Efficiency A factor used to convert the counting rate of a detector to the disintegration rate of the radioactive material counted. Since usage and factors involved vary considerably with different detectors, it is well to ascertain which factors (window transmissions, sensitive volume, energy dependence, etc.) are included in a stated efficiency. (RMB)

Effluent An outflow or discharge, as from a storm sewer, sanitary sewer, or sewage treatment plant. (FSL)

Effluent limitation Under the Clean Water Act, a restriction established by EPA on quantities, rates, and concentrations of chemical, physical, biological, and other constituents discharged from point sources, other than new sources, into navigable waters, the waters of the contiguous zone, or the ocean. 40 CFR §401.11(i). (DEW)

Effluent standard The maximum amount of specific pollutants permitted in a discharge from a sewage treatment plant or industrial treatment facility. (FSL)

Egregious A mega-fine policy structure implemented by OSHA whereby a separate penalty is assessed for each instance of a violation or, in some cases, for each employee exposed to an identified hazard. (MAG)

Elastic range The stress range in which a material will recover its original form when the force (or loading) is removed. *Elastic deformation* refers to dimensional changes occurring within the elastic range. *See also* Plastic range. (RMB)

Electrical safety The prevention of electrical hazards that can produce shock, heat, fire, or explosion. Overheating of electrical equipment is a common cause of fire. Arcing, which occurs when current flows through air between two conductors that are not in direct contact in the presence of an atmosphere containing combustible dust or flammable vapors, may cause an explosion. (SAN)

Electrochemistry The science that deals with the use of electrical energy to bring about a chemical reaction or with the generation of electrical energy by means of chemical action. (FSL)

Electrode Either terminal of an electrical apparatus. (RMB)

Electrodialysis A process that uses electrical current applied to permeable membranes to remove minerals from water. Often used to desalinize salty or brackish water. (FSL)

Electrolyte A chemical substance that breaks down into electronically charged particles (ions) when dissolved in water or melted. (FSL)

Electromagnetic radiation A traveling wave motion resulting from changing electric or magnetic fields. Familiar electromagnetic radiations range from X-rays and gamma rays, which are of short wavelength, through the ultraviolet, visible, and infrared regions, to radar and radio waves, which are of relatively long wavelength. All electromagnetic radiations travel with the velocity of light in a vacuum. (RMB)

Electrometer Electrostatic instrument for measuring the difference in potential between two points. Used to measure a change of electric potential of charged electrodes resulting from ionization produced by radiation. (FSL)

Electromotive force Potential difference across electrodes tending to produce an electric current. Abbreviation: emf. (RMB)

Electron A negatively charged particle that is a fundamental constituent of all atoms. A unit of negative electricity equal to 4.802×10^{-10} esu. Its rest mass is 0.000548 amu, which is approximately 1/1840 that of the hydrogen atom. It is not found in the atomic nucleus but rather in orbitals around the nucleus. (RMB)

Electron capture A mode of radioactive decay in which an orbital electron merges with a proton in the nucleus. The process is followed by emission of an electron or photon. Capture from a particular electron shell is designated as K-electron capture, L-electron capture, etc. (RMB)

Electron equilibrium The condition in which as many electrons are stopped as are produced in a slice of material when the material is under irradiation by photons. (RMB)

Electron volt A unit of energy equivalent to the amount of energy gained by an electron in passing through a potential difference of 1 volt. Larger multiple units of the electron volt are frequently used—e.g., keV (thousand electron volts), MeV (million electron volts), and BeV (billion electron volts). Equals 1.6×10^{-12} erg. Abbreviation: eV. (RMB)

Electrophile A chemical compound that accepts electrons in a chemical reaction. (RMB)

Electrophoresis The result of the application of an electrical field, on the movement of charged particles that have been suspended in different types of media (gel, paper, liquid). (RMB)

Electroscope Instrument for detecting the presence of electric charges by the deflection of charged bodies. (RMB)

Electrostatic field The region that surrounds an electric charge and in which another electric charge experiences a force. (RMB)

Electrostatic precipitator (ESP) A filtration system consisting of a high-potential electric field that is formed between discharge and collecting electrodes of opposite electrical charges. Ionization occurs as the air travels through the field and imparts an electrical charge to the particles, which are then attracted to the electrically charged collecting plate. The collecting plate can then be removed and cleansed of dust particles. Electrostatic precipitators are configured in either one-stage or two-stage systems, carrying up to 70,000 volts and 15,000 volts, respectively. (DEJ)

Electrostatic unit of charge (statcoulomb) The quantity of electric charge that, when placed in a vacuum 1 centimeter from an equal and like charge, will repel that charge with a force of 1 dyne. Abbreviation: esu. (RMB)

Element A pure substance that cannot be broken down into a simpler substance by a chemical change, but whose atoms will disintegrate into simpler particles through physical decomposition when exposed to drastic bombardment with high-energy particles. (RMB)

Eligible costs The construction costs for wastewater treatment works upon which Environmental Protection Agency grants are based. (FSL)

Elimination The removal of a chemical substance from the body by metabolism or excretion. (FSL)

Elutriation The process of separating lighter particles of a powder from heavier particles by means of an upward stream of fluid. (FSL)

Elutriator An air-sampling device that uses gravitational force to remove nonrespirable dust from the air sample. The dust is then collected on a filter for further analysis. This equipment is widely used in cotton-dust area air sampling. (DEJ)

Eluviation The process whereby soil material (clay, humus, etc.) is removed, as a suspension or solution, from a layer or layers of a soil. Usually the loss of material in solution is described by the term ''leaching.'' *See also* Alluvial; illuviation. (SWT)

Embryo An organism in an early stage of development. In humans, the developing organism from conception until the end of the second month. A primordial plant within a seed. (FSL)

Emergency (chemical) A situation created by an accidental release or spill of hazardous chemicals that poses a threat to the safety of workers, residents, the environment, or property. (FSL)

Emergency and destruct systems Systems developed for process safety to relieve pressure on vessels by providing a quick means of emptying the vessels. These measures include dump or quench tanks, incinerators, thermal oxidation units, scrubbers with both normal and emergency venting, and flare towers. *See also* Flare tower. (SAN)

Emergency egress Established routes or procedures for the evacuation of a structure, vehicle, or other facility in the event evacuation is necessary to protect life or limb. (FSL)

Emergency eyewash and shower OSHA requires that an emergency eyewash and shower be available for quick drenching or flushing of the eyes and body of any person who may be exposed to injurious corrosive materials. These facilities should be located within the work area, usually within 100 feet of the corrosive source. Eyewashes and showers should be inspected monthly and any water provided to them should be potable. (SAN)

Emergency plan A plan that includes written instructions, training, and drills for preparing for emergencies. The purpose of planning is to minimize the potential for harmful consequences in the event of a major incident. The plan must provide for the safety of personnel and the public, and must include a statement of priorities and procedures for emergency controls, hazard containment, personnel evacuations, and medical aid. It should deal with all types of emergencies that could occur, including fire and explosion, release of highly toxic chemicals, major chemical spills, acts of nature, power failure, sabotage (including bomb threats), and inclement weather. Supervisors and employees must be familiar with the plan and its procedures through instruction, discussion, and practice drills. Emergency information such as telephone numbers and exit routes must be clearly posted. (SAN)

Emergency Planning and Community Right-to-Know Act Part of the Superfund Amendments and Reauthorization Act of 1986; requires states and local areas to design and implement emergency response plans for hazardous-substance emergencies. Specified industries are also required to report releases into the environment of designated hazardous substances. These data are compiled and made available to the public through the Toxic Chemical Release Inventory. 42 USC §11001 *et seq. See also* Comprehensive Emergency Response Plan; Emergency Response Commission; local emergency planning committees; Toxic Chemical Release Inventory. (DEW)

Emergency Response Commission Required by the Emergency Planning and Community Right-to-Know Act, state commission responsible for appointing local emergency response planning committees and supervising and coordinating their activities under the act. *See also* Emergency Planning and Right-to-Know Act; local emergency planning committees. (DEW)

Emetic An agent that causes vomiting. (FSL)

Eminent domain Power of the government to take private property for public use; forced acquisition, with compensation paid to the land-owner. *See also* Taking. (DEW)

Emission Pollution discharged into the atmosphere from smokestacks, other vents, and surface areas of commercial or industrial facilities; from residential chimneys; and from motor vehicle, locomotive, or aircraft exhausts. (DEJ)

Emission inventory A listing, by source, of the amount of air pollutants discharged into the atmosphere of a community. It is used to establish emission standards. (FSL)

Emission peak The highest instantaneous release of any contaminant into the environment over a given time period, often 8 or 24 hours, usually expressed in a unit of concentration such as milligrams/cubic meter or parts per million. (DEJ)

Emission rate The quantity of contaminant released into the environment per unit time, usually expressed in grams per hour. The emission rate may be identified as either a peak emission rate or as a time-weighted average emission rate. (DEJ)

Emission standard The maximum amount of air-polluting discharge legally allowed from a single source, mobile or stationary. (FSL)

Emissions trading Environmental Protection Agency policy that allows a plant complex with several facilities to decrease pollution from some facilities while increasing it from others, so long as total results are equal to or better than previous conditions. Facilities where this is done are treated as if they exist in a bubble in which total emissions are averaged out. Complexes that reduce emissions substantially may "bank" their "credits" or sell them to other industries. (FSL)

Empirical Relying or based strictly on experiments and observation rather than theory or relying on practical experience without reference to scientific principles. (FSL)

Employee safety involvement That which is provided for and encouraged in the operation of an organization's health and safety program, including the decisions that affect employee safety and health. The employer consults with employees and/or their representatives concerning the development and implementation of the program. The plan provides for employee training in hazard recognition and accident investigation. Employees are involved in job health and safety and process hazard analyses to help restructure jobs and processes to prevent or control hazards. *See also* Accident investigation; job health and safety analyses; process hazard analyses. (SAN)

Emulsification The process of forming an emulsion. The function of a cleaning compound is to dislodge soil, often water-insoluble, and suspend soil particles for subsequent removal from surfaces. The suspension process of these water-insoluble materials through interaction with a soap is an example of emulsification. *See also* Emulsion. (HMB)

Emulsifier A surface-active agent that promotes the dispersion of one liquid in another, such as small fat globules in water. (HMB)

Emulsion A dispersion or suspension of one liquid in another, with the molecules of the two liquids immiscible or mutually antagonistic. An emulsion has three phases: one consists of suspended droplets, the second phase is the continuous phase (or the dispersion medium), and the third is the emulsifier. Cleaning compounds disperse soil deposits in the cleaning medium by creating an emulsion. *See also* Emulsification. (HMB)

Enclosure A fence, cage, or other boundary designed to keep organisms of a certain size or habitat within a designated area. (SMC)

Endangered species Under the Endangered Species Act, any species that is in danger of extinction throughout all or a significant portion of its range. This extinction may be due to the present or threatened destruction, modification, or curtailment of habitat or range, overutilization for commercial, recreational, scientific, or educational purposes, disease or predation, inadequacy of existing regulatory mechanisms, or other natural or human-related factors affecting the continued existence of the endangered species. 42 USC §1532(6) & 1533(a). *See also* Critical habitat; Endangered Species Act; threatened species. (DEW)

Endangered Species Act Federal law that provides a means whereby the ecosystems upon which endangered species and threatened species depend may be conserved, to provide a program for the conservation of such species, and to take such steps as may be appropriate to achieve the purposes of the relevant treaties and conventions to which the United States is a signatory. Specifically, the act provides for the listing of plants and animals as endangered or threatened, the preparation of plans for the recovery of those species and restrictions on their killing, collecting, and sale or purchase, and conservation of their habitats. Administered by the U.S. Fish and Wildlife Service of the Department of the Interior and the National Marine Fisheries Service of the Department of Commerce. 16 USC §1531 *et seq. See also* Endangered species; threatened species; critical habitat. (DEW)

Endangerment assessment A study conducted to determine the nature and extent of contamination at a site on the National Priorities List and the risks posed to public health or the environment. (FSL)

Endemic Refers to diseases or infectious agents in the human population within a given geographic area that are constantly present or usually prevalent. For example, malaria is endemic in various tropical regions of the world. (FSL)

Endoergic reaction Reaction that absorbs energy. (RMB)

Endosperm In seeds, the tissue containing stored food outside the embryo. (SMC)

Endospore A thick-walled structure formed within the cells of certain bacteria that allows the organism to withstand adverse environmental conditions, such as drying. (FSL)

Endothermic Refers to a reaction in which the products contain more energy than the reacting materials, and in which energy is therefore absorbed. *See also* Exothermic. (FSL)

Endotoxin A toxic substance associated with the outer membrane of some Gram-negative bacteria, such as *Brucella* and *Vibrio* species. In humans the presence of endotoxins may produce fever, diarrhea, and other biologic effects. Endotoxins are present in the cell but not in cell-free extracts of cultures of intact cells. These toxins are not proteins, but lipopolysaccharide complexes. They produce fever, severe diarrhea, and altered immunity states. Salmonellosis symptoms for example are caused by the effects of an endotoxin. (HMB)

Endotrophic Refers to organisms that receive nourishment from within other organisms, such as fungi that receive nourishment from plant roots. (SWT)

Endrin ($C_{12}H_8OCl_6$) One of the so-called ''hard'' or persistent chlorinated hydrocarbon insecticides that is known to be toxic to mammals and to bioaccumulate in the environment. The use of this material has been banned in the United States. (FSL)

Energy The capacity to do work. Energy (E) is the product of the force (F) acting on a body to produce motion, and the distance (s) through which the force acts, and the cosine of the angle (θ) between the direction of force and direction of motion. $E = Fs \cos \theta$. (RMB)

Energy dependence The characteristic response of a radiation detector to a given energy, as compared with some standard response. For gamma survey instruments, radium is often used as the standard. (RMB)

Energy-isolating device A mechanical device that physically prevents the transmission or release of energy. Such devices include: manually operated electrical circuit breakers, disconnect switches; manually operated switches (by means of which the conductors of a circuit can be disconnected from all ungrounded supply conductors, and no pole can be operated independently); line valves; and blocks. Push-buttons, selector switches, and other control-circuit-type devices are not considered energy-isolating devices. *See also* Lockout/tagout. (SAN)

Enforcement Action of the Environmental Protection Agency, the state, or local authorities to obtain compliance with environmental laws, rules, regulations, or agreements, and/or to exact penalties or criminal sanctions for violations. Enforcement procedures may vary, depending on the specific requirements of different environmental laws and related implementing regulatory requirements. (FSL)

Enforcement decision document (EDD) A document that provides an explanation to the public of the Environmental Protection Agency's selection of a cleanup alternative at an enforcement site on the National Priorities List; similar to a record of decision. (FSL)

Engineering controls Measures taken to prevent or minimize exposure to biological, chemical, or physical agents in the occupational environment. The measures can include improved ventilation, noise reduction, or substitution of a chemical with a lower order of toxicity for one that is highly hazardous. Engineering controls can include modification of the source, the transmission path, the amount of ventilation, or the quantity of contaminants released into the work area or the environment. (DEJ) (FSL)

Engulfment The situation that exists when a person or object is surrounded and effectively confined by a liquid or finely divided solid substance. *See also* Low-hazard permit space; permit-required confined space. (SAN)

Enrichment The addition of nutrients (e.g., nitrogen, phosphorus, carbon compounds) from sewage effluent or agricultural runoff to surface water. This process increases the growth potential for algae and aquatic plants. (FSL)

Entamoeba Parasitic amoeba that can affect human and animal species—e.g., *Entamoeba hystolytica*, which can cause dysentery and liver abscesses. (FSL)

Enteric Pertaining to the intestines. (HMB)

Enteritis Pertaining to an inflammation of some portion of the enteric system. Foodborne illnesses—e.g., staphylococcal intoxication—typically cause this type of condition. (HMB)

Entero- A combining form denoting relationship to the intestines. (HMB)

Enterobacteria Gram-negative, anaerobic, rod-shaped bacteria found in the intestinal tracts of humans and animals. A large proportion of hospital-acquired intestinal infections are due to enterobacteria, such as *Proteus, Klebisella*, and *Escherichia*. (FSL)

Enterobiasis An infection with the so-called ''pinworm'' (*Enterobius vermicularis*) found in the upper part of the large intestine in humans. Infestations with this parasite are common in children, who commonly complain of itching in the rectal area. (FSL)

Entisols Mineral soils with no distinct subsurface diagnostic horizons within one meter of the soil surface. (SWT)

Entomology The study of insects (Class Insecta, Phylum Arthropoda). (SMC)

Entrant An employee who is authorized by the employer to enter a permit-required confined space. Authorized entrants may rotate duties, serving as attendants if the permit program and the entry permit indicate. Any properly trained person with the authority to authorize entry by other persons may enter the permit space during the term of the permit, provided the attendant is informed of that entry. *See also* Attendant; entry permit; permit-required confined space. (SAN)

Entry permit In permitting systems, as applied to the occupational setting, a document, established by the employer, the content of which is based on the employer's hazard identification and evaluation for a specific confined space. It is the means by which an employer authorizes employees to enter a permit-required confined space. The entry permit defines the conditions under which the space may be entered, states the reason(s) for entering the space, and the anticipated hazards of the entry. For entries where the individual authorizing the entry does not assume direct charge of the entry, the permit lists the eligible attendants, entrants, and individuals who may be in charge of the entry. In entry permits, the length of time for which the permit may remain valid is established. *See also* Permit-required confined space; permitting systems. (SAN)

Envenomization The passage of a toxin from an arthropod to a human by a bite, sting, or contact with urticating hairs or vesicating fluid. (FSL)

Environment The sum of all external conditions affecting the life, development, and survival of an organism. (FSL)

Environmental assessment A written environmental analysis that is prepared pursuant to the National Environmental Policy Act, by a federal agency in anticipation of proposed legislation or federal action. The analysis briefly provides sufficient evidence for determining whether to prepare an environmental impact statement or a finding of no significant impact, and aids an agency's compliance with NEPA when no environmental impact statement is necessary. It also includes a brief discussion of the need for the proposal, of alternatives, and of the environmental impacts of the proposed action and alternatives, as well as a listing of agencies and persons consulted. 40 CFR Part 1508.9. *See also* Environmental impact statement; finding of no significant impact; National Environmental Policy Act. (DEW)

Environmental Audit As defined by the Environmental Protection Agency, ''. . . a systematic, documented periodic and objective review by regulated entities of facility operations and practices related to meeting environmental requirements . . . designed to: verify compliance with environmental requirements; evaluate the effectiveness of environmental management systems already in place; or assess risks from regulated and unregulated materials and practices.'' EPA, Environmental Auditing Policy Statement, 51 *Fed. Reg.* 25,004 (1986). (DEW)

Environmental engineer An individual who has received formal engineering training and applies that training to the prevention, control, and elimination of biologic, chemical, and physical hazards in the environment. The title ''environmental engineer'' has largely replaced the titles sanitary engineer and public health engineer. (FSL)

Environmental health The body of knowledge concerned with the prevention of disease through the control of biological, chemical, or physical agents in air, water, and food, and the control of environmental factors that may have an impact on the well-being of people. (FSL)

Environmental health specialist An individual who, by virtue of specialized formal training in biology, chemistry, microbiology, environmental health, industrial hygiene,or other specialized natural or physical sciences, designs and implements programs to prevent disease by protecting the air, water, food, and natural resources from contamination by chemical, biological, or physical agents. *See also* Registered sanitarian; sanitarian. (FSL)

Environmental impact statement (EIS) A report required of major federal actions significantly affecting the quality of the human environment, under the terms of the National Environmental Policy Act (NEPA). This report is, in effect, a detailed study of the proposed action and the environmental consequences of that action. Alternatives must be provided in the statement and a ''no action'' scenario must be presented. Specifically, the study must describe the environmental impact of the proposed action, any adverse environmental effects that cannot be avoided should the proposal be implemented, alternatives to the proposed action, the relationship between the local short-term uses of the human environment and the maintenance and enhancement of long-term productivity, and any irreversible and irretrievable commitments of resources that would be involved in implementation of the proposed action. 42 USC §4332(2)(C). *See also* National Environmental Policy Act. (FSL, DEW)

Environmentalist *See* Environmental health specialist; sanitarian. (FSL)

Environmental manipulation As applied to mosquito control, measures such as draining swamps, filling marshy areas, lowering or raising the level of impounded water, and other procedures to eliminate mosquito breeding areas. Other techniques, such as grass or brush removal and pasture rotation, can be used to reduce populations of ticks and chiggers. Environmental manipulation can include actions such as screening of doors and windows to prevent the entry of annoying insects. (FSL)

Environmental monitoring The measurement of contaminant concentrations in a workroom or in the environment, using a device that measures the air contamination concentration. (DEJ)

Environmental Pesticide Control Act of 1972 A revision of the Federal Insecticide, Fungicide, and Rodenticide Act of 1970, codified at 7 USC 136. (DEJ)

Environmental pollution The contamination of the environment (air, soil, or water) by deleterious materials that results from agricultural, industrial, or human activities. Environmental pollution exists in those situations in which natural resources (air, soil, water) cannot be used to their maximum without prior treatment, because of the presence of harmful biologic, chemical, or physical contaminants. (FSL)

Environmental Protection Agency (EPA) Independent federal agency charged with implementing and enforcing most environmental laws. (DEW)

Environmental quality standards The maximum allowable level or concentration of a pollutant permitted in environmental media (air, soil, water). The levels of pollutants are determined by the concentrations permitted under specific statutes—e.g., the Clean Water Act—and the regulations promulgated thereunder. (FSL)

Environmental Response Team Environmental Protection Agency experts located in Edison, NJ, and Cincinnati, OH, who can provide around-the-clock technical assistance to EPA regional offices and states during all types of emergencies involving hazardous wastes and spills of hazardous substances. (FSL)

Environmental sampling A battery of analytical techniques to measure gases, vapors, aerosols, noise, heat, cold, ionizing and non-ionizing radiation, and other stressors, either in the workplace, in exhaust stacks, or in the ambient environment. *See also* Area sampling; environmental monitoring; personal sampling. (DEJ)

Environment-driven zone (EDZ) The range of environmental heat stresses within which the body core temperature rises sharply with elevation in various climatic conditions. *See also* Prescriptive zone. (DEJ)

Enzootic Refers to the constant presence or usual prevalence of a disease or infectious agent in animal populations of a given geographic area. (FSL)

Enzyme An organic compound, frequently a protein, that accelerates (catalyzes) specific transformations of material, as in the digestion of foods. (FSL)

Eolian Refers to materials deposited by the wind—e.g., sand dunes, sandsheets, and loess. (SWT)

EPA identification number A 12-character code assigned by either EPA or the authorized state to each generator, transporter, and treatment, disposal, or storage facility. Facilities that are not generators but anticipate generation may also apply for and receive an EPA ID number. The first two characters are alphabetical and stand for the state in which the site is physically located. The third character can be either alphabetical or numeric. The remaining nine characters are always numeric. (FSL)

Epidemic The occurrence of cases that are of similar nature in human populations in a particular geographic area and that are clearly in excess of the usual incidence. **(1)** *Common-source epidemic:* An epidemic in which one human, animal, or specific vehicle has been the main means of transmitting the agent to the cases identified. **(2)** *Propagated-source epidemic:* An epidemic in which infections are transmitted from person to person or animal to animal in such a fashion that cases identified cannot be attributed to agents transmitted from a single source. (FSL)

Epidemiologist A person who applies epidemiologic principles and methods to the prevention and control of disease. (FSL)

Epidemiology The study of the distribution and determinants of disease causation in human populations. Epidemiology may be divided into the following classes. **(1)** *Descriptive:* describes the distribution of a disease, with comparisons of its frequency in different populations and in different segments of some populations. **(2)** *Analytic:* generates observational studies designed to examine hypotheses developed as a result of descriptive studies. **(3)** *Experimental:* studies humans or animal populations to test hypotheses that are not refuted by observational analytic studies. (FSL)

Epidermis The outer layer of cells of an animal or plant. (FSL)

Epigenetic Refers to the development of offspring as a result of the union of egg and sperm. (FSL)

Epilation (depilation) The temporary or permanent loss or removal of hair. (FSL)

Epipedon A surface horizon of soil that includes an upper portion that is darkened by organic matter. (FSL)

Epiphyte An organism that grows on another organism but is not parasitic on it; especially, a plant growing on another plant, but deriving no food or mineral nutrition from it. (SMC)

Episode (pollution) An air pollution incident in a given area caused by reaction of a concentration of atmospheric pollution with meteorological conditions that may result in a significant increase in illnesses or deaths. Although most commonly used in relation to air pollution, the term may also be used in connection with other kinds of environmental events, such as a massive water pollution situation. (FSL)

Epithelium Refers to cells that line all canals and surfaces that have communication with external air, and also to cells that are specialized for secretion in certain organs such as the liver and kidneys. (FSL)

Epizootic The occurrence of cases of similar nature in animal populations in a particular geographic area clearly in excess of normal incidence. Epizootics may result from common or propagated sources of infection. (FSL)

Epoch A unit of geologic time. Two or more epochs comprise a period. (FSL)

Epoxy A compound in which an oxygen atom is joined to two carbon atoms in a chain to form a bridge; a resin that polymerizes spontaneously when mixed with a diphenyl, forming a strong resistant adhesive, used in glues and coatings. Ethylene oxide and epichlorohydrin are well-known epoxy compounds. (FSL)

Equivalent diameter In sedimentation analysis, the diameter that is assigned to a nonspherical particle and that is numerically equal to the diameter of a spherical particle of the same density and velocity of fall. (SWT)

Equivalent method Any method of air pollution sampling and analysis that has been demonstrated to be, under specific conditions, an acceptable alternative to the normally used reference methods. (FSL)

Equivalent weight (soil colloid) The weight of a colloidal particle (clay or organic) that has a combining power equivalent to that of 1 gram-atomic weight of hydrogen. (SWT)

Eradicant fungicide *See* Fungicide. (FSL)

Erg A unit of work equal to the work done by a force of 1 dyne acting through a distance of 1 centimeter. (FSL)

Ergonomic factors *See* Environmental exposure. (DEJ)

Ergonomics A term that came into use after World War II, when a group of physical and biological scientists in the United Kingdom gathered to solve complex problems of the war effort. The term is derived from the Greek word for "force," combined with the word "economics." Ergonomics is a technique for consolidating disciplines and solving problems arising from work and the occupational environment. It is defined by the International Labor Organization as "the application of the human biological sciences in conjunction with the engineering sciences to achieve the optimum mutual adjustment of man and his work, the benefits being measured in terms of human efficiency and well-being." *See also* Anthropometry. (FSL)

Ergot A genus of the *Ascomycetes* fungus that infects cereals and grasses. The fungus is highly toxic and caused the disease known as holy fire during medieval times. The drug ergotamine is extracted from this fungus and is used to constrict blood vessels. (FSL)

Erosion The wearing away of the land surface by water, wind, or ice movement and by geological force. Erosion includes the following types. **(1)** *Accelerated:* Erosion that progresses much faster than natural erosion; can be caused by the actions of people or animals. **(2)** *Geological:* Erosion that is normal or natural and caused by geological processes acting over long geological periods. **(3)** *Gully:* Erosion that is caused by the accumulation of water in narrow channels and that, over a short period of time, will remove the soil from this narrow area to considerable depths. **(4)** *Interrill:* Removal of a fairly uniform layer of soil on a multitude of relatively small areas by splash, due to raindrop impact and film flow. **(5)** *Natural:* The erosion of the earth's surface by agents such as water, wind, ice, and other environmental conditions of climate, vegetation, etc., and not including the action of people. **(6)** *Normal:* Erosion that occurs on land used by people and that occurs at a rate that does not greatly exceed that of natural erosion. **(7)** *Rill:* A process that usually occurs on newly cultivated land and in which numerous small channels are formed. **(8)** *Sheet:* Erosion of soil by rainfall and surface runoff; includes rill and interrill erosion. **(9)** *Splash:* The detaching and movement of small soil particles by the impact of raindrops. **(10)** *Wind and Water:* Includes the following: (a) *saltation*, the bouncing or jumping action of soil particles 0.1 to 0.5 mm in diameter by wind, usually at a height less than 15 cm above the soil surface, or the bouncing or jumping action of mineral particles, including stones or gravel, effected by the energy of flowing water; (b) *surface creep*, the rolling of dislodged soil particles up to 51.0 mm in diameter by wind along the soil surface, or a slow movement of soil and rock debris that is not usually perceptible except through extended observation; (c) *suspension*, the movement of soil particles usually less than 0.1 mm in diameter through the air, usually at a height of more than 15 cm above the soil surface, for relatively long distances.

There are two equations which apply to the process of erosion. They are as follows.

- *Universal soil loss equation.* Allows prediction of A, the average soil loss in tons per acre per year, and is defined as $A = RKLSCP$, where R is the rainfall factor, K is the soil erodibility factor, L is the length of slope, S is the percent slope, C is the cropping and management factor, and P is the conservation practice factor.
- *Wind erosion equation.* Allows prediction of E, the average annual soil loss due to wind, in tons per acre per year, and is defined as $E = IKCLV$, where I is the soil erodibility factor, K is the soil ridge roughness factor, C is the local climatic, L is the field width, and V is the vegetative factor.

(SWT)

Erosional surface A land surface that is shaped by the erosive action of running water and that can also be influenced by ice, wind, or water. (SWT)

Erosion classes A grouping of erosion conditions that is based on the degree of erosion or on characteristic patterns of erosion. *See also* Erosion. (SWT)

Erosion index A measure of the erosive potential of a specific rainfall event. In the universal soil loss equation, it is defined as the product of two rainstorm characteristics: total kinetic energy of the storm times its maximum 30-minute intensity. (SWT)

Erosion pavement A layer of course fragments, such as sand or gravel, remaining on the ground after the removal of fine particles by a storm. (SWT)

Erosivity The potential ability of water, wind, gravity, etc., to cause erosion. (SWT)

Error *Type-one Error:* In hypothesis testing, the mistake of rejecting a true null hypothesis. *Type-two Error:* In hypothesis testing, the mistake of deciding in favor of a false null hypothesis. (VWP)

Erysipelas An inflammation of the skin, marked by red patches with sharp border lines. *See also* Cellulitis. (FSL)

Erythema An abnormal redness of the skin, due to distension of the capillaries with blood. It can be caused by many different agents—e.g., heat, certain drugs, ultraviolet rays, and ionizing radiation. (FSL)

Erythrocyte A red blood cell, which contains hemoglobin and transports oxygen to body tissues. *See also* Heme; hemoglobin. (FSL)

Erythromycin An antibiotic derived from strains of bacteria (*Streptomyces erythreus*) that is effective against a wide range of microbes. (FSL)

Escherichia A genus of rod-shaped Gram-negative bacteria found in the intestinal tract of humans and other warm-blooded animals. The organisms, which are nonpathogenic, include *Escherichia coli*, which is used as an indicator organism to determine the extent of pollution or water supply degradation. (FSL)

Ester A chemical compound resulting from the reaction of an alcohol and an acid, with removal of water. (FSL)

Estrogens Natural or synthetic hormones that stimulate the growth and activity of the reproductive system and the development of sex characteristics in mammals. These hormones are produced by ovaries, placenta, adrenal glands, and testes. (FSL)

Estuary An arm of the sea at the mouth of a river, in which the river current is met and influenced by the tides. Estuary ecosystems are strongly influenced by the variable strength and direction of the currents and by the fluctuating salinity. *See also* Tidal marsh. (SMC)

Ether A colorless, highly flammable, volatile chemical with a pungent odor, one type of which (diethyl ether) was previously used as an inhalation anesthetic. Following prolonged storage, some ether compounds form organic peroxides that are extremely shock-sensitive and make use or disposal of outdated material very risky. (FSL)

Ethnobiology The study of the use of or naming of nondomesticated plants, animals, and other organisms by a particular ethnic or cultural group of people. (SMC)

Ethnobotany The study of the use of or naming of plants by a particular ethnic or cultural group of people. Ethnobotanists sometimes focus on medicinal uses of plants in a search for chemically or medicinally active compounds. (SMC)

Ethology The study of animal behavior. (SMC)

Ethyl acrylate (CH_2:$CHCOOC_2H_5$) A foul-smelling gas produced in the manufacture of acrylates used in the plastic industry. The gas, which is highly toxic, can be removed from the air by the use of charcoal filters, but even a minuscule release can result in odor complaints. (FSL)

Ethylenediaminetetracetic acid (EDTA, $C_{10}H_{12}N_2Na_4O_8$) A white powder that is soluble in water and is used as a chelation agent in the treatment of lead poisoning. *See also* Lead. (FSL)

Ethylene dibromide (EDB, $BrCH_2CH_2Br$) A chemical used as an agricultural fumigant and in certain industrial processes. Extremely toxic and found to be a carcinogen in laboratory animals, EDB has been banned for most agricultural uses in the United States. (FSL)

Ethylene oxide [ETO, $(CH_2)_2O$] A colorless gas at room temperature that is among the highest-volume chemicals produced in the United States. Used in the manufacture of ethylene glycol, acrylonitrile, and rocket propellants, ETO is also used as a fungicide and sterilant. The compound is miscible with water and organic solvents, and in gaseous form has an ether-like odor detectable at concentrations of about 700 ppm. Human exposure should remain below levels of 1.0 ppm (PEL) (5.0 ppm STEL). Two major concerns are that ETO is a suspected human carcinogen and is extremely flammable; as little as 3% in air is flammable, with explosive properties if confined. Ethylene oxide has been used as a sterilizing agent in which microbial death follows first-order kinetics. It has particular application in laboratories and medical-care facilities, where it is used for the sterilization of biologically contaminated materials. The time required for sterilization depends on its concentration and temperature, the relative humidity, and the resistance of the material to penetration. The gas has been widely used to reduce microbial contamination and kill insects in various dried foods. (HMB)

Etiologic agents Infectious microorganisms, viruses, or parasitic agents capable of producing infection and/or disease in a susceptible host. The interstate transport of etiologic agents is regulated by the U.S. Public Health Service. The regulations cover the transport of etiologic agents in diagnostic specimens and biological products and set out packaging and labeling requirements for transport. Damaged packages or packages that are not received within 5 days of anticipated delivery must be reported to the Centers for Disease Control. 42 CFR §72. *See also* Biological products; diagnostic specimens. (DEW) (JHR)

Etiology The study or theory of the causation of disease; the sum of knowledge regarding causes. (FSL)

Eutrophication The slow aging process during which a lake, estuary, or bay evolves into a bog or marsh and eventually disappears. During the later stages of eutrophication the water body is choked by abundant plant life as the result of increased amounts of nutritive compounds such as those of nitrogen and phosphorus. Human activities can accelerate the process. *See also* Cultural eutrophication. (FSL)

Eutrophic lakes Shallow, murky bodies of water that have excessive concentrations of plant nutrients causing excessive algal production. (FSL)

Evacuation plan A written protocol that should be a part of an emergency planning effort. It should cover evacuation routes, both inside and outside plant routes, as well as primary and alternative routes. Designated safe areas for employee regrouping that are crosswind from the contaminant source and a safe distance from the danger zone should be identified to employees. Designated evacuation personnel to account for all individuals and to report on the medical condition of those present and the names of those missing should be noted. A communications system that consists of alarms and signals easily heard and understood should be in place, as well as lists of critical operations and personnel, as well as provisions to evacuate visitors and handicapped personnel as appropriate. *See also* Emergency planning. (SAN)

Evaporation ponds Areas in which sewage sludge is dumped and allowed to dewater. *See also* Lagoon, oxidation pond. (FSL)

Evaporation rate The rate at which a material will vaporize (evaporate), compared to the known rate of vaporization of a standard material. The designated standard material is usually normal-butyl acetate (NBUAC or *n*-BuAc), which has a vaporization rate designated as 1.0. Vaporization rates of other solvents or materials are then classified as: **(1)** fast evaporating, if the rate is greater than 3.0 (examples: methyl ethyl ketone = 3.8, acetone = 5.6, hexane = 8.3); **(2)** medium evaporating, if the rate is 0.8 to 3.0 (examples: 190-proof (95%) ethyl alcohol = 1.4, VM & P naphtha = 1.4, MIBK = 1.6); **(3)** slow evaporating, if the rate is less than 0.8 (examples: xylene = 0.6, isobutyl alcohol = 0.6, normal butyl alcohol = 0.4, water = 0.3, mineral spirits = 0.1). (DEJ)

Evaporites A class of soil materials including gypsum. (SWT)

Evapotranspiration The combined loss of water from a given area during a specified period of time, both from the soil surface and by transpiration from plants. Where absorption of water is limited, plant transpiration has been used to aid in the disposal of effluent from a subsurface sewage disposal system. The tile lines are placed near the top of the trench, where use can be made of the action of trees, shrubs, and grasses in the absorption and subsequent release into the atmosphere of appreciable quantities of moisture through the process of evapotranspiration. *See also* Soil absorption system. (SWT)

Event-tree analysis (ETA) A hazard analysis method used in industrial process safety analysis. An ETA is a decision-tree type of analysis that is constructed horizontally, starting with a single, initiating event and moving forward to allow prediction of the consequences of that event. A completed event tree depicts a process in several alternative states of failure, including that of the safety devices in the system. ETA does not identify multiple initiating events and interactions. *See also* Fault-tree analysis; process safety. (SAN)

Evidence Information presented to the court and upon which a decision must be made. In federal court, relevant evidence is "evidence having any tendency to make the existence of any fact that is of consequence to the determination of the action more probable or less probable than it would be without the evidence" (Federal Rules of Evidence Rule 401). (DEW)

Excavation Any cut, cavity, trench, or depression in an earth surface that has been formed by earth removal. Excavations must include a means of egress and exit. Specific construction methods and protective safeguards are also required depending on the particular soil classification. *See also* Distress; shoring; sloping; soil classification. (SAN)

Exceedance Violation of environmental protection standards by the exceeding of allowable limits or concentration levels. (FSL)

Excess flow valve A device designed to close when the fluid passing through it exceeds an established flow rate, as determined by a drop in pressure. (FSL)

Exchange capacity The total ionic charge of the adsorption complex active in the adsorption of ions. (SWT)

Exclosure A fence, cage, or other boundary designed to keep animals of a certain size or habitat out of a designated area. (SMC)

Exclusionary Refers to any form of zoning ordinance that tends to exclude specific classes of persons or businesses from a particular district or area. (FSL)

Exclusion zone An area at a hazardous-waste site or at an emergency-response scene where contamination does or could occur. The zone is to be demarcated by barrier tape, placards, or physical barriers. It can be subdivided into zones of different degrees of contamination, based on known or expected type and degree of hazards, for selection of appropriate levels of personal protection. *See also* Levels of protection. (MAG)

Excrement Feces or other waste matter discharged from the body of humans or animals. *See also* Sewage. (FSL)

Excretion The removal of a substance or its metabolites from the body in urine, feces, or expired air. (FSL)

Exempted aquifers Underground bodies of water defined in the Underground Injection Control program as aquifers that are sources of drinking water, although they are not being used as such, and that are exempted from regulations barring underground injection activities. (FSL)

Exempt solvents Specific organic compounds that are not subject to requirements of regulation because they have been deemed by the EPA to be of negligible photochemical reactivity. (FSL)

Exhaust gas circulation An air pollution control technique applied to internal combustion engines as a means of reducing emission pollutants. Most cars and trucks in the United States are equipped with such devices as a means of minimizing pollutants. (FSL)

Exoergic Refers to that which liberates energy. (FSL)

Exothermic When applied to reactions, describes those that produce substances that have less energy than the reacting materials, resulting in a release of energy during the reaction; refers to that which liberates energy, specifically as heat. (FSL)

Exotic A nonnative or introduced organism; refers especially to plants that are growing in but not indigenous to a region. (SMC)

Exotoxin A microbial toxin. Generally, exotoxins are heat-labile proteins produced by a specific microorganism and are not structural components of the cell. Exotoxins have specific and unique modes of action that differ from one species to another. A notable exception to the heat-sensitive characteristic is the heat-stable exotoxin produced by *Staphylococcus aureus*. These toxins are responsible for the symptoms of foodborne illness elicited by the microorganism producing them. They include a potent chemical formed and excreted by certain types of bacteria—e.g., the *Clostridium* genus, which includes the tetanus and botulism organisms. *See also* Botulism; staphylococcal food intoxication. (FSL) (HMB)

Expected value ($E(x)$) **(1)** (Of a continuous random variable) The average value taken on by the random variable. **(2)** (Of a discrete random variable) The sum of the products of the values the random variables with nonzero probability assume and their respective probabilities. Therefore:

$$E(x) = \sum_{i=1}^{k} x_i P(x_i)$$

where $i = 1, 2, \ldots, k$, and x_i = the values of the random variables; $P(x_i) \neq 0$. (VWP)

Experience rating A method of adjusting manual rates using a three-year history of the employer. Most worker's compensation insurance companies instead use manual rates to establish premiums based on a set rate for a specific industry and adjusted for the number of workers employed. This means that all employers in a given category have the same rate, regardless of their safety record (losses due to injury and illness claims). In contrast, with the experience-rating method, an employer with numerous losses (i.e., a poor safety record) will pay a higher premium for the same number of employees than an employer in the same industry with a good safety record. The method allows for a certain amount of loss (average losses determine basic rate). (SAN)

Experiment A planned activity in which results yield a set of data. (VWP)

Expert witness A witness qualified as an expert by knowledge, skill, experience, training, or education. Unlike other witnesses, an expert's testimony may be in the form of an opinion (Federal Rules of Evidence Rules 701–706). (DEW)

Expiration Exhaling of the lungs, caused by relaxation of the diaphragm and rib muscles, which causes a decreased chest cavity space, thus forcing air out through the trachea. (DEJ)

Expiratory reserve volume (ERV) The maximum amount of air that can be forcibly expired after a normal expiration. (DEJ)

Explosimeters *See* Combustible gas meter. (FSL)

Explosion A deflagration or a detonation. A deflagration is an exothermic reaction that expands rapidly from the burning gases to the unreacted material by conduction, convection, and radiation. A deflagration may not always produce sufficiently rapid increases in pressure to produce an explosion. A detonation is an exothermic reaction characterized by the presence of a shock wave that establishes and maintains the reaction. A detonation usually results in sufficiently rapid increases in pressure to produce an explosion. There are three primary hazards in explosions: blast waves, thermal effects, and scattering of fragments. *See also* Boiling liquid–expanding vapor explosions; condensed-phase detonations; dust explosions; runaway reactors. (SAN)

Explosive A chemical that causes an almost instantaneous release of pressure, gas, and heat when subjected to sudden shock, pressure, or high temperature. *See also* Explosion. (FSL)

Explosive classification As defined by the U.S. Department of Transportation, explosives are placed in the following classes: **(1)** *Class A:* those that possess detonating or other maximum hazard, such as dynamite or nitroglycerin; **(2)** *Class B:* those that pose a flammability hazard, such as propellant explosive or photographic flash powder; **(3)** *Class C:* those that contain Class A, Class B, or both as components, in limited quantities. (FSL)

Exponential growth phase *See* Log phase. (HMB)

Exposure **(1)** The amount of radiation or pollutant that represents a potential health threat to the living organisms in an environment. **(2)** The opportunity of a susceptible host to acquire an infection by either a direct or an indirect mode of transmission. An effective exposure is one in which the exposure actually results in infection. **(3)** The amount of biological, physical, or chemical agent that reaches a target population. **(4)** The means by which an organism comes in contact with a toxicant; such exposure can occur by the oral, dermal, or respiratory routes, or by inoculation. (FSL)

Exposure dose A measure of the X-radiation or gamma radiation at a certain place, based upon the ability of the radiation to produce ionization. The unit is the roentgen (R). (RMB)

Exposure dose rate The radiation exposure dose per unit time (R/unit time). (RMB)

External radiation Ionizing radiation in which the source is located outside the body and the radiation penetrates into the deeper tissues. (RMB)

Extraction procedure (EP) A two-step laboratory analysis of waste, involving **(1)** the extraction of certain compounds from the sample, and **(2)** analysis of the extract for the EP toxicity contaminants. A solid waste exhibits the characteristics of EP toxicity if test methods listed by EPA or equivalent methods reveal that the extract from a representative sample of the waste contains EPA-listed contaminants at specific concentrations.

In March 1990, EPA published a final, revised rule called the Toxicity Characteristic Rule. In essence, the rule supplanted the EP test procedures by addition of 25 new chemicals to the list of wastes having regulatory levels above which a waste becomes a characteristic hazardous waste. Chemical Abstract Service (CAS) numbers were added, as well as the chronic toxicity reference levels and the regulatory levels. A regulatory level is the concentration at or above which a chemical is defined as a hazardous waste. This new rule currently applies to 39 chemicals. *See also* Toxicity Characteristic Rule. (FSL)

Extraction procedure toxicity characteristics (EPTC) One of four existing hazardous waste characteristics (along with ignitability, corrosivity, and reactivity) for which the EPA has identified and developed regulation. The EPTC defines the toxicity of a waste by measurement of the potential for the toxic constituents in the waste to leach out and contaminate ground water at levels that are of health or environmental concern. In order to determine whether a waste exhibits the EPTC, constituents are extracted in a procedure that simulates the leaching action that occurs in municipal landfills. In 1990, the EPTC procedure was replaced by the toxicity characteristic leaching procedure. *See also* Comprehensive Environmental Response Compensation and Liability Act; hazardous waste; Resource Conservation and Recovery Act; toxicity characteristic leaching procedure. (FSL)

Extraneous material In a food product, any foreign matter associated with objectional conditions or practices in production, storage, or distribution. Examples include sand, soil, glass, rust, decomposed tissue, or other foreign substances. *See also* Filth. (HMB)

Extrapolation ionization chamber An ionization chamber with electrodes whose spacing can be adjusted and accurately determined, to permit extrapolation of its reading to zero chamber volume. (RMB)

Extremely hazardous substance Any of 406 chemicals identified by the Environmental Protection Agency on the basis of toxicity, and listed under SARA Title III. (FSL)

Extremities The hands, forearms, and, for certain safety restrictions, the head, feet, and ankles. Permissible radiation exposures for these regions are generally greater than for whole-body exposures because the regions contain less blood-forming material and have smaller volumes for energy absorption than do other body parts. (RMB)

Extrinsic factors Environmental factors that affect the growth and survival of microorganisms and that relate to the organisms' environment. Such factors include relative humidity, temperature, and oxygen availability. Control of these factors, singly or in combination, will affect the number and types of organisms present. *See also* Intrinsic Factors. (HMB)

Exudation The process that describes the separation of oily components from explosives during protracted storage, particularly at high temperatures. (FSL)

Fabric collector A dust collector that operates by removing particles from airstreams through straining, impingement, interception, diffusion, or electrostatic charge. Fabric collectors are usually configured as bags, tubes, pleated cartridges, or envelopes, and are often cleaned at regular intervals automatically. (DEJ)

Fabric filter A cloth device that catches dust particles from industrial emissions. (FSL)

Face velocity The average air velocity in the plane of an opening into an enclosure through which air moves, usually expressed in feet per minute or meters per second. A typical laboratory hood may have an average face velocity, measured in the plane of the sash opening, of 100 feet per minute. Average face velocities are often measured by determination of airflow at specific points along a grid covering the entire face opening. *See also* Biological safety cabinet; chemical hood. (DEJ)

Facility A building, equipment, structure, or other stationary item that is located on a single site, owned or operated by one person. *See also* person. (FSL)

Factorial If n is a positive integer, n factorial ($n!$) is $n(n - 1)(n - 2) \ldots 1$. Zero factorial is defined to be 1. (VWP)

Facultative anaerobe Microorganism that can multiply either in the presence or in the absence of oxygen. Such organisms are able to obtain energy either by respiration or by fermentation and do not require oxygen for biosynthesis. Use of the fermentation method implies that the organism synthesizes its adenosine triphosphate by substrate-level phosphorylation and disposes of its electrons by donation to organic molecules. (HMB)

Failure mode and effects analysis (FMEA) A hazard analysis method used in process safety analysis. FMEA starts with a diagram of an operation, and individually reviews all process components, such as transmitters, controllers, valves, pumps, and rotometers, that could fail and conceivably affect the safety of the operation. Each component is listed on a data tabulation sheet and analyzed for failure modes, failure causes, failure effects, and risk rating (probability of failure). The final step in the analysis is to develop a series of recommendations appropriate to risk management. FMEA does not directly evaluate worker interaction with equipment systems. Also known as failure modes, effects, and criticality analysis (FMECA). *See also* Checklist; event-tree analysis; fault-tree analysis; hazard analysis; HAZOPS; what-if. (SAN)

Failure modes, effects, and criticality analysis (FMECA) *See* Failure modes and effects analysis. (SAN)

Falling loads Such loads constitute a primary hazard to workers in materials-handling situations because loads are often elevated (in the case of forklifts) or free hanging (in the case of cranes and hoists). Falling loads can also be the result of the shifting or falling of a load, such as a stack of goods on pallets. (SAN)

Fallout The radioactive particulate matter that drops back to the earth's surface after having been lifted or blown into the atmosphere by a nuclear detonation. (RMB)

Fallout radiation That portion of total nuclear radiation associated with the deposition of radioactive particles. As such, it includes sources of the following: fission products, gamma rays, beta rays, alpha particles from weapon debris, and induced gamma and beta activity from neutron interaction in nearby materials. (RMB)

Fallow Cropland that is left unused in order to restore productivity through accumulation of water and nutrients. (SWT)

Fall protection A system used to halt an employee in a fall from a working level. It consists of an anchorage, connectors, and a safety belt or safety harness, and may include a lanyard, deceleration device, lifeline, or suitable combinations of these. Also called a personal fall arrest system. *See also* Anchorage; connector, deceleration device; lanyard; lifeline. (SAN)

Fall zone A fenced or guarded area that people at lower levels cannot enter because of the danger of falling debris. The guarding consists of overhead and side protection of the occupied area. The material and construction of the area must be able to withstand the weight (and momentum) of the types of objects against which it is designed to protect. (SAN)

False-negative result A test result that is reported to be negative, but the underlying test condition is actually positive. (VWP)

False-positive result A test result that is reported to be positive, but the underlying factor or condition is actually negative. (VWP)

Family The taxon in the classification hierarchy between order and genus. (SMC)

Family, soil A soil classification level intermediate between great soil group and soil series. (SWT)

Fan Landform built of stratified alluvium with or without debris-flow deposits. (SWT)

Farmer's lung disease A syndrome that consists initially of chills and fever, followed by impairment of lung function, and that is due to chronic exposure to moldy hay or other moldy organic material. The molds are typically thermophilic *Actinomycetes* that cause immunological reactions against the fungal antigens. They are most prevalent in cold and humid areas, or in situations in which hay has been improperly cured. *See also* Silo filler's disease. (DEJ)

Fatigue Tiredness that can lead to a decrease in physical performance and that can be due to subjective feelings, motivation problems, and/or deterioration of mental and physical abilities. Factors producing fatigue include monotony, poor illumination, poor climate, noise, excessive manual or mental work intensity, worry, conflict, illness, pain, and poor eating habits. Normally, fatigue will disappear after a rest period; in pathological cases, however, fatigue may continue. (DEJ)

Fault tree analysis (FTA) A hazard analysis method used in industrial process safety analysis to determine the cause of an unwanted event, such as a chemical release. FTA leads to a decision tree that is constructed vertically as a graphic representation (using event and logic symbols) for all possible sequences of events, starting with either the initial event or the final event. The logic symbols are known as *and* and *or* gates. The result resembles a tree with many branches, each branch containing a sequence of events leading to the top event. Event probabilities are calculated to determine the likelihood of the occurrence of an undesirable event. This technique is particularly useful in evaluating the effect of alternative actions in reducing the chances for an undesired event. *See also* Event-tree analysis; failure mode and events analysis; HAZOPS. (SAN)

Fauna A descriptive catalog of the animals of any geographical area or geological period. (SMC)

Favism A type of anemia caused by the ingestion of fava beans or inhalation of pollen from the fava plant. (FSL)

Favus A fungal infection that is caused by the organism *Trichophyton schoenleini* and that usually involves the scalp. (FSL)

F-distribution If two independent random samples are taken from two populations that are normally distributed with n_1 and n_2 respective sample sizes for the two populations, s_1^2 and s_2^2 respective variance for the two populations, and σ_1^2 and σ_2^2 respective population variances for the two populations, then the random variable F, given by

$$F = \frac{s_1^2}{s_2^2} \text{ (}F\text{ statistic)}$$

and has the F-distribution with $df = (n_1 - 1, n_2 - 1)$. The probabilities for the random variable F can be determined using areas under the F-curve with $df = (n_1 - 1, n_2 - 1)$. (VWP)

Feasibility study For the purposes of a cleanup action under the Comprehensive Environmental Response, Compensation, and Liability Act, a study undertaken by the lead agency to develop and evaluate options for remedial action. Studies emphasize data analysis and are generally performed concurrently and in an interactive fashion with the remedial investigation (RI), using data gathered during the RI. Also refers to a report that describes the results of the study. 40 CFR §300.5. *See also* Comprehensive Environmental Response; Compensation and Liability Act; remedial action; remedial investigation. (DEW)

Feasible, OSHA definition *See* Engineering controls. (DEJ)

Fecal coliforms Members of the coliform group of organisms that are of fecal origin. Elevated-temperature tests are used for the separation of organisms of the coliform group into those of fecal origin and nonfecal origin. Various tests have been employed with many modifications around the world to make this distinction. However, none can absolutely differentiate coliforms of fecal origin from those of nonfecal origin. Essentially, fecal coliforms are *Escherichia coli* isolates that can grow at 44.5°C. Food or water safety regulations may be based in part on a limit to the maximum number of fecal coliforms permitted. *See also* Coliforms; fecal streptococci; indicator organisms. (HMB)

Fecal streptococci Any of a number of bacteria of the genus *Streptococcus* that are of fecal origin. Fecal streptococci such as *Streptococcus faecalis*, *S. faecium*, *S. bovis*, and *S. avium*, belong to groups D and Q. These organisms are common inhabitants of animal and human intestines and therefore serve as indicators of fecal contamination. Whereas the coliform group is commonly used as an indicator of fecal contamination of foods or water in the United States, fecal streptococci are most commonly used as indicator microorganisms in European countries. They are used in testing for contamination of lakes, rivers, streams, and estuaries. Although the test for them is often of limited value, it may be used with coliform testing to identify the source of contamination. *See also* Coliforms; fecal coliforms; indicator organisms. (HMB)

Feces Waste discharged from the intestine, also known as stool or excrement. *See also* Excreta; sewage. (FSL)

Fecundity Capacity to produce offspring or reproduce. (SMC)

Federal Advisory Council for Occupational Safety and Health (FACOSH) A joint management–labor council that advises the Secretary of Labor on matters relating to the occupational safety and health of federal employees. (FSL)

Federal Coal Mine Health and Safety Act of 1969 A federal law governing occupational health and safety conditions in coal mines; a precursor to the 1977 Mine Safety and Health Act. *See also* Federal Mine Safety and Health Act. (DEJ)

Federal court Refers to any of the various courts of the federal judicial system, as opposed to the courts of the individual states. Federal courts consist primarily of district courts, courts of appeals, and the Supreme Court, and also include the bankruptcy court and the court of claims. The district courts have original jurisdiction over civil actions arising under the Constitution, laws, or treaties of the United States, and other civil actions where the matter in controversy involves more than $50,000 and there exists diversity of citizenship, meaning that the opposing parties are from different states or that a foreign state is implicated. The courts of appeals have jurisdiction over appeals from final decisions of the district courts and certain interlocutory decisions. The Supreme Court has jurisdiction over controversies between the states, or between the United States and a state. In addition, the Supreme Court may review cases from the highest court of a state when a federal statute is at issue or a state statute is challenged as being contrary to the Constitution, treaties, or laws of the United States, or particular claims are made thereunder. The process of appealing for review by the Supreme Court involves a writ of certiorari. (DEW)

Federal Emergency Management Agency (FEMA) The Federal agency accountable for emergency preparedness, mitigation, and response activities for all natural, human-caused, or nuclear emergencies. This agency also has the responsibility for administering certain training funds under the Superfund Amendments and Reauthorization Act (SARA) of 1986. (FSL)

Federal Food, Drug, and Cosmetic Act (FFDCA) A federal law originally passed in 1906 and modified by Congress by 1938 and numerous times thereafter. The law regulates the safety of foods, food additives, color additives, drugs, medical devices, cosmetics, pesticide residues, and poisonous or deleterious substances in covered items. The Delaney clause has been an important part of the earlier law. This clause states that food additives that cause cancer in humans or animals at any level shall not be considered safe and, therefore, that their use be prohibited. Regulations are primarily carried out through the prohibition, by the Food and Drug Administration, Public Health Service, and U.S. Department of Health and Human Services, of adulteration and misbranding. 21 USC §301 *et seq.*, 21 CFR Parts 1–1300. (DEW)

Federal Insecticide, Fungicide, and Rodenticide Act (1972) (FIFRA) Federal law that regulates the manufacture and use of pesticides as defined under the act. Sets out a process for the registration of pesticides with the Environmental Protection Agency. Registration includes the submission of data that demonstrate a product's safety and efficacy for specific uses. This law also regulates the labeling, packaging, and certified application of products. 7 USC §136 *et seq.*, 40 CFR Parts 162–180. (DEW)

Federal Mine Safety and Health Act of 1977 A federal law that governs occupational safety and health matters in mining. The Mine Safety and Health Administration under the Department of Labor enforces the legislation, with regulations found at 30 CFR Part 70. Also known as MSHA. (DEJ)

Federal Railroad Safety Act of 1970 Federal law that authorizes the Federal Railroad Administration within the Department of Transportation (DOT) to promulgate regulations for all areas of railroad safety; the Occupational Health and Safety Administration retains jurisdiction over safety and health conditions affecting workers in those working conditions for which DOT does not have standards. (DEJ)

Federal Register The official daily publication of the United States government, that provides a uniform system for publishing Presidential and federal agency documents. Contains certain Presidential documents of general applicability and legal effect, such as those required to be published by statute or regulation. Publication serves to provide official notice of a document's existence and content, and indicates that the document was duly issued, prescribed, or promulgated and is afforded judicial notice by a court. Regulations initially published in the *Federal Register* are codified in the Code of Federal Regulations. *See also* Code of Federal Regulations. (DEW)

Federal Tort Claims Act Federal law that allows the district courts to decide claims brought against the United States government for specified tort actions. Constitutes a waiver of sovereign immunity with respect to such claims. Does not allow suits for intentional torts or for acts within the "discretionary function or duty" of the federal government. 28 USC 2671–2680. (DEW)

Federal Water Pollution Control Act (FWPCA) *See* Clean Water Act. (FSL)

Feedlot An area in which large numbers of cattle are fattened for slaughter. Because of the concentration of highly nitrogenous manure, the runoff from these areas can create special stream pollution problems. Because of the organic material in it, this material is a suitable medium for breeding large numbers of flies, which can become a public health nuisance. (FSL)

Fen A wetland ecosystem characterized by peat formation and alkaline conditions. Fens are often situated around freshwater lakes and the upper parts of old estuaries. *See also* Bog. (SMC)

Fenestration The technique used to control natural light by regulating the type, size, location, and orientation of sources of such light by the installation of shades and awnings. (FSL)

Fentin A fungicide used to control plant diseases. It is an organic compound that contains tin, and that has been largely supplanted by other fungicides on the market. (FSL)

Feral Refers to animals, such as cats and dogs, that were formerly domesticated and that have reverted to the "wild" state and are capable of survival without assistance from humans. (FSL)

Fermentation The breakdown of organic substances by microorganisms, with a resulting release of energy. This process is used for the breakdown, by yeasts and bacteria, of carbohydrates that results in the formation of carbon dioxide, alcohol, and other organic materials. (FSL)

Ferran A cutan that has a high concentration of iron oxides. *See also* Cutan. (SWT)

Ferruginous bodies Bodies formed by fibers that have entered the lungs and that become coated with proteinaceous, iron-containing substances, possibly through contact with macrophages. Ferruginous bodies can be formed by any kind of durable fiber, including asbestos, fiberglass, and vegetable fibers of silicious origin. (DEJ)

Fertigation Fertilization of plants, using a solution of plant nutrients in irrigation water. (SWT)

Fertility, soil The ability of a soil to supply nutrients essential to plant growth. (SWT)

Fertilizer Material that has been applied to soils to provide nutrients for plant growth. Natural materials such as animal manure and compost are frequently used for this purpose, as well as chemical compounds including ammonia, other nitrogen compounds, and phosphates. Excessive use of these chemicals can result in contamination of water supplies and thus represents a human-health hazard. *See also* Eutrophication; methemoglobinemia. (FSL)

Fetus In humans, the product of conception from the end of the eighth week to the time of birth. (FSL)

Fiberglass A commercial, nonflammable fiber that is made from glass. It is used primarily for insulation. Fibers of the material can penetrate the skin and, when airborne, can affect the lungs. Fiberglass is resistant to most chemicals and solvents. (FSL)

Field moisture capacity The amount of water remaining in a soil after free drainage has occurred. (SWT)

Filariasis An infection with the filariae worm *Wuchereria bancrofti*, which is transmitted by the *Culex* mosquito or by mites or flies. *W. bancrofti* larvae invade the lymph tissues, which results in swelling of the tissues, inflammation, and pain. (FSL)

Filling Depositing dirt, mud, or other materials into aquatic areas to create land, usually for agricultural or commercial development purposes. (FSL)

Fill material For purposes of the Clean Water Act, any material used for the primary purpose of replacing an aquatic area with dry land or of changing the bottom elevation of a body of water. It does not include any pollutant discharged into the water primarily to dispose of waste. 33 CFR §323.2(e). *See also* Clean Water Act; wetlands; discharge of fill material. (DEW)

Film badge A pack of photographic film used for approximate measurement of radiation exposure, for personnel-monitoring purposes. The badge may contain two or three films of differing sensitivity, as well as a filter that shields part of the films from certain types of radiation. *See also* Photographic dosimetry. (RMB)

Film dosimeter A pack of photographic film and appropriate filters used for the detection of ionizing radiation exposure of the individual wearing the dosimeter. (RMB)

Film water (soil) A layer of water surrounding soil particles and varying in thickness from 1 to 100 molecular layers. Usually considered as the water that remains after drainage has occurred. (SWT)

Filth Any objectionable matter contributed by animal contamination, such as rodent, insect, or bird matter; or any other objectionable matter contributed by unsanitary conditions. Examples include, but are not limited to, extraneous materials such as insects or insect parts, rat or mouse hairs, rat, mouse, or insect excreta, feather fragments, and rodent urine. *See also* Contamination; extraneous material. (HMB)

Filtration As applied to water treatment and swimming-pool operations, the process of removing suspended matter from water as it passes through beds of porous material. The degree of removal will depend on the character and size of the filter media, and the thickness of the porous media, and the size and amount of the suspended solids. When a water source is characterized by a large amount of turbidity, a proportion of the water can first be removed by sedimentation, thus reducing the load on the filter. There are many types of filters, including: **(1)** slow sand filters, which operate at a rate of about 0.05 gallon per square foot of filtration area; **(2)** pressure sand filters, in which water is applied at a rate of about 2 gallons per square foot of filter area; **(3)** ceramic, porous stone, unglazed porcelain filters, which are typically small household filters that are attached to faucets and used for odor removal. (FSL)

Finding of no significant impact (FONSI) For purposes of the National Environmental Policy Act, a federal agency document that briefly presents the reasons why an action that is not categorically or otherwise excluded will not have a significant effect on the human environment and for which an environmental impact statement therefore need not be prepared. 40 CFR §1508.13. *See also* Categorical exclusion; environmental policy statement; National Environmental Policy Act. (DEW)

Fine Texture (soil) Texture characteristic of a soil that consists of or contains large quantities of fine fractions, particularly silt and clay. (SWT)

Fire The active principle of burning, characterized by the heat and light of combustion. Fires are distinguished by class as follows. **(1)** *Class A:* Involves cloth, paper, wood, and certain plastics. Such fires are most easily extinguished with water. **(2)** *Class B:* Involves flammable or combustible liquids, gases, greases, and similar materials. Common techniques for extinguishing fires in these materials include the exclusion of oxygen by the application of dry chemicals such as diammonium phosphate, sodium bicarbonate, and potassium carbonate. **(3)** *Class C:* Involve live electrical equipment. Such fires must be extinguished by the use of electrically nonconductive agents. Once the equipment has been deenergized, extinguishers suitable for class A or class B fires can be used. **(4)** *Class D:* Involves combustible metals such as magnesium, sodium, potassium, and titanium. Extinguishers used on these types of fires use various types of dry powders that are specific for the type of metal involved. Thus, for example, an extinguisher for a magnesium fire may not work on a titanium fire. For this reason, extinguishers for class D fires have name plates indicating the type of metal on which they should be used. (FSL)

Fire brigade A group of emergency response personnel engaged to function as a unit in the suppression of a fire. (FSL)

Firedamp An explosive mixture of air and methane (CH_4, marsh gas). The potential for explosion is greatest when the methane-air mixture is in the 8.5–9.5% range. (FSL)

Fire extinguisher A device specifically designed to quench, smother, or otherwise eliminate a fire. *See also* Fire; flammable. (FSL)

Fire extinguishing system (gaseous) A system that makes use of a fire-extinguishing chemical that is in the gaseous state at normal room temperature and pressure. Such agents have low viscosity, can expand or contract with changes in pressure and temperature, and have the ability to diffuse readily and to be distributed uniformly throughout an enclosure. Halon 1211 and Halon 1311 are examples. Computer rooms and other installations with electrical equipment typically have Halon systems because these agents will not damage the equipment. *See also* Fixed fire extinguishing system; predischarge employee alarm. (SAN)

Fire sprinkler systems Sprinkler systems designed for extinguishing fires and most commonly consisting of piping, heads, valves, and hangers for (piping) support. There are several classes of sprinkler systems, including wet-pipe, dry-pipe, pre-action, deluge, and combined systems. A wet-pipe system has water under pressure inside the piping at all times. A dry-pipe system has air or nitrogen under pressure inside the piping, for freeze protection; however, such a system is slower to deliver water than is a wet system because a valve separates the water from the piping. A pre-action system is a dry-pipe system with sensors that allow it to fill the piping with water faster. In deluge systems, all heads open at the same time. Combination systems usually consist of simple dry-pipe and pre-action sensors. *See also* Fire suppression systems. (SAN)

Fire suppression systems Systems that use chemical agents such as carbon dioxide, Halon, dry chemical, foam, or combustible metal agents for extinguishing fires. Their design depends on the material in the system. Carbon monoxide is used for class A and B fires. Halon is used to protect valuable electrical equipment such as that in switchgear rooms and computer rooms. Because of the potential displacement of oxygen, personnel must be evacuated before these systems are released. Dry chemicals and chemical foams are other materials often used in suppression systems. *See also* Aqueous film-forming foam; chemical foam; dry chemical; gaseous agent fire extinguishing system. (SAN)

Fire triangle The three components that are necessary for combustion to take place: fuel, oxygen, and a source of ignition. (FSL)

Fire wall A wall that is stable and durable enough to withstand the effects of the most severe fire exposure. (FSL)

Fire watch A person or persons assigned to guard against fire while hot work is being performed in an area where flammable or explosive contaminants are at potentially hazardous levels. The persons on fire watch must be instructed as to the specific anticipated fire hazards and how the firefighting equipment (usually #30 portable fire extinguishers, water hose, and/or fire blanket) provided is to be used. The persons on watch should remain for a sufficient period of time—e.g., 30 minutes—after completion of the hot work, to ensure that no possibility of fire exists. A productive fire watch consists of persons who participate in the hot work but are also responsible for extinguishing a fire if one occurs. A nonproductive fire watch is one that is simply on guard and has no other assigned duties. *See also* Hot work. (SAN)

Firing line The wires used to connect the electrical power source to an electric blasting cap. (FSL)

First-aid case For OSHA recording purposes, any one-time treatment, with subsequent observation, of minor scratches, burns, splinters, etc., that do not ordinarily require further medical care. Such treatment and observation are considered first aid even if provided by a physician or other professional personnel. First-aid cases do not have to be recorded on the OSHA log form 200. *See also* Lost workday case; OSHA Log of Occupational Injuries and Illness. (SAN)

First collision dose A term applied to neutron dose. It involves the total energy imparted to charged particles for only the first collision of a neutron within an absorbing medium. It is not assumed that the energy of the secondary particles is completely liberated within the area of interest. Thus, the first-collision dose may be smaller than the actual absorbed dose, depending on the radiation energy and the particles involved in the particular interaction. (RMB)

First draw The water that comes out when a tap is first opened. This water may have the highest level of lead contamination from plumbing materials. (FSL)

First quartile A number that is higher in value than at least one-fourth of the data and lower than at most three-fourths of the data. (VWP)

First responder Under the OSHA HAZWOPER standard, two levels of this concept are defined. **(1)** *Awareness Level:* Includes individuals who are likely to witness or discover a hazardous substance release and who have been trained to initiate an emergency response sequence by notifying the proper authorities of the release. At this level, individuals would take no further action beyond notifying the authorities of the release. **(2)** *Operations Level:* Includes individuals who respond to releases or potential releases of hazardous substances as part of the initial response to the site for the purpose of protecting nearby persons, property, or the environment from the effects of the release. These individuals are trained to respond in a defensive fashion without actually trying to stop the release. Their function is to contain the release from a safe distance, keep it from spreading, and prevent exposures. *See also* Hazardous Waste Operations and Emergency Response (HAZWOPER). (SAN)

Fisher exact test A test applied to a small sample of data to compare binomial proportions in two independent samples. (VWP)

Fissile Refers to cleavage, or splitting apart. In biology, the term refers to a form of asexual reproduction in which the parent organism divides into two or more equal parts that have identical characteristics. In nuclear energy, it refers to the splitting of the nucleus of an atom. (FSL)

Fission The process whereby the nucleus of a particular heavy element splits into (generally) two nuclei of lighter elements, with the release of substantial amounts of energy. The most important fissionable materials are uranium-235 and plutonium-239. (RMB)

Fission fragments The nuclei of elements produced by fission. (RMB)

Fission products A general term for the complex mixture of substances produced as a result of nuclear fission. A distinction should be made between these and direct fission products, or fission fragments formed by the actual splitting of heavy-element nuclei. As many as 80 different fission fragments result from the roughly

40 different modes of fission of a given nuclear species such as uranium-235 or plutonium-239. The fission fragments, generally being radioactive, immediately begin to decay, forming additional (daughter) products, with the result that the complex mixture of fission products so formed may contain about 200 different isotopes of over 30 elements. (RMB)

Fissure springs Springs that originate from breaks in rocks through which water passes. This type of spring may discharge water that has been contaminated by surface drainage from strata close to the surface. (FSL)

Fixation The process in a soil whereby certain chemical elements used for plant nutrition are converted from a soluble or exchangeable form to a much less soluble or a nonexchangeable form. Phosphate fixation is an example. (SWT)

Fixative A chemical, such as alcohol or formaldehyde, used for the preservation of biological materials. (FSL)

Fixed fire extinguishing system According to OSHA, a permanently installed system that either extinguishes or controls a fire at the location of the system. *See also* Sprinkler system; suppression system. (SAN)

Flame Manifestation of a chemical reaction or reaction product that is partially or entirely gaseous and that produces heat and light. (FSL)

Flammable Refers to a chemical that falls into one of the following categories. **(1)** *Flammable aerosol:* An aerosol that, when tested by the methods described in 16 CFR 1500.45, yields a flame projection exceeding 18 inches at full valve opening or a flashback (a flame extending back to the valve) at any degree of valve opening. **(2)** *Flammable gas:* A gas that, at ambient temperature and pressure, forms a flammable mixture with air at a concentration of less than or equal to 13% by volume; or a gas that, at ambient temperature and pressure, forms a range of flammable mixtures with air greater than 12% by volume, regardless of lower limit. **(3)** *Flammable liquid:* A liquid that has a flash point below 100°F (37.8°C), excepting any mixture that has components with flash points of 100°F (37.8°C) or higher that make up 99% or more of the total volume of the mixture. **(4)** *Flammable solid:* A solid, other than a blasting agent or explosive, that is liable to cause fire through friction, absorption of moisture, spontaneous chemical change, or retained heat from manufacturing or processing, or that can be ignited readily and that when reignited burns so vigorously and persistently as to create a serious hazard. A chemical shall be considered to be a flammable solid if, when tested by the method described in 16 CFR 1500.44, it ignites and burns with a self-sustained flame at a rate greater than one-tenth of an inch per second along its major axis. (FSL)

Flanged hood A barrier placed around the periphery of a chemical hood to reduce air turbulence and hood entry pressure loss by keeping the hood from drawing air from behind the hood face. *See also* Chemical hood. (DEJ)

Flare A pyrotechnic device that produces a single source of intense light or radiation. (FSL)

Flare tower A form of pressure relief system for chemical processes that allows gases and liquids to be vented from a vessel quickly while minimizing the amount released to the atmosphere. The flare tower should have: a knockout drum to separate liquid, flow backup alarms, pilot lights, a flame sensor, assist gas for emergency venting, excess-effluent alarms, an inert gas purge, a water seal between the stack and collection system to prevent flashback, and a proper stack height. The knockout drum should have a pressure relief device, liquid level indicator, steam or electrical tracing, automatic draining, and pressure rating. *See also* Process safety. (SAN)

Flash burn An inflammation of the lens of the eye due to excessive exposure to ultraviolet (UV) radiation from a welding arc. Ordinary glasses will not protect against high-intensity UV generated by welding arcs; therefore, special lens are required for both welders, helpers, and bystanders. *See also* Lens; shade numbers. (SAN)

Flashover The transmission of detonation from one cartridge to another in line. It is often called sympathetic detonation. *See also* Explosion. (FSL)

Flash point The minimum temperature at which a liquid gives off vapor in sufficient concentration to ignite when tested using the following testers: **(1)** *Tagliabue closed tester* (*see* American National Standards method of test for flash point by Tag closed tester 211.24–1979, ASTM D-56-79): For liquids with a viscosity of less than 45 Saybolt universal seconds (SUS) at 100°F (37.8°C) that do not contain suspended solids and do not have a tendency to form a surface film under test. **(2)** *Pensky-Martens closed tester* (*see* American National Standard method of test for flash point by Pensky-Martens closed tester, 211.7–1979, ASTM D-93-79): For liquids with a viscosity equal to or greater than SUS at 100°F (37.8°C) or that contain suspended solids, or that have a tendency to form a surface film under test. **(3)** *Setaflash closed tester* (*see* American National Standard method of test for flash point by Setaflash closed tester, 211–1979, ASTM D-3278-78). (FSL)

Fleming, Sir Alexander A Scottish bacteriologist (1881–1955) who discovered penicillin in 1938. (FSL)

Floc A clump of solids formed in sewage by biological or chemical action. *See also* Coagulation. (FSL)

Flocculation The process whereby the particles of a dispersion form larger clusters by contact and adhesion. This is one of the treatment procedures used in municipal water plants and swimming pools for the removal of suspended material clumps. (FSL)

Flood Unexpected aboveground accumulation of water that has been caused by melting snow, rapid runoff of surface water, and/or excessive rainfall. (FSL)

Flood plain Land that borders a stream and that is built of sediments deposited from overflow of the stream. A flood plain may be inundated when the stream is at flood stage. (SWT)

Floor sweep An exhaust fan/duct-work system designed to capture vapors that are heavier than air and that may collect along the floor in an area where chemicals are used or stored. (FSL)

Flora **(1)** A collective term applied to the plants or plant life of any particular region or epoch; **(2)** a descriptive catalog of the plants of any geographical area or geological period. (SMC)

Flow meter A gauge that shows the speed of wastewater moving through a treatment plant. Also used to measure the speed of liquids moving through various industrial processes. (FSL)

Flue The passageway within a fireplace, chimney, oven, or other structure containing a fire or combustion products from which spent gases can escape to the atmosphere. *See also* Flue gas. (FSL)

Flue gas Vented air emitted from a chimney after combustion in a burner. It can include nitrogen oxides, carbon oxides, water vapor, sulfur oxides, particles, and many chemical pollutants. (FSL)

Flue gas desulfurization A technology that uses a sorbent, usually lime or limestone, to remove sulfur dioxide from the gases produced by burning fossil fuels. Flue gas desulfurization is currently the state-of-the-art technology in use by major SO_2 emitters—e.g., power plants. (FSL)

Flume A natural or artificial channel that diverts water. (FSL)

Fluorescence Phenomenon involving the absorption of radiant energy by a substance and its reemission only during the period of radiation excitation, as visible or near-visible light. (RMB)

Fluorescent screen A sheet of material coated with a substance, such as calcium tungstate or zinc sulfide, that will emit visible light when irradiated with ionizing radiation. (RMB)

Fluoridation The process by which trace amounts of fluorine are added to water supplies for the purpose of reducing dental caries. Many water supplies contain levels of naturally occurring fluorine compounds above 1 ppm. Levels this high can cause mottling of the enamel of the teeth; consequently, the supplies must be adjusted to bring the fluoride concentration to the 1-ppm range. The fluoridation of community water supplies is a controversial practice, although most scientific evidence supports the practice as both effective in reducing dental caries and cost-beneficial. *See also* Fluorides; fluorosis. (FSL)

Fluorides Compounds of the lightest and most reactive of the halogens. They occur naturally and can also be released into the environment by a number of industrial processes, including the processing of phosphate from rocks. As pollutants, fluorides are harmful to plants and mammals. *See also* Fluorine; fluoridation. (FSL)

Fluorine A pale-yellowish-green gas that is more reactive than chlorine. Fluorine is an element that occurs naturally in combined form as fluorite and cryalite. *See also* Fluoridation; fluorides. (FSL)

Fluorite A mineral, also known as fluorspar, that has the formula CaF_2. This mineral, which is found in sedimentary rock, is used in the manufacture of hydrofluoric acid and in ceramics. (FSL)

Fluorocarbon (FC) Any of a number of organic compounds analogous to hydrocarbons in which one or more hydrogen atoms are replaced by fluorine. Once used in the United States as a propellant in aerosols, fluorocarbons are now primarily used in coolants and some industrial processes. FCs containing chlorine are called chlorofluorocarbons (CFCs). They are believed to be modifying the ozone layer in the stratosphere, thereby allowing more harmful solar radiation to reach the earth's surface. *See also* Chlorinated hydrocarbon. (FSL)

Fluorography (photofluorography) Photography of images produced on a fluorescent screen by X-radiation or gamma radiation. (RMB)

Fluoroscope A screen that is suitably mounted with respect to an X-ray tube for ease of observation and protection, and that is used for indirect visualization by means of X-rays of internal organs in the body or internal structures in apparatus or in masses of metal. (RMB)

Fluorosis Disease in ruminants and occasionally humans that is caused by the ingestion of excessive amounts of fluorine. It is characterized by mottling of the teeth and excessive bone thickening. Cattle contract the disease by consuming vegetation contaminated with fallout of fluorine from industrial plants. Humans are usually exposed by consuming water containing high levels of naturally occurring fluorine compounds. (FSL)

Fluorspar *See* Fluorite. (FSL)

Flush **(1)** To open a cold-water tap to clear out all the water that may have been contained for a long time in the pipes. In new homes, to flush a system means to send large volumes of water gushing through the unused pipes to remove loose particles of solder and flux. **(2)** To force large amounts of water through piping, tubing, or storage or process tanks, in order to clean them out. (FSL)

Flux The number of visible-light photons, gamma-ray photons, neutrons, particles, or energy crossing a unit surface area per unit time. The units of flux are number of particles (or energy, etc.) per square centimeter per second. (RMB)

Fly ash Noncombustible residual particles from the combustion process, carried by flue gas. (FSL)

Fly rock Those rocks ejected from blast areas by the force of an explosion. (FSL)

Foam A gas dispersed in a liquid. Foams are used as firefighting materials because of their ability to choke off the oxygen supply to combustible materials. *See also* Fire; fire extinguisher. (FSL)

Focal spot (X-rays) The part of the target of the X-ray tube that is struck by the main electron stream. (RMB)

Fog A term loosely applied to visible aerosols that are liquid; formation by condensation is sometimes implied. (DEJ)

Fogging Applying a pesticide by rapidly heating the liquid chemical so that it forms very fine droplets that resemble smoke or fog. It may be used to destroy mosquitoes, black flies, and similar pests. (FSL)

Foliar diagnosis An estimation of mineral nutrient deficiencies of plants, based on the chemical analysis of certain plant parts and the color and growth characteristics of the foliage of the plants. (SWT)

Fomite A subclass of vehicles, including inanimate objects such as articles of clothing, that can become contaminated and transmit pathogenic agents. (FSL)

Food additive An ingredient added to food to serve a specific function or to improve a specific property of the product. The use of a food additive may directly or indirectly affect the characteristics of the product. A material used in the production of containers or packages is also considered to be an additive if it may reasonably be expected to become a component or affect the characteristic of the food packaged in the container. *See also* Additive. (HMB)

Food and Agriculture Organization (FAO) Program within the United Nations devoted to the improvement of agricultural practices in developing countries. (FSL)

Foodborne disease Any disease that is transmitted through food. Such a disease might be caused by contamination of foods by bacteria, viruses, fungi, or parasites. Foods may also be contaminated with toxic chemicals or naturally occurring toxicants, or may cause a food-associated allergic reaction. Most common foodborne diseases include viral infections (e.g., hepatitis); bacterial diseases (e.g., salmonellosis, listeriosis, campylobacteriosis, *Clostridium perfringens* infections); bacterial intoxication (e.g., botulism, staphylococcal intoxication); diseases caused by naturally occurring toxicants (e.g., scombro toxin); and diseases caused by parasites (e.g., tapeworm). Suppression of foodborne diseases is largely successful because of good sanitation measures and food-handling practices in food production, processing, and sale, to the point of consumption. *See also specific diseases*, e.g., Campylobacteriosis; salmonellosis. (HMB)

Foodborne Disease Outbreak Defined as an incident in which two or more persons experience a similar illness, usually gastrointestinal, after ingestion of a common food, and in which epidemiologic analysis implicates the food as the source of the illness. By the Centers for Disease Control and Prevention's definition, exceptions to the above include botulism and chemical poisoning, a simple case of either of which constitutes an outbreak. (HMB)

Foodborne intoxication A type of foodborne illness characterized by the ingestion not of viable microorganisms, but rather of preformed exotoxin produced by the organisms. Examples of foodborne intoxications caused by bacteria include botulism, caused by the bacterium *Clostridium botulinum*, and staphylococcal intoxication, caused by the bacterium *Staphylococcus aureus*. It may be appropriate to include in this category poisoning by the ingestion of toxic chemicals, as, in its broad definition, to intoxicate simply means "to poison." However, it is more commonly found that foodborne intoxications arise from preformed microbial toxins, and toxic chemical ingestion is considered a form of chemical poisoning. Toxicants such as those found in tissues of certain plants and animals are often included in the injection category. (HMB)

Food chain A sequence of organisms, each of which uses the next-lower member of the sequence as a food source. (FSL)

Food-contact surfaces Those surfaces of equipment or utensils, used in either a food service or food-processing establishment, with which food normally comes in contact, and those surfaces onto which food may drain, drip, or splash. Regulations generally require that such surfaces be easily cleanable, smooth, and free of breaks, open seams, cracks, chips, pits, and similar imperfections, and be free of difficult-to-clean internal corners and crevices. All such equipment or utensil surfaces are required to be fabricated of material that is easily cleanable and durable under conditions of normal use and resistant to denting, buckling, pitting, chipping, and crazing. (HMB)

Food infection An infection (foodborne infection) that results from the ingestion of a contaminated food that contains sufficient numbers of viable infectious microorganisms to elicit the signs and symptoms of the disease eventually. Examples of common foodborne infections are salmonellosis, campylobacteriosis, listeriosis, hepatitis, and those caused by parasites. Large numbers of the organisms may survive in the gut, and proliferate, producing a toxin or some other condition resulting in the symptoms of the disease. Typically, foodborne infections exhibit longer incubation periods than do foodborne intoxications because of the time required for the infection to be established. *See also* Campylobacteriosis; listeriosis; salmonellosis. (HMB)

Food poisoning A broad term including foodborne illnesses caused by the ingestion of foods containing microbial toxins or chemical poisons. In either case, the toxin is produced in the food by the growth of the organisms or is present prior to consumption. Examples include botulism, staphylococcal food poisoning, and poisoning caused by chemical toxicants. *See also* Botulism; staphylococcal intoxication. (HMB)

Food-processing establishments Commercial establishments in which food is manufactured or packaged for human consumption. They do not include food-service establishments, retail food stores, or commissary operations. (HMB)

Food sanitation The opportunity for the contamination of foods occurs at numerous specific points from the time of food production in the field or on the farm to the point of consumption. Many food sanitation measures can be taken during this process to ensure a clean, safe, wholesome product, properly prepared for consumption. Examples include: thorough cooking of all products of animal origin—e.g., beef, pork, and poultry; production of dairy products from pasteurized milk only; the keeping of perishable foods at temperatures below 40°F or above 140°F (i.e., not in the temperature range within which most pathogens grow); the maintenance of frozen foods in a frozen state or the permitting of such foods to be thawed for only short periods of time; the thorough washing of raw fruits and vegetables; the adequate cleaning and sanitizing of equipment, dishes, or utensils; the taking of measures to ensure that food handlers are free of communicable disease; and the protection of appropriate areas and foods from contamination by wastes, sewerage, and insect or rodent infestation. (HMB)

Food service The preparation and serving of meals, short orders, sandwiches, frozen desserts, or other edible products, generally for sale to the public. (HMB)

Food-service establishment Any place where food is prepared and intended for individual-portion service. It includes any site at which individual portions are provided, regardless of whether consumption is on or off the premises and regardless of whether there is a charge for the food. This also includes delicatessen-type operations that prepare sandwiches intended for individual-portion service. Not included are private homes where food is prepared or served for individual family consumption, retail food stores, food vending machines, and supply vehicles. (HMB)

Food shield Device designed to prevent contamination of unpackaged food by a food-service customer. *See also* Sneeze shield. (HMB)

Food transport cabinet An enclosed cabinet capable of being transported and intended for the conveyance of foods. (HMB)

Food-warming table A device receiving foods that are at a temperature lower than serving temperature, and capable of heating them to the approximate serving temperature. (HMB)

Food zone The set of surfaces that normally come into contact with food, and those surfaces from which food may drain, drip, or splash back onto surfaces that normally come into contact with food. Often referred to as food contact surfaces. *See also* Food contact surface. (HMB)

Fool's gold *See* Iron pyrites. (FSL)

Foot-candle The illumination resulting from the uniform distribution of a flux of one lumen (lm) on a surface area of one square foot. One foot-candle equals one lumen per square foot. (DEJ)

Foot-lambert The unit of photometric brightness equal to the uniform brightness of a perfectly diffusing surface emitting or reflecting one lumen per square foot. (DEJ)

Footslope The relatively gently sloping, slightly concave to straight slope component of an erosional slope that is at the base of the backslope component. (SWT)

Force The push or pull that tends to impart motion to a body at rest, or to increase or diminish speed, or to change the direction of a body already in motion. (FSL)

Force majeure For insurance purposes, a superior or irresistible force. *See also* Act of God. (DEW)

Forest floor Dead vegetable or organic matter, including litter and unincorporated humus, on the mineral soil surface under forest vegetation. (SWT)

Forest soils Soils that have developed under forest vegetation. (SWT)

Formaldehyde A colorless, pungent, irritating gas, CH_2O, used chiefly as a disinfectant and preservative and in synthesizing other compounds and resins. (FSL)

Formulation The substance or mixture of substances that comprises all active and inert ingredients in a pesticide. (FSL)

Fossorial **(1)** Refers to organisms that live in the earth, especially animals that dig burrows; **(2)** adapted for or used in digging (e.g., fossorial limbs or moles). (SMC)

Fractionation Any of several processes, apart from radioactive decay, that result in change in the composition of radioactive debris. As a result of fractionation, delayed fallout generally contains relatively less strontium-90 and cesium-137 than does early fallout from a surface burst. (RMB)

Fragipan A natural subsurface horizon with high bulk density and/or high mechanical strength relative to the solum above, seemingly cemented when dry, but showing a moderate to weak brittleness when moist. The layer is low in organic matter, mottled, slowly or very slowly permeable to water, and usually shows occasional or frequent bleached cracks that form polygons. It may be found in profiles of either cultivated or virgin soils but not in calcareous material. (SWT)

Frass Excrement from insects. *See also* Excrement. (SMC)

Free air ionization chamber An ionization chamber in which a delimited beam of radiation passes between the electrodes without striking them or other internal parts of the equipment. The electric field is maintained perpendicular to the electrodes in the collecting region; as a result, the ionized volume can be accurately determined from the dimensions of the collecting electrode and the limiting diaphragm. This is the basic standard instrument for X-ray dosimetry within the range of 5 to 400 kVp. (RMB)

Free chlorine Chlorine as elemental chlorine, hypochlorite ion, and hypochlorous acid is referred to as free chlorine or free chlorine residual. Chlorine in water may be present in such an uncombined or free state or in combined form as chloramines. *See also* Chlorine; chlorine compounds; disinfectants; hypochlorites; hypochlorous acid; residual chlorine compounds. (HMB)

Freedom of Information Act (FOIA) Law that requires federal agencies to make their records available to the public upon request, subject to specific exemptions and privileges. 5 USC §552. (DEW)

Free-fall distance The vertical displacement of the fall arrest attachment point on an employee's safety belt or safety harness between onset of fall and just before the system begins to apply force to halt the fall. This distance excludes deceleration distance and lanyard and lifeline elongation, but includes any deceleration-device slide distance or self-retracting lifeline/lanyard extension before these devices operate and fall arrest forces occur. OSHA requires that the free-fall distance be kept to a minimum, and in no case exceed six feet. *See also* Fall protection; deceleration distance. (SAN)

Free field Field in which sound radiated from a source can be measured without influence from the test space. (DEJ)

Free-flowing explosives Commercial explosives, such as ammonium nitrate, that can be poured into boreholes and detonated. (FSL)

Freons The registered trade name for chemical compounds (hydrocarbons) in which most of the hydrogen atoms have been replaced by fluorine, chlorine, and/or bromine. *See also* Chlorofluorocarbons. (FSL)

Frequency **(1)** The number of cycles, revolutions, or vibrations completed per unit time. The frequency of sound, for example, describes the rate at which complete cycles of high- and low-pressure peaks are produced by the sound source. The unit of measurement is cycles per second or hertz (Hz). The normal human ear has a frequency range of 20 to 20,000 Hz at moderate sound pressure levels. **(2)** The number of times a data value occurs in a sample. (DEJ, VWP)

Frequency histogram A type of bar graph representing an entire set of data. This pictorial representation includes a title that identifies the population, a vertical scale that identifies the frequencies in the data classes, and a horizontal scale that identifies and provides a scale for the data values. (VWP)

Frequency polygon A line graph of class frequency plotted against class mark. It can be obtained by connecting midpoints of the tops of the rectangles in a frequency histogram. (VWP)

Fresh water Water that generally contains less than 1000 milligrams per liter of dissolved solids. (FSL)

Friable Refers to materials that have a tendency to crumble easily. Most often used to describe the condition that exists when asbestos fibers can potentially be released and become airborne and therefore present a respiratory hazard. (SWT)

Friction cone penetrometer An instrument used to measure the local friction of penetration resistance in soils. (SWT)

Frostbite The destruction of tissue, resulting from exposure to extreme cold or contact with extremely cold objects. *See also* Hypothermia. (FSL)

Frost heaving The process whereby soil is lifted due to the formation of ice lenses. (SWT)

Frost point *See* Dew point. (DEJ)

Frost wedging The process whereby freezing water infuses on rock pores and fractures the rock. (FSL)

Fruigivore An organism that feeds on fruits. (SMC)

F table Table that lists areas under the F-curve that use different significance levels (usually including 0.05, 0.025, and 0.01) and different degrees of freedom for the numerator and denominator of the F-statistic. *See also* F-distribution. (VWP)

F test A hypothesis test using an F-distribution. (VWP)

Fuel A substance capable of reacting with oxygen and oxidizers and producing heat in the process. (FSL)

Fuel Economy Standard The Corporate Average Fuel Economy Standard (CAFE), which went into effect in 1978. It was meant to enhance the national fuel conservation effort by slowing fuel consumption through a miles-per-gallon requirement for motor vehicles. (FSL)

Fuel gas A gas or gas mixture that is combustible. (FSL)

Fugitive emissions Emissions not caught by a capture system. (FSL)

Fuller's earth An argillaceous rock or clay with a high potential for the absorption of water, oils, and greases. (FSL)

Fulvic acid A mixture of organic substances remaining in solution upon acidification of a dilute alkali extract from the soil. (SWT)

Fume Small solid particles usually generated by condensation from the gaseous state of a metal or plastic after volatilization. Welding, for example, causes the volatilization of metals into a gas, followed by condensation upon contact with cooler air, creating small particles on the order of 0.1–1 μm in diameter. In popular usage, "fume" often refers to any type of air contaminant. (DEJ)

Fumigant A pesticide, such as picryl chloride or methyl bromide, that when applied in the gaseous state to a confined space, kills insects or rodents. *See also* Fumigation. (FSL)

Fumigation The process whereby gaseous pesticides are applied in an enclosed space. When used for the destruction of termites and other wood-destroying insects, this technique consists of sealing the affected areas and then applying pesticides, such as methyl bromide, in gaseous form to the structure. A similar technique is used to eliminate insects from stored grain crops. Rodents can be eradicated from burrows by the introduction of pesticides such as hydrogen cyanide. *See also* Fumigant. (FSL)

Fungicide Chemical developed for the control of fungi, particularly those affecting foods and ornamental crops. A few drugs are available to treat fungal infections in humans and animals. Some examples of fungicides include captan (dicarboximide), pentachlorophenol (a substituted aromatic), maneb (dithiocarbamate), benomyl (a nitrogen heterocycle), and methylmercury. *See also* Pesticides. (FSL)

Fungivore An organism that feeds on fungi. (SMC)

Fungology The study of fungi. (SMC)

Fungus A general term used to describe the diverse morphological forms of yeast, rust, mildew, and mold. Earlier classified as plants without chlorophyll, the fungi are now generally placed in the Kingdom Protista with algae (except blue-green types), the protozoans, and slime molds. Fungi are heterotrophs and obtain nourishment by absorption, usually from dead or decaying organic matter; they have rigid cell walls and reproduce by means of spores. Fungi are important as foods, in the fermentation process (in the development of alcohols), and in antibiotics, drugs, and antitoxins. Some—for example, the smuts—can infect grain and other plants. Others, such as *Histoplasma capsalatum*, are pathogenic for humans and can cause pulmonary disease. Fungi can be found in the soil, water, and air. (FSL)

Fuse A type of cord, consisting of a flexible fabric tube and a core of explosive, that is used in blasting and demolition work. (FSL)

Fusible plugs Such plugs are made of a metal, such as lead, that melts at a specified temperature, and can thereby create an opening in devices that contain them. Fusible plugs are used as pressure relief devices in boilers, compressed-gas cylinders, and other pressure vessels. The metal is selected so that the temperature at which the fusible plug will melt is well below that of the metal of which the vessel is constituted. *See also* Pressure Vessel. (SAN)

Fusion A nuclear reaction characterized by the joining together of light nuclei to form heavier nuclei, the energy for the reactions being provided by kinetic energy derived from violent thermal agitation of particles at very high temperatures. If the colliding particles are properly chosen and the agitation is violent enough, there will be a release of energy from the reaction. The energy of the stars is believed to be derived from such reactions. (RMB)

G

Galena Lead sulfide (PbS), which is the source of most of the world's production of lead used in storage batteries and as the gasoline additive tetraethyllead. In the 1920s, lead was shown to be responsible for a bizarre neurological disorder brought about by chronic worker exposure to leaded gasoline in refineries, and thus the term ''loony gas'' was coined. *See also* Lead poisoning. (FSL)

Galvanize To bind with zinc or place a coating of zinc on iron or steel to prevent corrosion. The process is carried out by immersing the iron or steel base metal in molten zinc or by depositing (electroplating) a solution of zinc sulfate on the surface. (FSL)

Galvanometer An instrument used for detecting and measuring an electric current. (FSL)

Gamete Any germ cell, such as an ovum, spermatozoan, or pollen cell. (FSL)

Gamma rays Electromagnetic radiations of high energy originating in atomic nuclei and accompanying many nuclear reactions—e.g., fission, radioactivity, and neutron capture. Physically, gamma rays are identical with X-rays of high energy, the only essential difference being that the X-rays do not originate from atomic nuclei but are produced in other ways—e.g., by slowing down (fast) electrons of high energy. *See also* Electromagnetic radiation. (RMB)

Ganglion A mass of human or animal tissue containing nerve cells (neurons). (FSL)

Gangplank A bridge between a platform and a vehicle or ship. Gangplanks must be sufficiently strong to carry the anticipated load, must be inspected regularly for defects, and must be repaired promptly. *See also* Dock plates. (SAN)

Gangrene An infection caused by one of the species of anaerobic *Clostridium* bacteria and causing the destruction of body tissue. The disease results from the death of body tissue, loss of blood supply, and, ultimately, putrefaction of the affected area. (FSL)

Garbage The putrescible and nonputrescible animal and vegetable wastes resulting from the handling, preparation, and consumption of foods. *See also* Compost; solid waste. (FSL)

Garnet A group of metals that have the general chemical composition $A_3B_2(SiO_4)_3$, found in a wide range of rocks and used in industrial processes because of their hard and abrasive qualities. (FSL)

Gas A thin fluid, like air, capable of indefinite expansion but convertible by compression and cold into a liquid and eventually a solid. Gases may be either elements (e.g., argon) or compounds (e.g., carbon dioxide). (FSL)

Gas amplification As applied to gas-ionization radiation-detecting instruments, the ratio of the charge collected to the charge produced by the initial ionizing event. (RMB)

Gas cap An accumulation of natural gas found above pooled oil in petroleum production areas. (FSL)

Gas chromatography A laboratory or portable analytical procedure involving a physical process for separating the components of complex mixtures. A gas chromatograph consists of a carrier gas supply, an injection system for the introduction of a gas or vaporizable sample, and a separation column. This column includes a stationary phase containing either inert material or an active substrate, a heater and oven assembly, a detector (such as a flame ionization detector, a photoionization detector, or an electron capture detector), and a recorder. Gas chromatography is widely used for analysis of solvents and other environmental contaminants. *See also* Chromatography. (DEJ)

Gas-forming bacteria Organisms that ferment lactose in foods or other carbohydrates, producing both acid and gas. Gas production may result in a food product that is unacceptable. The coliform bacteria responsible are known lactose-fermenters. *See also* Coliforms. (HMB)

Gasification Conversion of solid material such as coal into a gas for use as a fuel. (FSL)

Gas masks A full-face respirator equipped with an air-purifying cartridge or canister that removes contaminants and renders air breathable to the user. These cannot be used in oxygen-deficient atmospheres. The term "gas mask" is synonymous with respirators that are used by the armed forces to protect against chemical attack. *See also* Air-purifying respirator. (DEJ)

Gas pressure The force, generally designated in pounds per square inch (PSI), that is exerted by a gas on its surroundings. *See also* Vapor pressure. (FSL)

Gastric Pertaining to the stomach. (FSL)

Gastroenteritis Inflammation of the mucous membrane of the stomach and intestines. Gastroenteritis with acute onset may be caused by various bacteria and viruses. Many foodborne illnesses result in symptoms typical of gastroenteritis, or gastrointestinal syndrome. Symptoms vary and may include diarrhea, abdominal cramps, nausea, vomiting, fever, malaise, muscle ache, and fatigue. (HMB)

Gauntlet A cuff, collar, or portion of a glove above the hand intended to protect the lower portion of the arm from contamination. (FSL)

Gaussian distribution *See* Normal distribution. (VWP)

Geiger counter An electrical device that detects the presence of certain types of radioactivity. It consists of a needlelike electrode inside a hollow metallic cylinder filled with gas, which, when ionized by radiation, sets up a current in an electrical field. Named for its inventor, Hans Geiger (1882–1945). (RMB)

Geiger-Mueller Counter A refined version of a Geiger counter that has an amplifying system and is used for detecting and measuring radioactivity. *See also* Geiger counter. (FSL)

Geiger region In an ionization radiation detector, the operating voltage interval in which the charge collected per ionizing event is essentially independent of the number of primary ions produced in the initial ionizing event. (RMB)

Geiger threshold Minimum voltage at which a Geiger-Mueller tube will indicate the presence of ionizing events. (RMB)

Gel Material that has a jellylike appearance and that is formed when a colloidal suspension is allowed to stand. Gels resemble solids more than they do liquids. (FSL)

Gelamite The trade name for a specific type of semigelatinous explosive. *See also* Explosion. (FSL)

Gelatin A protein extract from collagen that is used in research, food processing, and the production of adhesives. *See also* Collagen. (FSL)

Gene A functional unit of heredity that occupies a specific location on a chromosome, is capable of reproducing itself exactly at each cell division, and can direct the formation of an enzyme or other protein. Genes consist of discrete segments of large DNA molecules that contain atoms in the correct sequence for coding the sequence of amino acids needed to form specific peptides. *See also* Deoxyribonucleic acid. (FSL)

Genera *See* Genus. (SMC)

General Duty Clause That part of the Occupational Safety and Health Act that requires employers to provide to each employee a place of employment free of recognized hazards that are likely to cause death or serious physical harm. Often applied by OSHA to cover hazards for which specific health or safety standards do not exist. (DEJ)

Generator Under the Resources Conservation and Recovery Act (RCRA), generators of solid waste must determine whether the waste is hazardous. If it is, the generator must follow EPA regulations covering the use of manifests, packaging, labeling, marking, and placarding of wastes for transport. Regulations also apply to the accumulation of hazardous wastes on site, the record-keeping and reporting requirements, and the export and import of hazardous wastes.

According to the Environmental Protection Agency, a small-quantity generator is one who produces in a calendar month: more than 100 kilograms (1 kg = 2.2 pounds) but less than 1000 kg of nonacutely hazardous waste; less than 100 kg of waste resulting from the cleanup of any residue or contaminated soil, water, or other debris from an acutely hazardous waste; or less than 1 kg of an acutely hazardous waste.

Generators who exceed these quantities are considered large-quantity generators. Generators who produce less than these quantities are considered conditionally exempt generators. That is, if they comply with certain conditions, they need not manage their wastes in accordance with the more stringent regulations established for small- and large-quantity generators of hazardous waste. In determining the quantity of waste produced, however, it is important to consider certain exemptions and exclusions for types of waste. These and other issues are described in the "Hazardous Waste Determination" section of the RCRA. 42 USC §6922, 40 CFR Part 262. *See also* Hazardous waste; manifest; Resource Conservation and Recovery Act; solid waste. (FSL, DEW)

Genetic engineering A process of inserting new genetic information into existing cells in order to modify one of the characteristics of an organism. (FSL)

Genome A total set of chromosomes derived from one parent; the haploid number of a gamete. Also, the total gene complement of a set of chromosomes found in a higher life form or the functionally similar but simpler linear arrangements found in bacteria and viruses. (FSL)

Genotoxic Refers to the ability of a chemical to affect adversely the genome of living cells, such that upon duplication of the affected cells, a mutagenic or carcinogenic event is expressed because of the alteration of the molecular structure of the genome. (FSL)

Genotoxin Chemical or radioactive substances known to cause or suspected of causing damage to the deoxyribonucleic acid (DNA) in individual cells, thus causing mutations or cancer. (FSL)

Gentian violet A purple dye, also known as methyl violet, with antibacterial, antifungal, and antihelminthic properties that is used as a stain in microscopy. (FSL)

Genus The taxon in the classification hierarchy between family and species. Plural: genera. *See also* Classification. (SMC)

Geode A hollow, round body in a rock that is lined with crystals projecting inward. (FSL)

Geomedicine The branch of medicine that involves the influence of climatic and environmental factors on human health. (FSL)

Geometric mean For a set of n numbers x_1, x_2, x_3, x_n, the nth root of the product of the numbers:

$$\sqrt[n]{x_1 \cdot x_2 \cdot x_3 \ldots x_n}$$

(VWP)

Geomorphology The science that deals with the evolution of the earth's surface. The science of landforms. The systematic examination of landforms and their interpretation as records of geological history. (SWT)

Geophagia A condition in which individuals consume dirt or other nonfood items. *See also* Pica. (FSL)

Geotrophism A response in which an organism turns toward gravity. The most common example is the turning of the roots of plants to follow gravity. (SMC)

Germ cells Cells whose function is to reproduce an organism. *See also* Gamete. (FSL)

Germicide Any substance that kills microbes; an agent that destroys pathogenic microorganisms. The term is synonymous with "bactericide" though broader in its range of meaning. It should be considered a layman's term and is somewhat dated. Detergents or other products that contain germicides are often labeled as having sanitizing or disinfecting properties. *See also* Bactericide; disinfectant. (HMB)

Germinal Pertaining to a primordium, the earliest tract of a structure within an embryo. In botany, relates to germination or development, bud or sprout. (FSL)

Gestation The act or period of carrying young in the uterus from conception to birth; pregnancy; includes embryonic and fetal life. (FSL)

Giardiasis An infection caused by the protozoan *Giardia lamblia*. The organism, which is spread by contaminated food and water and person-to-person contact, resides in the lumen portion of the small intestine. (FSL)

Gibberellins Plant hormones that control growth and development. They are responsible for stem elongation in some plants and play a role in the formation of fruit and flowers. (FSL)

Gilgai Microrelief structures of soils caused by expansion and contraction with changes in moisture. Soils that contain large amounts of clay that swells and shrinks considerably with wetting and drying can form gilgai. (SWT)

Girdling Cutting and removing a ring of bark and wood tissue from around a tree. This eventually kills the tree because the cambium and the vascular tissues (phloem and xylem) are removed, which prevents the movement of water and nutrients. (SMC)

Glacial Refers to a highly pure state of some acids, such as acetic acid; so designated because the freezing point is only slightly below room temperature. (FSL)

Glacial drift Transport or deposit of rock-debris by glaciers, either directly from the ice or from the melt water. (SWT)

Glacier A body of ice that originates on land by the melting, and freezing of snow and that shows evidence of current or past movement; a ''river of ice.'' (SWT)

Glaciofluvial deposits Material that has been moved by glaciers, and sorted and deposited by streams flowing from the melt water. (SWT)

Glanders A disease of horses. The disease can be transmitted to humans, however. This zoonotic infection, caused by the organism *Pseudomonas mallei* is characterized by fever, swelling of the glands beneath the lower jaw, and inflammation of the nasal mucous membranes, and is known in the chronic state as farcy. (FSL)

Glare The sensation produced by luminances within the visual field that are sufficiently greater than the luminances to which the eyes are adapted. This causes annoyance, discomfort, or loss in visual performance and visual ability, a concern especially for individuals who use video display terminals for extended periods of time. (DEJ)

Glass blower's cataract An opacity of the rear surface of the lens in the eye, caused by excessive exposure of the eyes to luminous radiation, primarily visible and infrared. A condition found in those occupationally exposed to furnaces or other hot devices over an extended period of time. Also known as heat cataract and blower's cataract. (DEJ)

Glazing The application of a smooth, shiny surface to pottery. Glazes typically contain lead and are a concern when applied to food containers because of the possibility of lead leaching into the food. *See also* Lead poisoning. (FSL)

Gley soil A soil developed in poor drainage conditions that result in the reduction of iron and other elements. Gley soils usually are gray and mottled. (SWT)

Gleyzation A soil-forming process that results in the development of a gley soil. (SWT)

Globulin Any of a group of proteins that are found in animal and vegetable tissues and that can be precipitated from serum (or plasma). Globulins are synthesized by the body during infections. Globulins also occur naturally on wheat and other cereal crops. (FSL)

Glutaraldehyde ($C_5H_8O_2$) Compound is used as a disinfectant for the control of bacteria, yeast, and fungi, particularly for disinfecting equipment or instruments that cannot be heat sterilized. It is also used as a fixative for electron microscopy. Known also as glutaral or pentanedial. *See also* Bacteriocide; disinfectant. (HMB)

Glycophytes Plants that do not grow well in soil solutions with high osmotic pressures. Nonhalophytic plants. (SWT)

Gneiss A metamorphic rock with alternate layers of dissimilar materials. (FSL)

Gnotobiotes Germ-free or formerly germ-free organisms in which the composition of associated microorganisms, if any are present, is fully defined. (FSL)

Goggle A tight-fitting device worn over the eyes to provide splash and/or impact protection. Shielded ventilation ports are typically provided on the sides to minimize fogging. Goggles are often used in combination with face shields in the laboratory. (DEJ)

Gold (Au) A metallic element that is bright yellow, soft, and acid-resistant and has an atomic weight of 196.967 and an atomic number of 79. Gold is found in hydrothermal veins and in placer deposits and is used in brazing alloys, electrical contact alloys, laboratory ware, dental alloys, jewelry, and specialized medical treatments—for example, gold leaf is used in surgery. (FSL)

Gonad A gamete-producing organ in animals; a testis or ovary. (FSL)

GOCO Government-owned/contractor-operated federal facility, owned by a government agency but with all or portions of the facility operated by private contractors, often defense contractors. (DEW)

GOGO Government-owned/government-operated federal facility. (DEW)

Goiania A community in Brazil where an abandoned teletherapy source was found in 1987 by junk dealers at a former medical clinic. The source was subsequently removed from its shield, taken home, and opened. As a result, many persons were exposed externally and internally to the soluble $^{137}CsCl$ in it. Several persons died and a large area of the city was contaminated. This accident is one of the largest involving radioactive material, probably exceeded only by the nuclear-reactor episode at Chernobyl in 1986. *See also* Chernobyl. (FSL)

Good Manufacturing Practices The Food and Drug Administration promulgated the first Good Manufacturing Practices (GMP) regulations, known as the umbrella GMPs, in 1969. These regulations deal with sanitation in manufacturing, packing, processing or holding facilities. The regulations have since been amended.

The practices established general requirements for sanitation in a food establishment. A sanitation section addresses areas including cleaning and sanitizing of equipment and utensils, storage and handling of clean equipment and utensils, maintenance of physical facilities, waste management, pest control, and proper storage and use of sanitizers, cleaning compounds, and pesticides. Specific GMPs supplement the umbrella standards. Each of them covers a specific industry or closely related group of foods. Critical aspects of the operations are addressed, including cleaning and sanitizing procedures, critical steps in the processing of the food, time-temperature relationships, storage conditions, and employee training. (HMB)

Goose pimples A skin condition marked by numerous small elevations around hair follicles, caused by the action of the arrectores pilorium (''raisers of hair'') muscles. These muscles raise the hair in order to trap a layer of air next to the skin. The air layer acts as an insulator and limits heat loss from the skin. (DEJ)

GOPO Government-owned/privately-operated facility that the government has leased, all or in part, to private operators for their profit. (DEW)

Grab sample **(1)** (Industrial hygiene) A type of air sample in which the air is admitted into a bag, vessel, or instrument instantaneously for subsequent analysis. The concentration of the contaminant is not integrated (averaged) over time, and the contaminant is usually a nonreactive gas that does not decompose over time. The grab sample procedure is to be distinguished from a time-weighted average (TWA) sample **(2)** (Environmental) A sample of environmental media (air, soil, or water) collected in a randomized fashion for analysis of contaminants. *See also* Time-weighted average. (DEJ)

Grain loading The rate at which particles are emitted from a pollution source. Measurement is made of the number of grains per cubic foot of gas emitted. (FSL)

Gram The basic unit of mass in the metric system; 1 gram is equal to 15.432 grains. (FSL)

Gram-atomic weight A mass in grams numerically equal to the atomic weight of an element. (FSL)

Gram-molecular weight (gram-mole) Mass in grams numerically equal to the molecular weight of a substance. (FSL)

Gram-negative bacterium Bacterium that reacts in a certain way to the conventional Gram staining procedure. A stained Gram-negative bacterium appears red when viewed microscopically. In Gram-negative bacteria, the alcohol solvent readily dissolves in and penetrates the outer layer of the cell, decolorizing it in the process. Safranin, a red counterstain, is then administered and absorbed by the cell, rendering the decolorized cell red. *See also* Gram-positive bacterium; Gram reaction; Gram stain. (HMB)

Gram-positive bacterium Bacterium that reacts in a certain way to the conventional Gram staining procedure. A stained Gram-positive bacterium appears blue when viewed microscopically. Gram-positive bacteria have thick cell walls that become dehydrated by the alcohol decolorization step in the staining procedure. This causes the pores in the walls to close, preventing the release of an insoluble crystal-violet–iodine complex formed inside the cell during the first two steps of the staining procedure. *See also* Gram-negative bacterium; Gram reaction; Gram stain. (HMB)

Gram reaction The reaction a bacterium has when subjected to the differential Gram stain procedure. Bacteria may be classified as either Gram-negative or Gram-positive. *See* Gram-negative bacterium; Gram-positive bacterium; Gram stain. (HMB)

Gram stain Stain used in one of the most important staining procedures in microbiology. On the basis of their reaction to the Gram stain, bacteria can be divided into two groups, Gram-positive and Gram-negative. This staining procedure is of considerable importance in bacterial taxonomy, and it also indicates fundamental differences in the cell wall structure of bacteria. The Gram stain procedure involves four steps: **(1)** staining with crystal violet, **(2)** treatment with an iodine solution, **(3)** decolorization with alcohol, and **(4)** counterstaining with safranin. *See also* Gram-negative bacterium; Gram-positive bacterium; Gram reaction. (HMB)

Grant A financial assistance mechanism whereby funds and/or direct assistance is provided to carry out approved activities. Generally used by the awarding agency when no substantial programmatic involvement with the recipient during performance of the activities is anticipated. (DEW)

Granule A natural soil aggregate that is relatively nonporous. (SWT)

Graphite Also known as black lead, this is a crystalline form of carbon found in nature. This material, which can also be manufactured from coke, is a component of "lead" writing pencils. It is used as a lubricant in paints and coatings, electrodes, bricks, motors, and generator brushes. In the natural powder form, graphite can be a fire risk and is considered a respirable dust with a TLV of 2.1 mg/m^3. (FSL)

Graunt, John (1620–1674) Considered to be the founder of vital-statistics systems. He described the use of numerical methods in his book *Natural and Political Observations on the Bills of Mortality*, which was published in 1662. (FSL)

Gravi's stain Stain used in a differentiating staining technique to segregate bacteria that retain the stain (Gram-positive) from those that do not (Gram-negative). Gram-positive organisms, such as *Staphylococci*, and Gram-negative ones, such as *Gonococcus*, have marked biochemical differences in their composition. *See also* Gram-negative bacterium; Gram-positive bacterium; Gram reaction; Gram stain. (FSL)

Gravitational water Water whose movement through soil is influenced by gravity. (SWT)

Gravity spring Type of spring that occurs where water percolating laterally through permeable material overlying nonpermeable strata comes to the surface. Such sources of water are typically low-discharge sources; however, when properly developed, they serve as satisfactory individual water supply systems. *See also* Spring. (FSL)

Gray water Domestic wastewater composed of washwater from sinks, and bathroom and laundry tubs. (FSL)

Grease trap A device designed and installed in a manner so as to separate and retain grease and other solids while permitting sewage or liquid wastes to discharge into a drainage system by gravity. Grease traps are frequently used in conjunction with sewer lines serving restaurants and other food-service establishments. (FSL)

Greenberg-Smith impinger An air-sampling device that has the capability of bubbling air, without special effort, through a liquid contained in the device to obtain intimate mixing of the air and the liquid. This instrument was used for analysis of airborne particulate concentrations many years ago and has been replaced by more-efficient devices. *See also* Impinger. (DEJ)

Greenhouse effect A warming of the earth's atmosphere, caused by a buildup of carbon dioxide or other trace gases; it is believed by many scientists that this buildup allows light from the sun's rays to heat the earth but prevents a counterbalancing loss of heat. (FSL)

Green manure Plant materials that are incorporated into the soil while still green for soil improvement. A crop grown for the purpose of plowing under while it is still green (e.g., rye grass) or shortly after maturation, in order to improve the soil. (SWT, SMC)

Grenz rays X-rays produced at voltages of 5 to 20 kilovoltage peak (KVP). (RMB)

Grinder pump A mechanical device that shreds solids and raises fluids to a higher elevation through pressure sewers. (FSL)

Grit Coarse nuisance dust particles that are larger than 75 microns in diameter. These particles, which may arise from industrial sources, are not normally a threat to human health via inhalation, because particles larger than 5 microns do not penetrate the alveoli in the lungs. (FSL)

Grit chamber A water pollution device or preseparator designed to remove grit, which is defined as being particles 0.2 mm in diameter with a specific gravity of 2.7 or greater. A variety of systems are designed to accomplish this separation. Many incorporate diffused aeration, which keeps organic matter in suspension while causing the grit to settle out. Reduced flow velocity is necessary (horizontal velocity at about 1.0 ft./sec.). Settled grit is removed from the chamber by conveyor, pump, or elevator systems. Grit may be defined as heavy particulate material, such as sand, glass, metal shavings, and coffee grounds, in wastewater, that will settle from the wastewater when flow velocity is reduced. Grit can be abrasive to pipes and pumps and may clog pipes by deposition, or settle in tanks and digesters. Also, this material is not biodegradable and therefore must be removed from the waste stream. (HMB)

Groin A small jetty extending from a shoreline to prevent beach erosion. *See also* Revetment; riprap. (FSL)

Gross alpha particle activity Total activity due to emission of alpha particles. Used as the screening measurement for radioactivity generally due to naturally occurring radionuclides. Activity is commonly measured in picocuries. (RMB)

Gross beta particle activity Total activity due to emission of beta particles. Used as the screening measurement for radioactivity from human-made radionuclides because the decay products of fission of such nuclides are typically beta particles and gamma-ray emitters. Activity is commonly measured in picocuries. (RMB)

Ground As applied to electricity, a conducting body, such as the earth or an object connected to the earth, whose potential is zero and to which can be connected an electric circuit or a device, such as an iron pipe or metal stake, that makes such a connection. Both the equipment-grounding conductor and the grounding electrode conductor should be connected to the grounded circuit conductor on the supply side of the service-disconnecting means, or on the supply side of the system-disconnecting means or overcurrent devices if the system is separately derived. A conductor used for grounding fixed or movable equipment should have the capacity to conduct safely any fault current that may be imposed on it. *See also* Bonding. (FSL, SAN)

Ground cover Plants, such as ivy, grown to keep soil from eroding. (FSL)

Ground fault circuit interrupter (GFCI) A safety device that senses when electrical current passes to ground through any path other than the proper conductor. These units trip, thus stopping current flow in the circuit and through any person receiving a ground fault shock. The devices are required by the electrical codes in many municipalities. According to OSHA, all 120-volt, single-phase, 15- and 20-ampere receptacle outlets that are on construction sites, that are not a part of the permanent wiring of the structure, and that are in use by employees must have approved GFCIs for personnel protection. (FSL, SAN)

Ground water Any water occurring naturally in underground formations. Under these conditions, water pressure is greater than atmospheric pressure. The term also refers to the supply of fresh water that is found beneath the earth's surface (usually in aquifers) and that is often used for supplying wells and springs. Because ground water is a major source of drinking water, there is growing concern regarding areas in which leaching agricultural or industrial pollutants or substances from leaking underground storage tanks are contaminating ground water. *See also* Surface water. (FSL, SWT)

Growth rings Woody tissue formed each year in woody vascular plants. The newest wood is in the outermost (most recent) rings of a branch or stem and the oldest wood is in the innermost rings. In temperate regions the rings can be used to tell the age of a tree, in studies of dendrochronology, because the rings of each year are distinct. During the growing season cell growth in the vascular region is rapid and the resulting wood is light-colored. In the winter the cells continue to divide but expand little, and the resulting wood is dark. This growth pattern creates rings of dark wood, each representing one year's growth. *See also* Dendrochronology. (SMC)

Guard A device or shield placed over or in association with a piece of machinery, such as a drill, grinder, or saw, to protect the operator from physical harm. (FSL)

Guar Gum The ground portion of the Indian plant *Cyanogensis tetragonolaba* that is added to commercial powder explosives to protect them from the influx of water. (FSL)

Gully A channel caused by erosion and concentrated but intermittent flow of water. (SWT)

Guncotton A product, containing nitrocellulose, that is used as an explosive in military munitions. (FSL)

Gust Wind or turbulence that is close to the ground and that has been caused by buildings or other obstacles that prevent the direct flow of air. (FSL)

Guttation The exudation, or seepage, of fluids from plants, due to root pressure. (SMC)

Guyon tunnel syndrome A condition that results from pressure on the ulnar nerve as a result of compression of the Guyon tunnel in the wrist. This condition is associated with tasks requiring prolonged flexing and extending of the wrist along with repeated pressure on the base of the palm (just above the fingers). Workers performing bricklaying, soldering, and hammering have been shown to be prone to this disorder. *See also* Cumulative trauma disorder. (SAN)

Gypsum Hydrated calcium sulfate ($CaSO_4$), which is used in the production of wallboard, plaster, and sulfuric acid. There are several forms of gypsum, including alabaster and desert rose. (FSL)

Gypsum requirement In soil, the amount of gypsum required to reduce the amount of exchangeable sodium to an acceptable level. *See also* Salic horizon. (SWT)

Gyttja Sedimentary peat consisting mainly of plant and animal matter that has precipitated from standing water. (SWT)

H

Haber, Fritz German inventor (1868–1934) who perfected the process for synthesizing ammonia from atmospheric nitrogen and hydrogen. In this process, these gases are passed through a catalyst bed containing osmium. (FSL)

Habitat The place where a population (e.g., human, animal, plant, microorganism) lives and its surroundings, both living and nonliving. (FSL)

Hair restraint Any device placed on the head to control contact with hair or to control loss of hair to a worker's immediate environment. Generally, this is required apparel for individuals working in a foodhandling or food-service environment. *See also* Snood. (HMB)

Half-life (Biological) The time required for the body to eliminate, by natural biological means, half of the material taken into it. (RMB)

Half-life, effective The time in which the quantity of a radioactive isotope in the body will decrease to half as a result of both radioactive decay and biological elimination. It is related to radioactive half-life by the equation:

$$T = \frac{T_b \times T_r}{T_b + T_r}$$

where T is effective half-life, T_b is the time required for biological elimination, and T_r is radioactive half-life. (RMB)

Half-life (radiation) The time required for the activity of a given radioactive isotope to decrease to half of its initial value, due to radioactive decay. The half-life is a characteristic property of each radioactive isotope and is independent of its amount or condition. (RMB)

Half-thickness The thickness of any given absorber that will reduce the intensity of a beam of radiation to one-half its initial value. It may be expressed in units of thickness or of mass per unit area. *See also* Half-life (radiation). (RMB)

Half-value layer (half-thickness) The thickness of any particular material necessary to reduce the intensity of an X-ray or gamma-ray beam to one-half its original value. (RMB)

Halide meter An instrument used for the direct measurement of halogenated hydrocarbons. This instrument, however, lacks the capability to distinguish between different halogenated hydrocarbon compounds. (DEJ)

Halite An evaporite mineral, also known as rock salt, that flows under the pressure of overburden, leading to a process known as halinokinesis and thus playing a role in the internal movement of salt domes. (FSL)

Halodine The boundary between two masses of water with different levels of salinity. (FSL)

Halogen Collective term for any of the five very active nonmetallic chemical elements fluorine, chlorine, bromine, iodine, and astatine. *See also* Disinfectant; freon. (FSL)

Halogenation The process whereby chlorine, fluorine, bromine, or iodine is used for disinfection purposes. *See also* Chlorine disinfection; water purification. (FSL)

Halomorphic soil Saline and alkaline soil that is formed under imperfect drainage in arid regions. (SWT)

Halons Bromine-containing compounds with long atmospheric lifetimes whose breakdown in the stratosphere causes depletion of ozone. Halons are used in firefighting. (FSL)

Halophytic vegetation Plants that tolerate or require a saline environment. (SWT)

Hamilton, Alice (1869–1970) Physician, author, and investigator who conducted numerous studies associated with occupational hazards, including many involving lead poisoning. Dr. Hamilton, who was the first woman faculty member at Harvard University (1919), wrote an autobiography entitled *Exploring the Dangerous Trades*. (FSL)

Hammermill A high-speed machine that has hammers and cutters to crush, grind, chip, or shred solid wastes. (FSL)

Hand-washing facility Employees of food-service facilities are required by regulation to wash their hands and the exposed portions of their arms with soap and warm water before starting work, during work as often as necessary to keep them clean, and after eating, smoking, drinking, or using the toilet. A hand-washing facility separate from dish or utensil washing sinks is required for new food-service establishments by most food-service regulations, and should be added to those undergoing renovation. Current OSHA regulations require that hand-washing facilities be accessible to workers in most industrial settings. (HMB)

Hangfire The detonation of an explosive charge at some point after its normally designated firing time. (FSL)

Hapten An antigen that is incapable by itself of causing the production of antibodies but is capable of combining with specific antibodies. (FSL)

Hard hat Helmets or devices worn on the head to protect workers from impact and penetration by falling and flying objects and from limited electric shock and burn. According to the ANSI standard Z89.1-1081, there are three classes of helmets. **(1)** *Class A:* Intended to reduce the force of impact of falling objects and to reduce the danger of contact with exposed low-voltage conductors. Representative sample shells are proof-tested at 2200 volts (phase to ground; note that is not intended to be an indication of the voltage at which the headgear protects the wearer). **(2)** *Class B:* Intended to reduce the force of impact of falling objects and to reduce the danger of contact with exposed high-voltage conductors. Representative sample shells are proof-tested at 20,000 volts (phase to ground; not intended to be an indication of the voltage at which the headgear protects the wearer.) **(3)** *Class C:* Intended to reduce the force of impact of falling objects. This class offers no electrical protection. (DEJ)

Hardness A relative specification of the quality or penetrating power of X-rays. In general, the shorter the wavelength, the harder the radiation. (RMB)

Hardpan A soil layer in the A or B horizon that has been hardened by the cementation of soil particles with organic matter or materials such as silica, sesquioxides, or calcium carbonate. The hardness does not change with moisture and pieces do not slake in water. (SWT)

Hard water A term ordinarily applied to water that has a high soap-consuming potential or water in which a lather cannot be produced without the use of excessive amounts of soap. Hardness is caused by metallic ions, such as calcium and magnesium ions. The presence of iron, manganese, copper, barium, and lead ions can also cause slight increases in hardness. The principal adverse effects of water hardness are: **(1)** the consumption or neutralization of soap; **(2)** the clogging of skin, discoloration of porcelain fixtures, and staining of fabrics; **(3)** the increasing of difficulty in textile and paper manufacture, canning, and other industrial processes; and **(4)** formation of scale in boilers, resulting in high heat transfer losses and danger of boiler failure. *See also* Water softening; scale. (FSL)

Harnesses Devices worn to secure workers in high places and prevent falls. Harnesses are also used to lower workers and for rescue purposes. Also known as body belts, chest harnesses, and suspension belts. (DEJ)

Haverhill fever An acute febrile disease first described in Haverhill, Massachusetts, in 1926 and caused by the organism *Streptobacillus moniliformis*. The disease is transmitted by the bite of an infected rat. (FSL)

Hazard A risky, perilous, or dangerous condition that could result in the exposure of individuals to unnecessary health risks. Defined relative to its use in a hazard analysis critical control point (HACCP) application and in the food-service industry, it is any biological, chemical, or physical property that may cause an unacceptable consumer health risk. Hazards can include unacceptable contamination, unacceptable growth or survival of microorganisms of concern in regard to safety or spoilage, and/or unacceptable production or persistence in foods of products of microbial metabolism—e.g., toxins and enzymes. Chemical hazards are those such as result from the presence in food of cleaning/sanitizing compounds, pesticides, lubricants, adhesives, or other materials used in a processing environment. Physical aspects include conditions in the surrounding environment, such as heat, cold, and moisture, and also the nature of packaging materials. *See also* Hazard analysis critical control point (HACCP). (HMB)

Hazard analysis An evaluation of all procedures concerned with the production, distribution, and use of raw materials and food products, conducted in order to: **(1)** identify potentially hazardous raw materials and foods that may contain poisonous substances, pathogens, or large numbers of spoilage organisms and conditions that support microbial growth; **(2)** identify specific points and potential sources of contamination through observation of each step in the food chain; **(3)** determine the potential for microorganisms to survive or multiply during production, distribution, processing, storage, and preparation for consumption; and **(4)** assess the risks and the severity of hazards identified. *See also* Hazard analysis critical control point (HACCP). (HMB)

Hazard analysis critical control point (HACCP) A systematic approach that can be used in food production, processing, or food service as a means of assuring food safety. The same principles can be applied to assure food safety in the home. The HACCP approach consists of several elements, including: **(1)** assessment of the hazards and risks associated with growing and harvesting a food, with its raw materials and ingredients, and with processing, manufacturing, marketing, preparation, and consumption of the product; **(2)** determination of critical control points (operations or conditions for which control measures can be asserted) required to control the identified hazards; **(3)** establishment of specific critical limits or criteria that must be met for each identified critical control point so that it can be determined whether a particular operation is under control; **(4)** establishment of monitoring procedures to make such determinations; **(5)** the taking of appropriate corrective measures when the monitoring system indicates that a critical control point is not in compliance with established limits; **(6)** establishment of a record-keeping system to document HACCP procedures; and **(7)** establishment of procedures for verifying that the HACCP system is working.

The HACCP concept is applied to the causes of food spoilage and foodborne illness. It is a continuous process because problem detection and resolution are ongoing. It is systematic and complete because it covers all operations, procedures, ingredients, and product usage. By directing efforts toward the control of key factors that affect food safety, the HACCP approach can assure that higher levels of food safety and quality are maintained. *See also* Critical control point; hazard; hazard analysis; sensitive ingredient. (HMB)

Hazard and operability study (HAZOP) HAZOP is a formal structure method of investigating each element of a system for ways in which important parameters can deviate from the intended design conditions and create hazards and operability problems. This method can identify complex failure scenarios that involve multiple independent events. Teams of individuals from a wide variety of backgrounds are involved in the process. The hazard and operability problems are typically determined by a study of the piping and instrumental diagrams (or plant model) by a team of personnel who critically analyze effects of potential problems arising in each pipeline and each vessel of the operation. An assessment is made, weighing the consequences, causes, and protection requirements needed, and recommendations for improvements are made. *See also* Checklist; event-tree analysis; fault-tree analysis; failure modes and effects analysis; hazard analysis; process safety; what-if. (SAN)

Hazard Classification System for Particulates *See* Engineering controls. (DEJ)

Hazard Communication Standard Standard (29 CFR 1910.1200) promulgated on May 23, 1985, by the Occupational Safety and Health Administration (OSHA) and applicable to all employees who might be exposed to chemicals. Although originally intended to apply to the manufacturing sector only, the scope of the standard was expanded in March 1989, to include the nonmanufacturing environment as well. This standard, often referred to as the "employee-right-to-know" regulation, is now applicable to all workplace settings involving chemicals. In summary, the standard requires the preparation of a chemical inventory and a written hazard communication plan, training, labeling of containers containing chemicals, and accessibility to material safety data sheets (MSDS) for those chemicals in the workplace. *See also* Chemical hygiene plan; Material safety data sheets. (FSL)

Hazardous air pollutants Specified pollutants listed pursuant to the Clean Air Act for which the Environmental Protection Agency must set emissions standards. 42 USC §7412. *See also* Clean Air Act; NESHAPS; pollutants. (DEW)

Hazardous chemical According to OSHA (29 CFR, 1910.1200), any chemical that is a health hazard or physical hazard and for which there is statistically significant evidence, based on at least one study conducted in accordance with established scientific principles, that acute or chronic health effects may occur in exposed individuals. Includes chemicals that are carcinogens, toxic or highly toxic agents, reproductive toxins, irritants, corrosives, sensitizers, hepatotoxins, nephrotoxins, neurotoxins, agents that act on the hematopoietic systems, and chemicals that damage the lungs, skin, eyes, or mucous membranes. *See also* Hazardous waste. (FSL)

Hazardous (classified) locations Refers to the National Fire Protection Association classification for areas that is based on the hazards of the chemicals present and the type of area. Electrical and other potential spark-producing equipment must meet certain criteria to be used in hazardous locations. The locations are classified as follows:

Class I: Locations in which flammable gases or vapors are or may be present in the air in quantities sufficient to produce an explosive or ignitable mixture. Chemical atmospheres for Class I locations include: A, acetylene; B, hydrogen; C, ethyl ether; D, gasoline, naphtha, alcohols, acetone, lacquer, and benzene.

Division 1: Locations in which ignitable concentrations of flammable gases or vapors may exist under normal operating conditions; or in which ignitable concentrations of such gases or vapors may exist frequently because of repair or maintenance operations or because of leakage; or in which the breakdown or faulty operation of equipment or processes might release ignitable concentrations of flammable gases or vapors, and might also cause simultaneous failure of electric equipment.

Division 2: Locations in which volatile flammable liquids or flammable gases are handled, processed, or used, but in which they are normally confined within closed containers or closed systems from which they can escape only in the case of an accidental rupture or breakdown of such containers or systems or through abnormal operation of equipment; or a location in which ignitable concentrations of gases or vapors are normally prevented from building up by positive mechanical ventilation, and that might become hazardous through failure or abnormal operations of the ventilating equipment; or a location adjacent to a Class I, Division 1 location, and to which ignitable concentrations of gases or vapors might occasionally be communicated unless this is prevented by adequate positive-pressure ventilation from a source of clean air and effective safeguards against ventilation failure are provided.

Class II: Locations that are hazardous because of the presence of combustible dust. Groupings for contaminant atmospheres in Class II are: E, metal dust; F, carbon black, coal dust, and coke dust; G, grain dust, flour dust, starch dust, organic dust.

Division 1: Locations that have combustible dusts that are or may be in suspension in the air under normal operating conditions, in quantities sufficient to produce explosive or ignitable mixtures; or where mechanical failure or abnormal operation of machinery or equipment might cause such explosive or ignitable mixtures to be produced, and might also provide a source of ignition through simultaneous failure of electric equipment, operation of protection devices, or from other causes; or in which combustible dusts of an electrically conductive nature may be present.

Division 2: Locations that have combustible dusts that are not normally in suspension in the air in quantities sufficient to produce explosive or ignitable mixtures, and where dust accumulations are normally insufficient to interfere with the normal operation of electrical equipment or other apparatus; or that have dust that may be in suspension in the air as a result of infrequent malfunctioning or handling of processing equipment; the dust accumulations resulting, therefore, may be ignitable by the abnormal operation or failure of electrical equipment or other apparatus.

Class III: Locations that are hazardous because of the presence of easily ignitable fibers or flyings but in which such fibers or flyings are not likely to be in suspension in the air in quantities sufficient to produce ignitable mixtures.

Division 1: Locations in which easily ignitable fibers or materials producing combustible flyings are handled, manufactured, or used.

Division 2: Locations in which easily ignitable fibers are stored or handled, except in process of manufacture. Examples are baled waste, cocoa fiber, cotton, excelsior, hemp, thistle, jute, kapok, oakum, sisal, Spanish moss, and synthetic fibers. (SAN)

Hazardous material As defined by the Transportation Safety Act of 1974 (PL93-633, Section 103-Title I), a substance or material in a quantity and form that may pose an unreasonable risk to health and safety or property when transported in commerce. The following categories of hazardous materials are defined: Class 1, explosives; Class 2, gases; Class 3, flammable liquids; Class 4, flammable solids; Class 5.1, oxidizers; Class 5.2, organic peroxides; Class 6, poisons; Class 7, radioactive materials; Class 8, corrosives; Class 9, miscellaneous.

Further, an additional category ("other regulated materials," ORM-D) is included in the regulations. It includes consumer commodities that, although otherwise subject to the regulations, present a limited hazard during transportation due to their form, quantity, and packaging. *See also* Hazardous waste. (FSL)

Hazardous materials response team *See* HAZMAT. (FSL)

Hazardous materials specialists As defined by OSHA HAZWOPER standard, individuals who respond with and provide support to hazardous materials technicians. Their duties parallel those of the hazardous materials technician; however, those duties require a more directed or specific knowledge of the various substances they may be called upon to contain. The hazardous materials specialist would also act as the site liaison official with Federal, state, local, and other government authorities in regard to site activities. *See also* Hazardous materials technicians; Hazardous Waste Operations and Emergency Response (HAZWOPER). (SAN)

Hazardous materials technicians As defined by the OSHA HAZWOPER standard, individuals who respond to releases or potential releases for the purpose of containing the release. They assume a more aggressive role than a first responder at the operations level, in that they will approach the point of release in order to plug, patch, or otherwise stop the release of a hazardous substance. *See also* First responder; hazardous materials specialists; Hazardous Waste Operations and Emergency Response (HAZWOPER). (SAN)

Hazardous Materials Transportation Act An Act of Congress authorizing the Secretary of Transportation to develop, issue, and enforce any safety aspect of the transportation of materials found to be hazardous. The act applies to rail, air, water, and highway transportation. The law regulates the packaging, labeling, and shipping of hazardous materials. 49 USC §1801 *et seq.*, 49 CFR Parts 106, 107, 171–179. (DEW)

Hazardous ranking system The method used by the EPA to evaluate the relative potential of hazardous-substance release to cause health or safety problems or ecological or environmental damage, and particularly to evaluate sites for inclusion on the National Priorities List. 42 USC §9605(c), 40 CFR §300.5. *See also* Comprehensive Environmental Response; Compensation and Liability Act; national priorities list. (DEW)

Hazardous substance As defined by CERCLA, a substance listed or defined by specified sections of federal law or regulation, including the Clean Air Act, Clean Water Act, Solid Waste Disposal Act, Resource Conservation Recovery Act, and Toxic Substances Control Act. Excludes petroleum and petroleum fractions not listed in the specified sections of law, and natural gas. 42 USC §9601(14). For purposes of occupational safety and health standards for hazardous-waste operations and emergency response, includes: CERCLA-defined hazardous substances, in addition to any biological agent and other disease-causing agents, that, after release into the environment and upon exposure, ingestion, inhalation, or assimilation into any person, either directly from the environment or indirectly by ingestion through food chains, will or may reasonably be anticipated to cause death, disease, behavioral abnormalities, cancer, genetic mutation, physiological malfunctions (including malfunctions in reproduction), or physical deformations in such persons or their offspring; any substance listed by the U.S. Department of Transportation as a hazardous material under the Hazardous Materials Transportation Act; and any specified hazardous waste under RCRA. 29 CFR §1910.120. *See also* Clean Air Act; Comprehensive Environmental Response, Compensation, and Liability Act; hazard communication standard; Hazardous Materials Transportation Act; Solid Waste Disposal Act. (DEW)

Hazardous Substances Act Federal law authorizing the Consumer Product Safety Commission to regulate a hazardous substance in interstate commerce, by imposing packaging and labeling requirements to protect public health and safety, or, in the case of hazardous substances intended for use by children, accessible to children, or intended for household use, to ban or prohibit distribution. 15 USC 1261–1274, 16 CFR, Subchapter C, Part 1500, *et seq.* (DEJ)

Hazardous waste As defined by the Resource Conservation and Recovery Act, and enforced by EPA, a solid waste (or combination of solid wastes) that because of its quantity, concentration, or physical, chemical, or infectious characteristics may cause, or significantly contribute to, an increase in mortality or an increase in serious irreversible or incapacitating reversible illness; or pose a substantial present or potential hazard to human health or the environment when improperly treated, stored, transported, disposed of, or otherwise managed. Regulations further specify characteristics of hazardous wastes (ignitability, corrosivity, reactivity, and toxicity) and list covered waste streams and commercial chemicals. The definition excludes household wastes and other specified wastes. Includes any mixture of a listed hazardous waste with a nonhazardous waste, and mixtures of characteristic hazardous waste with nonhazardous waste unless the mixture no longer exhibits the hazardous characteristic. Also includes any solid waste derived from the treatment, storage, or disposal of a hazardous waste. 42 USC §6902(5), 40 CFR Part 261.

The characteristic of ignitability is met if a representative sample of the waste has any of the following characteristics: liquid with a flash point of less than 60°C (140°F) (e.g., acetone, toluene, xylene); nonliquid with the capability of spontaneous and sustained combustion; ignitable compressed gas (e.g., hydrogen, acetylene).

The characteristic of corrosivity is met if a representative sample is: aqueous and exhibits a pH of less than 2 or greater than 12.5 (e.g., sodium hydroxide, nitric acid), or is a liquid and will corrode steel at a rate of $\frac{1}{4}$ inch per year at a temperature of 55°C (130°F).

A chemical has the characteristic of reactivity if it has any of the following properties: it is normally unstable and undergoes violent change without detonating; it has a violent reaction with water (e.g., sodium metal); it forms mixtures with water that are potentially explosive; upon mixture with water it generates toxic gases, vapors, or fumes; it contains cyanide or sulfide and will generate toxic gases, vapors, or fumes at a pH of between 2 and 12.5; it is capable of detonation if heated under confinement or subjected to a strong initiating source (e.g., picric acid with $<10\%$ water).

A chemical waste exhibits the characteristic of toxicity if, using the toxicity characteristic leaching procedure (TCLP) or equivalent test methods, the extract from a representative sample of the waste contains concentrations greater than a set milligram-per-liter concentration of certain metals such as arsenic or mercury, a pesticide such as lindane or 2,4-D, or a solvent such as benzene, methyl ethyl ketone, or chloroform.

In addition to the establishment of a chemical as a hazardous waste based on ignitability, corrosivity, reactivity, and toxicity, a chemical may be listed as a hazardous waste if: a small dose will cause human mortality; it has an oral LD_{50} of less than 50 mg/kg; it has an inhalation LC_{50} of less than 2 mg/L; it has a dermal LD_{50} of less than 200 mg/kg. *See also* Hazardous material; Resource Conservation and Recovery Act; solid waste; toxicity characteristic leaching procedure. (DEW, FSL)

Hazardous Waste Operations and Emergency Response (HAZWOPER) OSHA standard (29 CFR 1910.120) that regulates the safety and health provisions of employees involved in cleanup operations at uncontrolled hazardous waste sites being abated under government mandate, in certain hazardous waste treatment, storage, and disposal operations conducted under the Resource Conservation and Recovery Act of 1976, as amended, and in any emergency response to incidents involving hazardous substances. The standard provides for employee protection during initial site characterization and analysis, monitoring activities, materials handling activities, and training and emergency response.

The standard contains the requirement that, during initial site entry, personal protective equipment be selected and used that will provide protection to a level of exposure below permissible exposure limits and published exposure levels for known or suggested hazardous substances and health hazards, and that will provide protection against other known and suspected hazards identified during the preliminary site evaluation. Four levels of protection (A, B, C, D) have been established and guidelines for selected personal protective equipment have been developed. (See table.)

<table>
<tr><th>EPA* Level</th><th>Respiratory Protection</th><th>Protective Clothing</th><th>Hand and Foot Protection</th><th>Additional Protection</th></tr>
<tr><td>A</td><td rowspan="2">NIOSH-approved positive-pressure full-facepiece self-contained breathing apparatus (SCBA)
OR
NIOSH-approved positive-pressure, supplied-air respirator with escape SCBA (minimum 5-minute duration)</td><td>Totally encapsulating chemical protective suit specifically designed to resist permeation by chemicals that are encountered</td><td rowspan="3">Gloves: outer and inner chemical-resistant gloves

Boots: chemical-resistant, with steel toe and shank.</td><td>Coveralls
Long underwear
Hard hat
Two-way radio communications system</td></tr>
<tr><td>B</td><td rowspan="2">Hooded chemical-resistant clothing made of materials resistant to the chemicals encountered (overalls and long-sleeved jacket; coveralls: one- or two-piece chemical splash suit: disposable chemical-resistant overalls)</td><td>above plus:
Faceshield
Boot covers (disposable, chemical-resistant)</td></tr>
<tr><td>C</td><td>NIOSH-approved full-facepiece or half-mask air-purifying respirators.</td><td>above plus:
Escape mask</td></tr>
<tr><td>D</td><td></td><td>Coveralls</td><td>Boots: chemical-resistant, with steel toe and shank</td><td>above plus:
Safety glasses or splash goggles
Gloves</td></tr>
</table>

*Appendix B. OSHA 29 CFR. Part 1910.120: EPA Guidelines may differ.

Further, the standard contains specific requirements for the extent of training that must be provided for workers on waste disposal sites. For example, general site workers (equipment operators, laborers, and supervisory personnel) engaged in removal or other activities that expose or potentially expose them to hazardous substances and health hazards shall receive a minimum of 40 hours of instruction offsite and a minimum of 3 days' actual field experience under the direct supervision of a trained, experienced supervisor. Training requirements for other types of workers, depending upon the types of tasks to be performed, are specified in the standard. *See also* First responder; hazardous materials specialists; hazardous materials technicians. (FSL)

Hazards analysis The procedures involved in: **(1)** identifying potential sources of release of hazardous materials from fixed facilities or transportation accidents; **(2)** determining the vulnerability of a geographical area to a release of hazardous materials; and **(3)** comparing hazards to determine which presents the greater or lesser risks to a community. (FSL)

Hazards identification The providing of information on which facilities have extremely hazardous substances, what the chemical substances are, and the quantity of them at each facility. The process also provides information on how the chemicals are stored and whether they can sustain high temperatures. (FSL)

HAZMAT Hazardous materials response team. An organized group of employees, designated as such by the employer, who are expected to perform work to handle and control actual or potential leaks or spills of hazardous substances, and whose duties may require possible exposure to the substances. Team members perform responses to releases or potential releases of hazardous substances for the purposes of control or stabilization of the incidents. 29 CFR §1910.120. *See also* Hazardous substance. (FSL)

HAZOP *See* Hazard and Operability Study. (SAN)

Health As defined by the World Health Organization, a state of complete physical, mental, and social well-being, and not merely the absence of disease or infirmity. (FSL)

Health education The process by which individuals and groups of people learn to promote, maintain, or restore health. (FSL)

Health physics That branch of radiological science dealing with the protection of personnel from harmful effects of ionizing radiation. (FSL)

Hearing conservation A medical surveillance program required by OSHA regulations and involving regular testing of an individual's hearing ability, as well as training, provision of protective devices, and monitoring of the work environment. Hearing conservation program regulations are found at 29 CFR 1910.95(c). These programs are required whenever an individual is exposed to noise in the workplace greater than 85 dBA, measured as an 8-hour time-weighted average. (DEJ)

Hearing impairment Loss of the ability to hear, either partially or completely. Types of hearing loss include conductive impairment, sensorineural impairment, mixed impairment (both conductive and sensorineural), central impairment, and psychogenic impairment. *See also* Noise. (DEJ)

Hearsay A statement made out of court that is offered as evidence to prove that the matter asserted in the statement is true. Generally not admissible as evidence except as provided for in specific exceptions (Federal Rules of Evidence 801–806). (DEW)

Heat The energy associated with a mass because of the random motions of its molecules. *See also* Heat stress. (FSL)

Heat cataract *See* Glass blower's cataract. (DEJ)

Heat cramps A condition related to work and/or exercise in hot environments that causes painful muscle spasms. The cramps are due to heavy sweating and consumption of large amounts of water without adequate salt intake and inadequate exercise–rest balance. (DEJ)

Heat exchanger Device for transferring heat from one fluid or body to another for the purpose of heating or cooling. (FSL)

Heat exhaustion A condition caused by high-temperature work environments and exertion, marked by mild elevation in body temperature, pallor, weak pulse, pale complexion, dizziness, fainting, profuse sweating, headache, low blood pressure, and cool, moist skin. The condition can be dangerous and immediate medical attention is usually required. (DEJ)

Heat island effect The presence of a dome of elevated temperatures over an urban area, caused by structural and pavement heat fluxes and pollutant emissions from the area below. (FSL)

Heat of combustion The maximum amount of heat that can be released by the total combustion of a measured amount of combustible material, defined according to specific parameters. The caloric equivalent of the total combustion energy of a given substance. The heat of formation is often determined by first determining the heat of combustion. *See also* Heat of formation; National Fire Protection Association. (FSL)

Heat of explosion The amount of heat liberated during explosive decomposition. *See also* Explosion. (FSL)

Heat of formation The heat resulting from the formation of one gram-molecule of a compound from its constituent elements. *See also* Heat of combustion. (FSL)

Heat of fusion The heat given off by a liquid freezing to a solid or gained by a solid melting to a liquid without a change in temperature. (FSL)

Heat of solution The amount of heat absorbed or released when a substance is dissolved in a solvent. (FSL)

Heat stress Thermal stress upon the body from the environment, including heat stroke, heat cramps, and heat exhaustion, caused by the body's inability to rid itself of excessive heat. *See also* Engineering controls; heat cramps; heat exhaustion; heat stroke. (DEJ)

Heat stroke A serious condition marked by a rapid rise in body temperature, hot dry skin, mental confusion, loss of consciousness, convulsions, coma, and the absence of sweating. The condition is often caused by excessive physical exertion in hot environments by unacclimatized individuals, dehydration, and recent alcohol intake. Individuals suffering from heat stroke must receive immediate medical treatment. (DEJ)

Heat syncope A condition marked by fainting while standing erect and immobile in hot environments, and caused by pooling of blood in dilated vessels of the skin and the lower part of the body. The individual affected should be removed to a cool environment and medical treatment should be obtained. (DEJ)

Heavy metal As used in environmental regulations, this term includes arsenic, barium, cadmium, chromium, lead, mercury, and silver. Most of these materials do not rapidly break down in the body or the environment and thus can exert toxic effects because of their cumulative or residual properties. (FSL)

Heavy water Deuterium oxide, or water in which the hydrogen has been replaced by deuterium (D_2O). (FSL)

Heinrich, H. W. (1886–1962) A pioneer in the field of safety who wrote a book entitled *Industrial Accident Prevention*. Heinrich established the theory that preventive measures should focus on unsafe acts and unsafe behaviors. He established the 300 : 29 : 1 ratio that suggests that for every group of 330 accidents of the same kind, 300 result in no injuries (are near misses), 29 produce minor injuries, and 1 results in a major, lost-time injury. *See also* Domino theory; unsafe act; unsafe condition. (SAN)

Helium (He) An element with an atomic weight of 4.0026 and atomic number 2 that is an inert gas occurring in the atmosphere in trace concentrations. Helium has numerous commercial, scientific, and medical uses. (FSL)

Hematite One of the most important iron ores. It occurs in igneous rocks and in hydrothermal veins, has a cherry-red streak, and is used as a pigment and in anticorrosion paints. (FSL)

Heme The nonprotein, iron-containing part of the hemoglobin molecule. It carries oxygen and accounts for the color of the blood. *See also* Erythrocyte; hemoglobin. (FSL)

Hemoglobin The red pigment matter in the red blood corpuscles of vertebrates that is a protein yielding heme and globin on hydrolysis. The hemoglobin carries oxygen from the lungs to the tissues and carbon dioxide from the tissues to the lungs. *See also* Heme; hemolysis. (FSL)

Hemolysis The alteration, dissolution, or destruction of red blood cells in such a manner that hemoglobin is released into the medium in which the cells are suspended. *See also* Hemoglobin. (FSL)

Hemotoxin Any substance that causes destruction of red blood cells; usually refers to substances of biologic origin in contrast to chemicals. (FSL)

HEPA filter *See* High-efficiency particulate air filter. (DEJ)

Heparin A complex anticoagulant agent first isolated from canine liver and now known to be a constituent of many types of tissue, including liver and lung. It appears to decrease the rate of interaction between thrombin and fibrogen, thus slowing clot formation. *See also* Warfarin. (FSL)

Hepatitis B virus (HBV) A virus that causes inflammation of the liver, a condition that can occasionally also be caused by toxic agents. (FSL)

Hepatotoxic Refers to an agent that produces damage to the liver; pertaining to damage to the liver. The chemical carbon tetrachloride (CCl_4) has been shown to have this property. (FSL)

Heptachlor ($C_{10}H_7Cl_7$) A organochlorine DDT derivative that breaks down in the soil to heptachlor epoxide, a compound more toxic than the parent material. It was formerly used as an insecticide, but all its uses have been eliminated in the United States because of its persistence and tendency to bioaccumulate in the environment. (FSL)

Herb A vascular plant lacking a persistent woody stem. (SMC)

Herbicide Chemicals used to destroy unwanted plants. There are a number of different categories of these chemicals, including the chlorophenoxy compounds (2,4-D and 2,4,5-T), the bypyridyls (paraquat and diquat), the carbamates (e.g., propam), the amides (e.g., propanil), and the organic arsenicals (e.g., MSMA). *See also* Pesticides. (FSL)

Herbivore An organism that feeds on plant tissue. (SMC)

Herd immunity Resistance of a group to the introduction and spread of an infectious agent. Such resistance is based on the immunity of a high proportion of individual members of the group and on the uniform distribution of the immunes within the group. (FSL)

Heredity Transmission of characteristics and traits from parent to offspring. (FSL)

Heroin [$C_{17}H_{17}No(C_2H_3O_2)_2$] Product, also known as diacetylmorphine, resulting from the alkylation of morphine. Heroin is a narcotic alkaloid, has addiction potential, and is strictly regulated. (FSL)

Herpetology The study of reptiles and amphibians (Class Reptilia and Class Amphibia, Phylum Chordata). (SMC)

Hertz *See* Frequency. (DEJ)

Heterogeneous reactor A nuclear reactor in which the fissionable material and the moderator (if used) are combined in a mixture. This mixture is represented either by a solution of fuel in the moderator or by discrete particles having dimensions small in comparison with the neutron mean free path. (RMB)

Heterotroph An organism that must obtain food energy (organic compounds and energy for metabolic processes) by ingesting other organic material. Predators are heterotrophs that consume other animals; herbivores eat plants; detritivores eat dead organic material. The energy for all heterotrophs ultimately comes from the action of autotrophs, or producers. *See also* Detritivore; fungivore; herbivore; insectivore; parasite; predator. (SMC)

Hexachlorophene [$(C_6HCl_3OH)_2CH_2$] A germicidal compound frequently incorporated in soaps and handwashing compounds. Studies have shown it can be toxic and in some cases fatal when repeatedly applied to the skin. It is no longer commonly used in nurseries because of its potential neurotoxicity. (FSL)

High-density polyethylene A material that is used to produce plastic bottles and other products and that produces toxic fumes when burned. (FSL)

High-efficiency particulate air filter (HEPA) A filter capable of removing 99.97% of all particles with a mean aerodynamic diameter of 0.3 micron. Often used to filter air in air-purifying respirators, vacuum systems, exhaust systems, and fans. HEPA filters are incorporated into biological safety cabinets. Also known as high-efficiency particulate arrestor filters and high-efficiency particulate absolute filters. *See also* Biological safety cabinet. (DEJ)

High-level radioactive wastes (HLW) Irradiated reactor fuel, liquid wastes resulting from the operation of the first-cycle solvent extraction system, or the equivalent, and the concentrated wastes from subsequent extraction cycles, or the equivalent, in a facility for reprocessing irradiated reactor fuel and solids into which such liquid wastes have been converted. *See also* Low-level radioactive waste. (FSL)

Highly toxic Refers to a chemical that has any of the following median lethal doses (LD_{50}) or lethal concentrations (LC_{50}): **(1)** a median lethal dose of 50 mg or less per kg of body weight when administered orally to albino rats weighing between 200 and 300 grams each; **(2)** a medial lethal dose of 200 mg per kg of body weight when administered by continuous contact for 24 hours (or less if death occurs within 24 hours) with the bare skin of albino rabbits weighing between 2 and 3 kg each; **(3)** a median lethal concentration in air of 200 parts per million by volume or less of gas or vapor, or 2 mg per liter or less of mist, fume, or dust, when administered by continuous inhalation for 1 hour (or less if death occurs within 1 hour) to albino rats weighing 200 and 300 grams each. (FSL)

Hinge The median of the top or bottom half of a ranked data set. It is used in a box plot display. (VWP)

Histogram *See* Frequency histogram. (VWP)

Histolysis Process whereby tissue is broken down. (FSL)

Histopathology The study of the cytologic and histologic structure of abnormal or diseased tissue. *See also* Pathology. (FSL)

Histoplasmosis A systemic fungal disease caused by the organism *Histoplasma capsulatium*. The organism is dustborne and the disease results from the inhalation of contaminated particles. (FSL)

Histosol An order of soils that are characterized by their high organic content. Bog soils may be included in this order. (SWT)

HIV *See* Human immunodeficiency virus; acquired immunodeficiency syndrome. (FSL)

Hoist A device intended to raise and lower a suspended or supported load. A traction hoist is a type of hoisting machine that does not accumulate a suspension wire rope on the hoisting drum or sheaf, and is designed to raise and lower a suspended load by the application of friction forces between the suspension wire rope and the drum or sheaf. A winding drum hoist accumulates the suspension wire rope on the hoisting drum. (SAN)

Holding pond A pond or reservoir, usually made of earth, built to store polluted runoff. *See also* Oxidation pond; sewage lagoon. (FSL)

Holdout A restraint device that is a machine-guarding mechanism that includes attachments such that, when anchored and adjusted, it prevents the operator's hands from entering the point of operation. (SAN)

Hole watch A designated individual on duty at the entrance to a confined space occupied by workers. The watch person should be in audible or visual contact with the workers inside at all times. In the event of an emergency, it is this person's responsibility to summon help or, if lifelines are in use, to attend them. If properly trained, the watch may don a self-contained breathing apparatus (SCBA) and enter the space to assist injured workers. Otherwise, the watch cannot enter the confined space until help arrives. *See also* Attendant; fire watch; permit-required confined space. (SAN)

Hologram A three-dimensional image created by lasers, with one laser beam used to illuminate the object being photographed, and another split off from the first and bounced to a plate using a mirror, and thus producing an interference pattern between the two beams. (DEJ)

Homeostasis The tendency to maintain a state of normal stability within an organism by coordinated responses; automatic adjustment of organ systems for environmental changes. (FSL)

Homogeneous radiation A beam or flux consisting of radiation of the same kind and energy. (RMB)

Homogeneous reactor A nuclear reactor in which the fissionable material and the moderator (if used) are combined in a mixture such that an effectively homogeneous medium is presented to the neutron. This mixture is represented either by a solution of fuel in the moderator or by discrete particles having dimensions small in comparison with the neutron mean free path. *See also* Heterogeneous reactor. (RMB)

Homogenization The process of breaking up fat globules to such an extent that after quiescent storage no separation of fat occurs, as in the homogenization of milk. Homogenizers are used in dairy operations, for example, on products of high fat concentration or on ultra-high-temperature pasteurized products (UHT). This processing sequence makes the cleaning and sanitizing of the homogenizer of extra importance in a plant sanitation program. (HMB)

Homologous Similar or alike in certain critical attributes. In chemistry this term is used to refer to members of a single chemical series that differ by fixed increments. In genetics, it refers to chromosomes or chromosome parts that are identical with respect to their genetic loci. (FSL)

Hood A shaped inlet designed to capture contaminated air and conduct it into an exhaust duct and/or exhaust fan. Differently shaped hoods have varying efficiencies and pressure losses. External hoods are designed to capture an air contaminant generated some distance in front of the hood face. Internal hoods are designed to capture air contaminants generated inside an enclosure behind the hood face, as in a laboratory hood. *See also* Biological safety cabinet; chemical hood. (DEJ)

Hookworm Parasitic worm that infests people and causes debilitation. Major infestations can cause anemia and retardation of mental and physical development. Adult hookworms feed on blood and tissue from the wall of the intestine. Eggs pass out in the feces and undergo a period of development in soil, and the larva enter a new host by burrowing through the skin. *Ancyclostoma duodenale* and *Necator americanus* are two of the species of hookworms of concern in the United States. (FSL)

Hopcalite A mixture of manganese dioxide and other oxides used as a catalytic agent in some carbon monoxide monitors and also for sampling mercury vapor in air. (DEJ)

Hormone A chemical substance (e.g., cortisol) found in one organ or part of the body and carried in the blood to another part. Hormones can alter the function, and sometimes the structure, of one or more organs. (FSL)

Hornblende A calcium magnesium aluminosilicate mineral. (FSL)

Horse latitudes The regions at approximately 30°N latitude and 30°S latitude that are characterized by descending air masses. Air that is heated (and rises) at the equator cools adiabatically as it rises and moves north or south, and then sinks at the horse latitudes. (SMC)

Host An organism, simple or complex and including humans, that is capable of being infected by a specific agent. Primary (definitive) hosts are those in which the parasite attains maturity or passes through its sexual stage. Secondary (intermediate) hosts are those in which the parasite exists in the larval stage or other asexual stage. Accidental (dead-end) hosts are those that have no role in the propagation or transmission of a particular infectious agent; they include humans or other organisms, including birds, and arthropods, that under natural conditions provide subsistence or lodgment to an infectious agent. (FSL, HMB)

Host range The extent of the variety of species susceptible to infection by a specific kind of agent. (FSL)

Hot A colloquial term applied to chemicals and meaning highly reactive or highly toxic. *See also* Highly toxic. (FSL)

Hot-food storage equipment An enclosed electrically heated device that, when preheated and connected to a power source, is intended to receive and to hold food at not less than 140°F (60°C). (HMB)

Hot spot The region in a contaminated area in which the level of radiative contamination is somewhat greater than in neighboring regions in the area. *See also* Contamination. (RMB)

Hot tap A procedure used in repair, maintenance, and service activities that involves welding on a piece of pressurized equipment (pipelines, vessels, or tanks), in order to install connections or appurtenances. The technique is commonly used to replace or add sections of pipeline without the interruption of service for air, gas, water, steam, and petrochemicals. In general this practice should be discouraged unless it is demonstrated that continuity of service is essential and shutdown of the system is impractical. When carried out, documented procedures must be followed and special equipment that will provide effective protection for employees must be provided. *See also* Hot work. (SAN)

Hot-wire anemometer A device, also known as a thermal anemometer, used to measure air velocity by the cooling effect of moving air over a heated element. Its use is based on the principle that the amount of heat removed by an airstream is related to the velocity of the airstream, so long as a correction factor is applied to account for the temperature of the airstream. Hot wires are also used to support combustion in combustible-gas meters. (DEJ)

Hot work Applies to the workplace or portion of the work site where operations are conducted involving use of flame-producing or spark-producing equipment such as grinders, or where welding, burning, or brazing, capable of igniting flammable vapors or gases, takes place. A permit and a fire watch are generally required for most types of hot work. The permit certifies that the fire prevention and protection requirements have been implemented prior to the beginning of the hot-work operations; it also indicates the dates authorized for hot work, and identifies the equipment or facility on which the work is to be done. The permit must be kept on file until completion of the hot-work operations. *See also* Fire watch; permitting systems. (SAN)

Hot zone *See* Exclusion zone. (FSL)

Housekeeping Practices and procedures initiated in the work environment and elsewhere to prevent and/or remove the buildup of trash, debris, refuse, and other materials. OSHA requires that combustible scrap and debris be removed at regular intervals. Containers should be provided for the collection and separation of waste, trash, oily and used rags, and other refuse. Garbage and other waste must be disposed of at regular intervals. (SAN)

Human immunodeficiency virus (HIV) The virus that causes acquired immunodeficiency syndrome (AIDS). *See also* Acquired immunodeficiency syndrome. (FSL)

Humic acid The dark-colored fraction of soil humus that can be extracted with dilute alkali and precipitated by acidification to pH 1–2. (SWT)

Humidity The amount of moistness or dampness in the air. *See also* Relative humidity. (FSL)

Humification **(1)** The conversion of organic residues through biological activity, microbial synthesis, and chemical reaction. **(2)** The breakdown of organic matter in the soil by microbial organisms. (FSL, SWT)

Humus **(1)** The fraction of soil organic matter, peat, or compost that is formed during the biological decomposition of organic matter. **(2)** Decomposed organic matter in the soil that aids the texture of the soil and holds water, thus preventing nutrients from being lost. (FSL, SWT)

Hydatid disease The infection of people by the larval stages of taenid cestodes of the genus *Echinococcus*. This species ranks high as a cause of morbidity. The disease is also known as sensu lato. (FSL)

Hydrargyria Chronic mercury poisoning. (FSL)

Hydration The process of absorbing or combining with water; the chemical addition of water to a compound. (FSL)

Hydraulic conductivity An expression of the ability of a liquid, such as water, to flow through a soil in response to a given potential gradient. (SWT)

Hydrazine (H_2NNH_2) Chemical intermediate resulting from the production of explosives and photographic chemicals that has been used as a rocket fuel. Hydrazine is toxic by ingestion, inhalation, and skin absorption and is a strong irritant to the skin and eyes. It is designated a carcinogen by OSHA and has a TLV of 0.1 ppm in air. (FSL)

Hydric **(1)** Referring to or containing hydrogen; **(2)** wet, referring to systems with some visible surface water. (SMC)

Hydrocarbon A compound composed solely of the two elements hydrogen and carbon. (FSL)

Hydrofluoric acid (HF) Among the most corrosive and strongest acids known. It reacts with water and a variety of materials. It is used in refineries and for etching glass and cleaning masonry. It is highly corrosive to skin and mucous membranes. (FSL)

Hydrogen (H) A highly flammable and explosive gaseous element with an atomic weight of 1.00797 and atomic number 1. In addition to being used as a carrier gas in chromatography, it is used in the synthesis of ammonia, hydrochloric acid, and methane. (FSL)

Hydrogenation The addition of hydrogen to a gaseous substance by the use of gaseous hydrogen combined with a catalyst. (FSL)

Hydrogen cyanide (HCN) Also known as hydrocyanic acid and prussic acid, a highly poisonous gas used as a burrow gassing agent in areas with large rodent populations. It has also been used as a fumigant in buildings with rodent infestations. Its use for these purposes has been largely eliminated because of its toxicity. (FSL)

Hydrogenic soil Soil developed under the influence of water standing for considerable periods of time; usually formed in cold humid regions. (SWT)

Hydrogen peroxide (H_2O_2) An unstable, reactive, colorless liquid with a variety of uses such as that of bleaching agent, antiseptic, and oxidant for rocket fuels. A strong oxidizing agent that at high concentrations can be a dangerous fire and explosion risk. (FSL)

Hydrogen sulfide (H_2S) A toxic, foul-smelling gas, the inhalation of which in sufficient concentrations can cause headaches and death. It is used in the purification of hydrochloric acid and sulfuric acid, as a laboratory reagent in numerous chemical processes, and in the manufacture of some types of cloth. It has a TLV of 10 ppm in air. (FSL)

Hydrologic cycle The cycle that begins when water is precipitated from the atmosphere, continues when the water evaporates back into the atmosphere, and ends when the water is recycled as precipitation. (SWT)

Hydrology The study of the distribution and movement of water. It has the following subcategories. **(1)** *Agrohydrology:* The study of water and soil solution movement and distribution to and from the root zone in agricultural land. **(2)** *Surface hydrology:* The study of the distribution and movement of water on the soil surface. **(3)** *Soil water hydrology:* The study of the distribution and movement of soil solution in the soil profile. **(4)** *Ground water hydrology:* The study of movement of the soil solution in the saturated zone of the soil profile. (SWT)

Hydrolysis The formation of an acid and a base from a salt by the ionic dissociation of water. Also occurs by the decomposition of organic compounds by interaction with water. (FSL)

Hydromorphic soils Soils formed under conditions of poor drainage in marshes, swamps, etc. (SWT)

Hydrophilic Refers to a material that rapidly absorbs water—e.g., silica gel. (FSL)

Hydrophobia **(1)** Fear of water. **(2)** An obsolete term for the disease rabies. (FSL)

Hydrophobic soils Soils that have developed a water-repellent surface due to dense mycelial mats. (SWT)

Hydroponics The cultivation of plants in a solution of water and dissolved nutrients rather than in soil. Sometimes also called aquiculture or aquaculture. (SMC)

Hydrostatic The pressure exerted by a fluid—e.g., by a rising water table. (FSL)

Hydrostatic pressure Pressure created by water at rest equally at any point within a confined area. (FSL)

Hydrostatic testing The procedure that involves submerging in water a tank designed to hold pressurized liquids or gases and pressurizing it with air to 150% of the design pressure. Portable fire extinguishers and air tanks for self-contained breathing apparatus must be hydrostatically tested at regularly specified intervals. Pressure vessels must be hydrostatically tested in order to qualify for a rating by the American Society of Mechanical Engineers (ASME). In some cases, pneumatic testing is used rather than hydrostatic testing. The month and year in which the test was performed is marked either on the tank or on a label attached to the tank. *See also* Pneumatic testing. (SAN)

Hydrothermal The generic term that refers to any geologic process involving heated or superheated water. (FSL)

Hygiene Named for the Greek god Hygeia, this is the science of health and the preservation of well-being. *See also* Health. (FSL)

Hygrometer An instrument used for the detection of atmospheric moisture. (FSL)

Hygroscopic Refers to substances that absorb water from the atmosphere. (FSL)

Hymenolepsis A genus of tapeworms, including *H. nana*, that has been found in humans. (FSL)

Hyperbarism A condition resulting from exposure to atmospheric pressure that exceeds the pressure within the body. (FSL)

Hyperkinesis A high state of motility of an organism or a muscle. (FSL)

Hyperphagia Voracious eating and the consumption of edible and nonedible substances past the point of normal satiation. *See also* Aphagia; pica. (FSL)

Hyperplasia An unusual increase in the number of cells in an organism, such as in animal tumors and plant cells. (FSL)

Hyperthermia A marked sustained increase in body temperature due to the inability of the body to dissipate excessive heat generated through metabolic activity. The buildup of heat may be due to the body's attempt to rid itself of a bacterial or viral infection or to an increase in physical activity coupled with the wearing of clothing that incumbers heat transfer under varying environmental conditions. Severe cellular damage and death can result if the condition is not treated promptly. (MAG)

Hypochlorinators Pumps that are used to inject chlorine solutions into water. When properly maintained, they provide a reliable method for the application of chlorine for disinfection purposes. There are several types of hypochlorinators, including the following. **(1)** *Suction feeders:* Those in which chlorine is siphoned from the supply container by the suction created by an operating water pump. **(2)** *Positive-displacement feeders:* Those in which a piston or diaphragm pump injects the solution. **(3)** *Aspirator feeders:* Those that operate on the hydraulic principle that uses a vacuum created when water flows through a venturi tube or perpendicular to a nozzle; the vacuum created draws the chlorine from a container into the chlorinator unit, where it is mixed with water and injected into the water system. **(4)** *Gaseous feeders:* Those that may use chlorine as the disinfectant when large quantities of water require disinfection; in these cases, the chlorine gas is fed directly into the system by various metering devices. **(5)** *Tablet hypochlorinators:* Devices that consist of special pot feeders into which calcium hypochlorite tablets are placed; small jets of water are injected into the lower portion of the tablet bed, the slow dissolution of the tablets providing a continuous source of disinfectant. (FSL)

Hypochlorite A chlorine compound used as a sanitizer in the food-processing or food-service industry. Sodium and calcium hypochlorite are the important commercially available hypochlorites and are the most active of various chlorine compounds that might be used as sanitizers. They are routinely used in aqueous solutions to reduce microbial loads at low concentrations with short exposure times. A 90% reduction in numbers of organisms might be expected in less than ten seconds at relatively low levels of free available chlorine. A typical concentration for use on a precleaned surface would be 50 ppm with an exposure of 1.0 min., assuming a temperature of about 75°F. *See also* Chlorine compounds; sanitizer. (HMB)

Hypochlorous acid (HOCl) A chemical compound used as a disinfectant and bleaching agent. *See also* Chlorine. (FSL)

Hypothalamus The portion of the brain that controls body temperature and produces hormones that affect the pituitary gland. (FSL)

Hypothermia Loss of body heat and decreased temperature due to extensive exposure to cold. (FSL)

Hypothesis An unproven assertion or statement, based on available information, that commonly deals with the identity of an etiologic agent, a source of infection, or a mode of transmission. Its role is to provide a rational basis for further investigation. (FSL)

Hypothesis (biostatistical) (alternative H_a or H_1) A supposition about a population parameter that contradicts the null hypothesis. (VWP)

Hypothesis, null (biostatistical) (H_o) A supposition about a population parameter that is to be tested. (VWP)

Hypothesis test Procedure that allows the tester to accept or reject hypotheses or to determine whether observed samples differ significantly from expected results. (VWP)

I

IAG Interagency agreement. (DEW)

Ichthyology The study of fishes (Class Pisces, Phylum Chordata). (SMC)

Igneous rocks Rocks that are derived from the magma deep in the earth. They include granite and other coarsely crystalline rocks, dense rocks such as those that occur in dikes and sills, basalt and other lava rocks, and cinders, turf, and other fragmental volcanic materials. (FSL)

Ignitable Capable of burning or causing a fire. (FSL)

Igniter A device, usually a propellant or pyrotechnic, used to initiate burning. (FSL)

Ignition The introduction of some external spark, flame, or glowing object that initiates self-sustained combustion. The ignition temperature of a substance is the minimal temperature to which it must be heated in order to ignite. (FSL)

Ignition sources Sources that can initiate combustion. These include: open flames; lightning; smoking materials; cutting and welding devices; hot surfaces; frictional heat; static, electrical, and mechanical sparks; spontaneous-ignition-producing sources, including heat-producing chemical reactions; and radiant heat. *See also* Hot work. (SAN)

Illness A condition of pronounced deviation from the normal health state; sickness. Illness can be the result of disease or injury. *See also* Morbidity. (FSL)

Illumination The density of light flux incident upon a surface. (DEJ)

Illuvial horizon A soil layer or horizon that is formed by material from an overlying layer that has been precipitated from solution or deposited from solution. A layer of accumulation. (SWT)

Illuviation The deposition of soil material removed from an upper horizon. (SWT)

Imhoff tank A component of a sewage treatment system in which the sewage solids are combined with anaerobic biological materials. The incoming sewage enters an upper chamber and the solids settle through slots in a lower digestion chamber, with sludge removed automatically. The tank used in this process was named for its inventor, Paul Imhoff (1876–1965). (FSL)

Immature soil A soil whose horizons are poorly developed, due to its short exposure time to soil-forming processes. (SWT)

Immediately dangerous to life or health (IDLH) Describes concentration of a hazardous agent that will produce irreversible health effects or death, following a 30-minute exposure. Concentrations above the lower flammable limit are considered to be IDLH. Limits for chemical substances have been established by the National Institute for Occupational Safety and Health (NIOSH). Used as a "level of concern." *See also* Level of concern. (DEJ)

Imminent danger Refers to a work situation in which there is an immediate danger of death or serious physical harm. An OSHA inspector may close a facility if a court order is obtained, and can also levy a substantial fine. With a restraining order from a court, an MSHA inspector may close a mine if the possibility of imminent danger to the workers exists. (DEJ)

Immiscible Refers to substances that are not capable of being mixed or those that can be mixed only with great difficulty. (FSL)

Immobilization The conversion of an element from an inorganic form to an organic form in tissue, in a process that renders the element unavailable to other organisms. (SWT)

Immune individual A person with specific protective antibodies or cellular immunity as a result of previous infection or immunization. (FSL)

Immune serum globulin A sterile solution of globulins that contains those antibodies normally present in adult human blood. (FSL)

Immunity The power that an organism possesses to resist and overcome infection. *See also* Resistance. (FSL)

Immunodeficient Lacking in the ability to produce antibodies in response to an antigen. *See also* Human immunodeficiency virus. (FSL)

Impeachment The presentation of evidence at trial or other attack on the credibility of a witness. (DEW)

Impeded drainage A condition that hinders movement of water under the influence of gravity through soils. (SWT)

Impingement The physical process of bringing matter into contact. **(1)** In air pollution abatement technology, the adherence of dust particles in a collector to prevent discharge to the environment. **(2)** In industrial hygiene, a process for the collection of particulate matter in which a particle-containing gas is directed against a wetted glass plate, causing the particles to be retained by the liquid. (FSL)

Implode To burst inward. (FSL)

Impoundment Confinement of a body of water or sludge by a dam, dike, floodgate, or other barrier. (FSL)

Inapparent infection Infection without recognizable clinical signs or symptoms. Such a condition may occur at any time during the course of a disease, dependent upon the nature of the illness. *See also* Asymptomatic. (HMB)

Incendiary A material that is primarily used to start fires. *See also* Ignition. (FSL)

Inceptisols Soil orders that have one or more diagnostic horizons that have been formed quickly without significant illuviation or eluviation, or extreme weathering. (SWT)

Incidence Number of new cases of a disease occurring within a particular population during a specified period of time. *See also* Prevalence. (FSL)

Incidence rate For OSHA record-keeping purposes, the number of injuries, illnesses, or lost workdays related to a common exposure base of 100 full-time workers. The common exposure base enables one to make accurate inter-industry comparisons, trend analyses over time, or comparisons among firms regardless of size. The rate is calculated as follows:

$$\text{Incidence rate} = N/EH \times 200{,}000$$

where N is the number of injuries and/or illnesses or lost workdays, EH is the total hours worked by all employees during the calendar year, and 200,000 is the base for 100 full-time equivalent workers (working 40 hours per week, 50 weeks per year). *See also* Lost workday case; OSHA Log of Occupational Injuries and Illnesses; severity rate. (SAN)

Incident Occurrence of an equipment problem that may result in harm or extensive environmental damage in a situation in which no injuries or illnesses occur to employees. *See also* Accident; incident investigation. (SAN)

Incident commander The ranking on-scene official of the jurisdiction within which a hazardous-materials incident (spill, fire, etc.) has occurred. (FSL)

Incident command system (ICS) An organized approach to control and manage operations effectively during a hazardous-materials emergency response. The individual in charge of the ICS is the senior official responding to the incident. A command post from which the incident commander manages the incident is established. The incident commander can delegate responsibility for performing various tasks to subordinate officers. All communications are routed through the command post. *See also* HAZWOPER; on-scene incident commander. (SAN)

Incident investigation standard The OSHA process safety standard, 1910.119, requiring investigation of every incident that results in, or could reasonably have resulted in, a major accident in the work place. Incident investigations should be initiated as promptly as possible, but no later than 48 hours following the incident. An incident investigation team must be established and should consist of persons knowledgeable about the process involved and other appropriate specialties. At the conclusion of the investigation, a report must be prepared that includes: the date of the incident; the date the investigation began; a description of the incident; a description of the factors that contributed to the incident; and any recommendations resulting from the investigation. The report is to be reviewed with all operating, maintenance, and other personnel whose work assignments are within the facility in which the incident occurred. (SAN)

Incineration **(1)** A controlled burning of certain types of materials that are changed into gases and produce residues that are relatively free of combustible materials. **(2)** A treatment technology involving destruction of waste by controlled burning at high temperatures—e.g., the burning of sludge to remove water and reduce the remaining residues to a safe, nonburnable ash that can be disposed of safely on land, in some waters, or in underground locations. (FSL)

Incineration at sea Disposal of waste by burning at sea on specially designed incinerator vessels. (FSL)

Incinerator Any device intended or used for the reduction or destruction of solid, liquid, or gaseous waste by burning. (FSL)

Incipient-stage fire A fire that is in the initial stage and that can be controlled or extinguished by portable fire extinguishers or Class II standpipe or small-hose systems, without the need for protective clothing or breathing apparatus. *See also* Interior structural firefighting; small-hose systems; standpipe systems. (SAN)

Incontestable clause A provision that states that after the lapse of a specified time, an insurer cannot dispute a policy on the grounds of misrepresentation, fraud, or similar wrongful conduct on the part of the insured. (DEW)

Incubation The growth and development of microorganisms. In the laboratory, the providing of the proper conditions for such growth. *See also* Incubation period. (FSL)

Incubation period The time interval between effective exposure of a susceptible host to an agent (infection) and onset of clinical signs and symptoms of disease in that host. Incubation periods vary with specific etiologic agent, and knowledge of such periods may serve as an important tool in the determination of the causative agents of specific outbreaks under investigation. (FSL, HMB)

Indemnify To agree to secure or reimburse in case of loss or damages. (DEW)

Independent events Events for which the occurrence or nonoccurrence of one event does not affect the occurrence or nonoccurrence of the other. (VWP)

Indeterminate Not of a limited, definite growth or size; continuing to grow as long as there is space, nutrients, or other resources available. (SMC)

Index case The first case among a number of similar cases that are epidemiologically related. Index cases are often identified as sources of contamination or infection. (FSL)

Indicating thermometer A nonrecording thermometer that allows the user to measure the temperature, generally on the Fahrenheit scale. The most common and versatile type of indicating food thermometer, the bimetallic type, has a metal stem at least five inches long, the temperature-sensing area of which is the lower two inches, which makes the thermometer appropriate for submersion in foods. Mercury-filled or glass thermometers should not be used to measure food temperatures. *See also* Recording thermometer. (HMB)

Indicator organisms Any of two groups of microorganisms normally used as indicators of fecal contamination in foods or in the measurement of effectiveness of wastewater treatment. The two groups are the coliform group and the fecal streptococci. The indicator organisms are chosen because they are present in relatively high numbers in feces, persist in the environment for relatively extended periods, and are easily cultivated. Their presence indicates the possible presence of fecal contamination. *See also* Coliforms; fecal coliforms; fecal streptococci. (HMB)

Indicator plants Plants that survive under specific soil or site conditions such as pH or soil moisture. (SWT)

Indicator species A species that is used as an indicator of the quality of a habitat. Certain species are more sensitive than others to changes in environmental conditions such as temperature, pollution, acidity, and salinity. The presence or absence of an indicator species can be used to assess the environmental conditions of a habitat without direct measurement. (SMC)

Indictment Formal accusation of a crime, presented to the grand jury by a prosecuting attorney. Also presented to the accused to inform him or her of the crimes charged. (DEW)

Indirect discharge Introduction of pollutants from a nondomestic source into a publicly owned waste treatment system. Indirect dischargers can be commercial or industrial facilities whose wastes go into the local sewers. (FSL)

Individual water supply system A single system of piping, pumps, tanks, or other facilities utilizing a source of ground or surface water to supply a single property. (FSL)

Indole acetic acid A growth-regulating hormone (auxin) produced by many plants. (FSL)

Indoor air The breathing air inside a habitable structure or conveyance. (FSL)

Indoor air pollution The presence of chemical, physical, or biological contaminants in indoor air in concentrations that could have an adverse effect on human health. (FSL)

Indoor climate Temperature, humidity, lighting, and noise levels in a habitable structure or conveyance. Some aspects of indoor climate can affect indoor air pollution. (FSL)

Induced radioactivity Radioactivity produced in certain materials as a result of nuclear reactions, particularly the capture of neutrons, that involve the formation of unstable (radioactive) nuclei. The activity induced by neutrons from a nuclear explosion in materials containing sodium, manganese, silicon, or aluminum may be significant. (RMB)

Inductively coupled plasma emission spectroscopy (ICPES) A method typically used for analysis of many heavy metals simultaneously. This method employs an inductively coupled, argon-supported plasma as an emission source. A liquid sample is converted into a fine aerosol and introduced into the plasma, where it is reduced to its atomic state in the hot ionized gas. This gives rise to spectral transitions at wavelengths characteristic of the metallic elements in the sample. (DEJ)

Induration The hardening of sediments through the action of heat, pressure, or cementation. (FSL)

Industrial hygiene The art and science of anticipating, recognizing, evaluating, and controlling occupational and environmental health hazards in the workplace and the surrounding community. (DEJ)

Industrial sanitation The specialized application of community environmental health services, including the control of the spread of infection or other insults to employee health not related specifically to the manufacturing process. (DEJ)

Industrial waste Solid waste generated by manufacturing or industrial processes that is not hazardous waste. Such wastes include, but are not limited to, wastes resulting from the following manufacturing processes: electric power generation, iron and steel manufacturing, nonferrous metals manufacturing/foundries, plastics and resins manufacturing, textile manufacturing, and water treatment. The wastes also include those involved in the production or use of: fertilizer/agricultural chemicals, food and related products/by-products, inorganic chemicals, leather and leather products, organic chemicals, pulp and paper, rubber and miscellaneous plastic products, and stone, glass, clay, concrete products, and transportation equipment. *See also* Hazardous waste. (FSL)

Inerted atmosphere The atmosphere of a confined space that has been made nonflammable, nonexplosive or otherwise chemically nonreactive. This can be done by displacing or diluting the original atmosphere with steam or gas (nitrogen, helium, carbon dioxide) that is nonreactive with respect to that space. This is a common practice carried out in order to minimize the potential for explosion when underground storage tanks and moving tank trucks that contain flammables are removed. (SAN)

Inert gas Nonreactive gas such as krypton, argon, helium, or neon. (FSL)

Inertial separator A device that uses centrifugal force to separate waste particles. (FSL)

Inert-waste landfill A disposal site accepting only wastes that will not, or are not likely to, produce leachate of environmental concern. Such wastes are limited to dirt and dirtlike products, concrete, rock, bricks, yard trimmings, stumps, tree limbs, and leaves. (FSL)

Infected person A person who harbors an infectious agent, whether or not the infection is accompanied by disease. An infectious person is one from whom the infectious agent can be naturally acquired by exposed susceptibles. (FSL)

Infection The entry and multiplication of an infectious agent that occurs in the body tissues of a human being or animal and that results in cellular injury. Infection is not synonymous with *infectious disease*. Examples of infections associated with foods and food sanitation include salmonellosis, listeriosis, and vibriosis. **(1)** *Inapparent infection.* An infection resulting in no perceptible clinical signs and symptoms. **(2)** *Apparent infection.* An infection resulting in clinical signs and symptoms (disease). *See also* Listeriosis; salmonellosis; vibriosis. (FSL) (HMB)

Infectious agent An organism, usually a microorganism, but including helminths, that is capable of producing infection or infectious disease. Any organism (such as a virus or a bacteria) that is capable of being communicated by invasion and multiplication in body tissues and is capable of causing disease or of having an adverse health impact in humans. *See also* Disease; infectious disease. (FSL)

Infectious disease A disease of humans or animals, resulting from an invasion of the body by pathogenic agents and the reaction of the tissues to these agents and/or the toxins they produce. (FSL)

Infectivity The ability of an agent to infect a host. (FSL)

Inferential statistics Methods used to determine conditions under which conclusions about a population may be drawn from an analysis of a sample of the population. (VWP)

Infestation The lodgment, development, and reproduction of arthropods, such as ticks, mites, or fleas, on the surface of the body, in clothing, or in dwellings. Infested articles or premises are those that harbor arthropods or rodents.

Effective measures that are intended to minimize the presence of insects, flies, rodents, and other pests should be utilized and are usually required by food-service or food-processing sanitation regulations. Control measures usually include elimination of food and water sources, prevention of entry into the establishment, and elimination of harborages. Chemical controls—e.g., insecticides, baits, and repellents—or nonchemical means, such as insect light traps, may be applied. The use of such measures is an essential aspect of sanitation in the food industry, as insects are aesthetically displeasing and may be carriers of disease organisms to people. *See also* Arthropods. (FSL,HMB)

Infiltration **(1)** The movement of water through the ground surface into subsurface soil or from the soil into sewer or other pipes through defective joints, connections, or manhole walls. **(2)** A technique whereby large volumes of wastewater are applied to land and allowed to penetrate the surface and percolate through the underlying soil. *See also* Percolation. (FSL)

Inflammation Normal tissue response to cellular injury or foreign material, characterized by dilation of small blood vessels (capillaries) and mobilization of defense cells (blood and tissue leukocytes and phagocytes). *See also* Leukocyte; phagocyte. (FSL)

Inflorescence Flower or flower cluster. (SMC)

Inflow Entry of extraneous rain water into a sewer system from sources other than infiltration, such as from basement drains, manholes, storm drains, and street washing. (FSL)

Influent Water, wastewater, or other liquid flowing into a reservoir, basin, or treatment plant. (FSL)

Ingestant A substance capable of entering the body through the mouth or digestive system. (FSL)

Infrared Electromagnetic radiations of wavelength between the longest visible red (7000 Å or 7×10^{-4} millimeter) and about 1 millimeter. *See also* Electromagnetic radiation. (RMB)

Infrared gas analyzer A real-time air-sampling device that measures absorbance of inorganic and organic gases and vapors. Many gases and vapors have characteristic infrared spectra that absorb infrared radiation over a range of wavelengths, the values of which can be converted into characteristic graphs. Because absorbance is proportional to concentration, and specific gases absorb infrared energy at designated frequencies, the concentrations of many gases can be measured rapidly in ambient air. (DEJ)

Infrared spectrum The portion of the electromagnetic spectrum that includes wavelengths from 0.78 micron to approximately 300 microns—i.e., those larger than the wavelengths of visible light and shorter than those of microwaves. (FSL)

Inhalable dust *See* Respirable dust. (DEJ)

Inhalation The process of drawing air or other substances into the lungs. (FSL)

Inherent filtration (X-rays) The filtration introduced by the wall of an X-ray tube and any permanent tube enclosure; to be distinguished from added primary and secondary filtration. (RMB)

Inhibitors Chemicals that may be added in small quantities to unstable materials to prevent a vigorous reaction. Substances that have the capability to interfere with chemical reactions, growth, or various types of biologic activity. (FSL)

Initial nuclear radiation Nuclear radiation (essentially neutrons and gamma rays) emitted from the fireball and the cloud column during the first minute after a nuclear explosion. The time limit of one minute is set, somewhat arbitrarily, as that required for the source of the radiations (fission products in the atomic cloud) to reach the earth's surface. *See also* Residual nuclear radiation. (RMB)

Initiation **(1)** The process whereby explosive charges are detonated. **(2)** The first step in carcinogenesis; the subtle alteration of cells by exposure to a carcinogenic agent such that they potentially form a tumor upon subsequent exposure to a promoting agent. (FSL)

Initiator A chemical whose introduction into the body may be the first step in carcinogenesis, a process that may involve changes in the genetic material in affected cells. (FSL)

Injection well A well into which fluids are injected for purposes such as waste disposal, improvement of the recovery of crude oil, and solution mining. (FSL)

Injection zone A geological formation, group of formations, or part of a formation receiving fluids through a well. (FSL)

Injunction Generally, a court order forbidding the defendant in the action to perform some action. In federal practice, a preliminary injunction shall be issued with notice to the adverse party; FRCP Rule 65(a). *See also* Temporary restraining order. (DEW)

Injury Physical harm or damage to a person or property. *See also* Illness. (FSL)

Inlet-type connection A type of cross-connection used for filling a receptacle open to the atmosphere. The connection may be below a rim or submerged, such as an inlet to a dishwashing tank. *See also* Backflow; backsiphonage; cross-connection; submerged inlet. (HMB)

Innocent-landowner defense Legal defense under Superfund for persons who unknowingly purchase contaminated property after exercising "due diligence" to uncover contamination prior to acquisition. *See also* Due diligence. (FSL)

Inoculum **(1)** Bacterium placed in compost to start biological action. **(2)** A medium containing organisms that is introduced into media or cultures of organisms. (FSL)

Inorganic chemicals Chemical substances of mineral origin, not basically of carbon structure. (FSL)

Insecticides Chemical products (pesticides) designed to kill various kinds of insects. These chemicals, which are subject to the regulatory provisions of the Federal Insecticide, Fungicide, and Rodenticide Act (FIFRA), are usually classified on the basis of their mode of entry into the body. They include: stomach poisons, which must be ingested in order to cause death; contact poisons, which penetrate the body wall; fumigants, which are chemical vapors that penetrate the breathing pores (spiracles) and body surface; and desiccants, which are dusts that scratch, abrade, or absorb the fatty, water-resistant outer layer of the exoskeleton, causing death by dehydration. Insecticides may also be classified as to the particular stage of development of the insect upon which the chemical acts. For example, adulticides kill the insect in the adult state, larvicides the immature states, and ovicides the eggs. Insecticides can also be classified on the basis of their chemical structure—e.g., chlorinated hydrocarbons (such as DDT), organophosphates (such as malathion), carbamates (such as Sevin), and botanicals (such as pyrethrums). *See also* Insect infestation; pesticide. (FSL)

Insectivore An organism that feeds on insects. (SMC)

In situ As used in the management of hazardous waste, treatment of contaminated areas without excavation or other removal, such as the treatment of soils through biodegradation of the contaminants on site. (FSL)

Inspiration The process of drawing air into the lungs. (FSL)

Inspiratory reserve volume (IRV) The maximum volume of air that can be forcibly expired following a normal inspiration. (DEJ)

Institute of Makers of Explosives (IME) A trade organization composed of the leading producers of commercial explosive materials in the United States. The goal of the IME is to promote the safe manufacture, transportation, storage, and use of explosive materials. (FSL)

Insulin A sulfur-containing hormone produced by the islets of Langerhans in the pancreas of vertebrates. This hormone stimulates the conversion of glucose to glycogen and fat. A deficiency in the hormone results in excess blood sugar and causes the condition diabetes mellitus. (FSL)

Integral absorbed dose The energy imparted to matter by ionizing particles in a region of interest. The unit is the gram-rad and is equal to 100 ergs. (RMB)

Integral dose (volume dose) A measure of the total energy absorbed by a patient or any object during exposure to radiation. According to British usage, the integral dose of X-rays or gamma rays is expressed in gram-roentgens. (RMB)

Integrated neutron flux The product of neutron flux and time, expressed in units of neutrons per square centimeter. It is a measure of neutron exposure dose. *See also* Dose; neutron flux. (RMB)

Integrated pest management Employment and integration of a variety of methods to control insects or other pests. Such integrated control methods should be both environmentally sound and more effective than application of a single pest control procedure. Various combinations of compatible approaches that do not attempt eradication but rather aim to keep pest levels below that which would cause significant economic loss have been successful. Pesticide application may be a part of this approach, but an integrated approach employs more natural controls. Among those methods that might be used are general sanitation, crop rotation methods, development of resistant host plants, the use of pathogens and parasites of the pests to be controlled, the use of juvenile hormones to control insect behavior, the use of natural enemies of the pest, and the introduction of sterile males into the population. None of the above methods alone is effective; however, in combination, this approach is generally safe and ecologically sound. (HMB)

Integrating circuit An electronic circuit that records at any time an average value for a number of events occurring per unit time; or an electrical circuit that records the total number of ions collected in a given time. (RMB)

Integration The process by which small bits of information, represented by a variable, are added with respect to time or another variable. (FSL)

Intensity (radiation) The energy of any radiation incident upon (or flowing through) a unit area, perpendicular to the radiation beam, in a unit of time. The intensity of thermal radiation is generally expressed in calories per square centimeter per second falling on a given surface at any specified instant. As applied to nuclear radiation, the term "intensity" is sometimes used, rather loosely, to express the exposure dose rate at a given location—e.g., in roentgens (or milliroentgens) per hour. (RMB)

Interaction of two events A situation wherein two events occur simultaneously and interact to produce a given effect. (VWP)

Interagency testing committee A committee created by the Toxic Substances Control Act and consisting of representatives of various federal agencies. The committee makes recommendations to the Environmental Protection Agency for chemical substances to be tested under Section 4 of the act. 15 USC §2603(e). *See also* Toxic Substances Control Act. (DEW)

Interceptor sewers Large sewer lines that, in a combined system, control the flow of the sewage to the treatment plant. In a storm, they allow some of the sewage to flow directly into a receiving stream, thus preventing an overload by a sudden surge of water into the sewers. They are also used in separate systems to collect the flows from main and trunk sewers and carry them to treatment plants. (FSL)

Interconnection A physical connection between two potable water-supply systems. *See also* Backflow; cross-connection. (FSL)

Interference phenomenon Temporary resistance by humans to infections by viruses, induced by an existing virus infection and attributable in part to the protein interferon. *See also* Interferon. (FSL)

Interferon Low-molecular-weight protein produced by cells infected with viruses. Interferon has the property of blocking viral infection of healthy cells and suppressing viral multiplication in cells already infected; interferon is active against a wide range of viruses. (FSL)

Intergrade A soil that possesses distinct characteristics of two or more great soil groups. (SWT)

Interim status Under the Resource Conservation and Recovery Act, the status of facilities that were in existence in 1980 that treat, store, or dispose of hazardous waste, and that are allowed to continue operating upon notification and filing of a Part A application and compliance with applicable regulations. 42 USC §6925(e), 40 CFR Parts 265, 270. (DEW)

Interior structural firefighting The physical activity involved with fire suppression, rescue, or both inside buildings or enclosed structures in a fire situation beyond the incipient stage. *See also* Incipient-stage fire. (SAN)

Interlocks Devices used in industrial settings to ensure that operations or motions occur in proper sequence. An interlocked mechanical power-press barrier guard, for example, is a barrier attached to the press frame and interlocked so that the press stroke cannot be started normally unless the guard itself, or its hinged or movable sections, enclose the point of operation. Interlocks are also used on electrical systems and lasers. (SAN)

Interlocutory decision Decision that is made during the course of a suit and that decides specific matters but not the final disposition of the controversy. (DEW)

Internal combustion engine An engine powered by the combustion of fuel in an enclosed space, the power from which is used to produce motion. The high temperatures involved create the production of pollutants such as sulfur oxides, oxides of nitrogen, carbon dioxide, and carbon monoxide. (FSL)

Internal conversion A mode of radioactive decay in which the gamma rays from excited nuclei cause the ejection of orbital electrons from an atom. The ratio of the number of internal-conversion electrons to the number of gamma quanta emitted in the deexcitation of the nucleus is called the conversion ratio. (RMB)

Internal radiation Nuclear radiation (alpha and beta particles and gamma radiation) resulting from radioactive substances in the body. Important sources are iodine-131 in the thyroid gland, and strontium-90 and plutonium-239 in bone. (RMB)

International Agency for Research on Cancer (IARC) An agency of the United Nations established by the World Health Organization. It evaluates the carcinogenicity of chemicals and publishes reports classifying chemicals according to their potential as human carcinogens. (FSL)

International Atomic Energy Agency (IAEA) A United Nations organization headquartered in Vienna, Austria, that is concerned with the scientific uses of radioisotopes and other agents of atomic energy. (RMB)

International Bank for Reconstruction and Development (IBRD) Also referred to as the World Bank, this organization was founded following the 1945 meeting of the United Nations Monetary and Financial Conference at Bretton Woods, New Hampshire. The aim of the bank is to stabilize currencies and bring about monetary reforms in member countries. (FSL)

Interrogatory Refers to written questions that are answered in writing and under oath for purposes of discovery. (DEW)

Interstate carrier water supply Federally regulated sources of water for drinking and sanitary use on planes, buses, trains, and ships operating in more than one state. (FSL)

Interstitial monitoring The continuous surveillance of the space between the walls of an underground storage tank. (FSL)

Interval estimation The specification of a range within which a parameter's value is likely to fall. (VWP)

Interval scale data Cardinal data with zero placed on the scale arbitrarily. (VWP)

Intrinsically safe Refers to a condition of equipment and associated wiring in which any spark or thermal effect, produced either normally or in specified fault conditions, is incapable, under certain prescribed conditions, of causing ignition of a mixture of flammable or combustible material in its most easily ignitable concentration. (SAN)

Intrinsic factors Environmental factors that affect the growth and survival of microorganisms may be considered either intrinsic or extrinsic. Intrinsic factors relate to characteristics of the menstruum of growth and can be manipulated to control growth rates or survival. These include availability of specific nutrients, pH, water activity, competition, inhibitors, or oxidation-reduction potential. *See also* Extrinsic factors. (HMB)

Invasiveness The ability of a microorganism to enter the body and to spread throughout tissue. Such dissemination of microorganisms may or may not result in infection or disease. (FSL)

Inversion An atmospheric condition in which a layer of warm air prevents the rise of cooler air and traps such air beneath it. This prevents the rise of pollutants that might otherwise be dispersed and can cause an air pollution episode. *See also* Donora. (FSL)

Investigations Studies conducted to identify causal factors associated with injuries, illnesses, or exposures to harmful agents. (FSL)

In vitro Refers to an experiment or procedure that is observable within a test tube, other laboratory equipment, or an artificial environment. (FSL)

In vivo Occurring within the living body of a plant or animal. In-vivo tests are those laboratory experiments carried out on live animals or human volunteers. (FSL)

Iodine A chemical element with atomic number 53 and atomic weight 126.904 used as a disinfectant in food-service operations and water treatment, and as a general-purpose germicide. Iodine is toxic by ingestion and inhalation and is a strong irritant to eyes and skin; TLV is 0.1 ppm in air. *See also* Disinfection. (FSL)

Iodine compound A type of halogen sanitizer. Elemental iodine is only slightly water-soluble. Mixed with certain nonionic wetting agents, it is readily soluble and more effectively used. The types of iodine compounds include iodine in aqueous solution, alcohol–iodine solutions, and iodophors. For bacterial action the active ingredients are either elemental iodine or hypoiodous acid. (HMB)

Iodophor Term applied to an iodine compound—i.e., elemental iodine, in combination with nonionic surface active agents—that is a commercially available sanitizer. Such water-soluble complexes are the most common form of iodine in use and have the greatest bactericidal properties when used under acidic conditions. Iodophors may be available as detergent-sanitizer preparations as well. The concentration of free available iodine determines the bactericidal activity of the iodophor. In use, iodophors usually have a low pH, which enhances their bactericidal quality. Typically a sanitizing solution of iodophor is used at a concentration of 25 ppm. *See also* Chemical sanitizer; iodine compound; sanitizer. (HMB)

Ion An atom or chemical radical (group of chemically combined atoms) bearing a positive or negative electrical charge caused by a deficiency or excess of electrons. (FSL)

Ion chromatography A laboratory analytical technique utilizing liquid chromatography and ion exchange. In the process ions of the same charge are exchanged, usually between an aqueous solution and an immobilized phase with many exchange sites. Different substances move through the phase at different rates, permitting separation and detection. *See also* Chromatography. (DEJ)

Ion exchange A chemical process involving the reversible interchange of ions between a solution and a particular solid material, such as an ion-exchange resin consisting of a matrix of insoluble material interspersed with fixed ions of opposite charge. The ion-exchange process is used in many of the water-softening devices on the market. *See also* Water softening; zeolite. (FSL)

Ionization The separation of a normally electrically neutral atom or molecule into electrically charged components. The term is also employed to describe the degree or extent to which this separation occurs. Ionization is the removal of an electron (negative charge) from the atom or molecule, either directly or indirectly, leaving a positively charged ion. The separated electron and ion are referred to as an ion pair. *See also* Ionizing radiation. (RMB)

Ionization chamber A device consisting of two oppositely charged plates used to measure radioactivity. (RMB)

Ionization density Number of ion pairs per unit volume. (RMB)

Ionization path (track) The trail of ion pairs produced by an ionizing particle in its passage through matter. (RMB)

Ionization potential The potential necessary to separate one electron from an atom, resulting in the formation of an ion pair. (RMB)

Ionizing event An event in which an ion pair is produced. (RMB)

Ionizing radiation Electromagnetic radiation (X-ray or gamma-ray photons or quanta) or corpuscular radiation (alpha particles, beta particles, electrons, positrons, protons, neutrons, or heavy particles) capable of producing ions by direct or secondary processes. Any electromagnetic or particulate radiation capable of producing ions, directly or indirectly, in its passage through matter. (RMB)

Ion pair Two particles of opposite charge, usually the electron and positive atomic or molecular residue resulting after the interaction of radiation with the orbital electrons of an atom. (RMB)

Iron (Fe) An element that has an atomic weight of 55.847 and an atomic number of 26. Iron is very reactive, a strong reducing agent, and oxidizes readily in moist air. A major source of exposure to iron (particularly to iron oxide fumes) of workers (TLV, 5 mg/m^3) can occur during welding operations. Iron dust and fine particles suspended in air are flammable and an explosion risk. Iron is a constituent of hemoglobin and necessary for plant and animal life. (FSL)

Iron pan A soil horizon that has become indurated primarily with iron as a cementing agent. (SWT)

Iron pyrite *See* Pyrite. (FSL)

Irradiation Exposure to radiation. (RMB)

Irrigation The application of water to crops to promote growth. (SWT)

Irritability The capacity of organisms to respond to stimuli. (FSL)

Irritation A reaction of tissues to an injury that results in an inflammation; the response or reaction by tissues to the application of a stimulus. (FSL)

Ischemia A condition in which there is an insufficient amount of blood to a part of the body, due to a functional constriction or blockage of a blood vessel. An extended lack of blood flow will result in tissue damage to the affected part of the body. (MAG)

Isobars Elements having the same mass number but different atomic numbers. (RMB)

Isodose chart A chart showing the distribution of radiation in a medium by means of lines or surfaces drawn through points receiving equal doses. Isodose charts have been determined for beams of X-rays traversing the body, for radium applicators used for intracavity therapy, and for working areas in which X-rays of radioactive isotopes are employed. (RMB)

Isodose curve A curve depicting points of identical radiation dosage in an area or medium. (RMB)

Isodose lines An imprecise term applied to imaginary lines on the ground (or on a water surface) or lines drawn on a map that join points in a radioactive field at which the total accumulated radiation dosage is the same; sometimes used erroneously in place of the term "isointensity lines." (RMB)

Isointensity lines Imaginary lines on the ground (or on a water surface) or lines drawn on a map that join points in a radioactive field that have the same radiation intensity at a given time. The lines are constructed on a map by adjusting to a common time readings of radiation intensity that have been taken at various points and different times. (RMB)

Isokinetic sampling An air-sampling technique used to measure particulates and other contaminants in exhaust stacks. The flow rate of the sampled gas stream entering the inlet nozzle of the sampling probe is made equal to the velocity of the stack effluent at the specific point in the stack cross-section at which the sample effluent is collected, in order to minimize turbulence and particle momentum effects. When the velocity of stack gas is higher than that at the sampling probe, fewer small particles and more large particles are collected, resulting in a nonrepresentative sample. Similarly, when the velocity of the stack gas is lower than that at the sampling probe, more small particles and fewer large particles are captured. (DEJ)

Isolation **(1)** (Physical) The separation of a permit space from unwanted forms of energy that could be a serious hazard to entrants. Isolation is usually accomplished by such means as blanking or blinding; removal or misalignment of pipe sections or spool pieces; use of double block and bleed; and use of lock and/or tagout. **(2)** (Biological) The physical separation, for the period of communicability, of infected persons or animals from those that are not infected, in such places and under such conditions as will prevent the direct or indirect transmission of the infectious agent from those infected to those who may be susceptible or who may spread the agent to others. Many categories of isolation, depending upon the disease and its communicability, have been established. (SAN, FSL)

Isomers **(1)** (Chemical) Chemical compounds with the same chemical formula but with different properties, due to the different arrangement of atoms within the molecule. **(2)** (Radiation) Nuclides that have the same number of neutrons and protons but that are capable of existing, for a measurable time, in different quantum states with different energies and radioactive properties. Commonly, the isomer of higher energy decays to the one with lower energy by the process of isomeric transition. *See also* Isomeric transition. (RMB)

Isomeric transition The process by which a nuclide decays to an isomeric nuclide (i.e., one of the same mass number and atomic number) of lower quantum energy. Isomeric transitions, often abbreviated I.T., proceed by gamma-ray and/or internal-conversion-electron emission. (RMB)

Isometric work Static effort in which little if any movement is accomplished, but energy is expended. The pressing of the hands together is an example of an isometric exercise. (DEJ)

Isomorphous substitution Replacement of one atom in a crystal structure by another of similar size but not necessarily the same valence, without disrupting or seriously changing the structure. (SWT)

Isopleth A line connecting points on a graph or diagram that have equal or corresponding values. Isopleths are used to display gradients or plume dispersal patterns for chemicals or in the development of noise maps. (DEJ)

Isotone One of several different nuclides having the same number of neutrons in their nuclei. (RMB)

Isotope dilution analysis A method of chemical analysis for a component of a mixture based on the addition to the mixture of a known amount of labeled component of known specific activity, followed by isolation of a quantity of the component and measurement of its specific activity. (RMB)

Isotope effect The effect of the difference in the mass between isotopes of an element on the rate and/or equilibria of chemical transformations. (RMB)

Isotopes Forms of the same element having nearly identical chemical properties but differing in their atomic masses (due to different numbers of neutrons in their respective nuclei) and in nuclear properties such as radioactivity and fission. For example, hydrogen has three isotopes, with masses of 1, 2, and 3 atomic mass units. ^{2}H and ^{3}H are commonly called deuterium and tritium, respectively. The first two of the three isotopes are stable (nonradioactive), but the third (tritium) is radioactive. Other examples are the common isotopes of uranium with masses of 235 and 238 units, which are both radioactive, emitting alpha particles, but have different half-lives. (RMB)

Isotope separation Process in which a mixture of isotopes of an element is separated into its component isotopes or in which the abundance of isotopes in such a mixture is changed. (RMB)

Isotopic carrier *See* Carrier.

Itai-itai So-called ''ouch-ouch'' disease, caused by cadmium poisoning and leading to bone deterioration. The disease derives the name itai-itai from the Japanese because of several large outbreaks in that country associated with the consumption of contaminated shellfish. *See also* Cadmium. (FSL)

Iteroparous Refers to species in which individuals may reproduce repeatedly in their lifetime. Examples are perennial plants, humans, bears, deer, and horses. (SMC)

J

Jenner, Edward (1749–1823) English physician who demonstrated in 1796 that infection with cowpox could prevent the disease smallpox. His observations were published in *An Inquiry into the Causes and Effects of the Variolae Vaccine, a Disease Discovered in Some of the Western Counties of England, Particularly Gloucestershire, and Known by the Name of Cowpox.* (FSL)

Jetted well A type of well constructed by forcing water down a riser pipe and releasing water from a special washing point. The well point and pipe are then lowered as material is loosened by the jetting action. An outside protective casing may be installed to the depth necessary to provide protection against the possible entry of contaminated surface water. (FSL)

Job health and safety analysis (JHSA) A technique used to identify potential sources of traumatic injury and environmental exposures in the workplace. Jobs that are determined to be serious risks to health and safety are the first targets for analysis. The process involves: **(1)** determining the job to be analyzed; **(2)** reducing the job to a sequence of steps; **(3)** determining key factors related to each job step; and **(4)** performing an efficiency check. A job analysis is best prepared by actual observations of a worker or workers doing the job. Also called job safety analysis (JSA). (SAN)

Job safety analysis (JSA) *See* Job health and safety analysis. (SAN)

Joint A crack in rocks, along which little or no movement has taken place. (FSL)

Joint and several liability A situation in which multiple parties are liable to a plaintiff and one or more of them is sued for the entire amount of compensation. The burden of demonstrating the divisibility of the injury is on the defendants. (DEW)

Joint Commission on the Accreditation of Hospitals (JCAH) A private, nonprofit organization composed of representatives from the American College of Physicians, the American College of Surgeons, and the American Hospital Association. The primary purpose of the JCAH is to promote high standards of medical care. (FSL)

Joule The amount of heat energy provided by one watt flowing for one second. (RMB)

Judicial notice Recognition by a court of the existence and truth of certain facts without their being proven by evidence, through verification of them by accepted sources or by general acceptance of their truth. (DEW)

Jurisdiction Generally, the authority of the court to hear and decide a case. Refers to the court's authority over the subject matter of dispute and the persons or parties to the suit, and its ability to render the type of judgment required. (DEW)

Juvenile hormones Chemicals that control the larval metamorphosis of insects, named because they induce the retention of insects' juvenile characteristics and prevent maturation. In the male wild silk moth they are produced by a small gland called the corpora allata. (FSL)

Juvenile water Water from magmatic (derived from liquid of molten rock) aquifers. (FSL)

K

K Symbol for the population size or density of a population at equilibrium. Also known as carrying capacity. (SMC)

Kala azar Disease caused by an infection with the protozoan *Leishmania donovoni*. The vectors for this disease are sandflies of the genus *Phlebotomus*. (FSL)

Kaolin A fine white clay also known as China clay and used in the manufacture of porcelain, as a filler in textiles, paper, and cosmetics, and in medications. It is composed mainly of kaolinite (50% alumina, and a high proportion of silica, plus impurities and water). (SWT)

Kaposi's sarcoma An opportunistic neoplasm associated with acquired immunodeficiency syndrome (AIDS). *See also* Acquired immunodeficiency syndrome; human immunodeficiency virus. (FSL)

K-electron capture The process wherein an electron in the K shell of an atom is captured by the nucleus during a nuclear reaction. In the process, a characteristic X-ray is emitted. (RMB)

Kepone ($C_{10}H_{10}O$) An insecticide and fungicide also known as chlordecone and used for control of fire ants. It can cause excitability, tremors, skin rash, opsoclonus (rapid irregular nonrhythmic movements of the eyes due to disorders of the brain stem or cerebellum), ataxia, hallucinations, irritability, cerebral edema, anovulation, spermatogenic arrest, low sperm motility, cancer of the liver, teratogenesis, nerve and muscle damage, and weight loss in humans, and can also cause testicular atrophy in animals. The "kepone incident" refers to an incident at a company that produced the pesticide in a Hopewell, Virginia, renovated garage where dust was poorly controlled. The incident resulted in the severe poisoning of many workers, pollution of the James River, contamination of the neighborhood, and the ultimate increased regulation of pesticides and working conditions. *See also* Fungicide; insecticide; pesticide. (DEJ)

Kerosene An oily liquid with a strong odor, derived from the distillation of petroleum. This chemical, which is toxic by inhalation, is used as jet engine fuel, rocket fuel, tractor fuel, and solvent. Kerosene is a moderate fire risk; its explosive limits in air range from 0.7% to 5%. (FSL)

Ketone A class of liquid organic compounds that are derived by the oxidation of secondary alcohols. They are used as solvents in paints and explosives. Acetone, or dimethyl ketone, is the least complex ketone. (FSL)

keV An abbreviation for kilo-electron-volt. The symbol for 1000 electron volts. (RMB)

Kilo Prefix indication for 1000, as in kilogram, kilocurie, or kilometer. (RMB)

Kilovolt The unit of electrical potential equal to 1000 volts. Abbreviation: kV. (RMB)

Kilovolts peak The crest value of the potential wave in kilovolts. When only one-half the wave is used, the crest value is to be measured on that half of the wave. Abbreviation: kV_p. (RMB)

Kinesiology The study of the movement of the human muscular system, often employed in ergonomics to prevent musculo-skeletal disorders among workers performing highly repetitive tasks. *See also* Ergonomics. (DEJ)

Kinetic energy The energy that a body possesses by virtue of its mass and velocity, or the energy of motion. The equation for kinetic energy is $K.E. = \frac{1}{2}mv^2$, where m is mass and v is velocity. *See also* Energy. (RMB)

Kinetic energy released in material (KERMA) A term for the quantity that represents the kinetic energy transferred to charged particles by the uncharged particles per unit mass of the irradiated medium. This is the same as ''first collision dose,'' a term used in neutron dosimetry. (RMB)

Kinetic rate coefficient A number that describes the rate at which a water constituent, such as biochemical oxygen or dissolved oxygen, increases or decreases. (FSL)

Kingdom The largest, most inclusive group in taxonomic classification. There are five kingdoms in the classification scheme proposed by Whittaker and Margulis in 1978. They are Animalia, Plantae, Fungi, Protista, and Monera. (SMC)

Koch's postulates Often called Koch's law, which states that in order for a specific microorganism to be considered responsible as a cause of a disease, the following conditions must be met: **(1)** the organism must be present in all cases of the disease; **(2)** the organism must be cultivated and isolated in pure culture; **(3)** inoculum from the pure culture must produce the disease in animals; **(4)** the organism must be observed and must be cultivated from experimentally infected animals. (FSL)

Konimeter A device used to measure airborne dust concentrations through a particle-count analytical procedure. The device was used in South Africa decades ago, but is no longer in widespread use. (DEJ)

Krebs, Sir Hans Adolph (1900–1981) The German-born biochemist who won the Nobel Prize for medicine and pharmacology in 1953 for his discovery of the citric acid cycle in carbohydrate metabolism. (FSL)

Kurtores (biostatistics) The degree of peakedness of a distribution, usually taken relative to the normal distribution. (VWP)

Kuru Disease in humans that is caused by a virus that affects the central nervous system and that is transmissible to subhuman primates. (FSL)

Kwashiokor Nutritional disease caused by protein deprivation and highly prevalent in developing countries. (FSL)

Kwell Commercial name for a shampoo that contains the gamma isomer of benzene hexachloride (lindane), which is used for delousing. *See also* Infestation. (FSL)

Kyphosis A posture of the lumbar spine, caused by bending forward over a workbench or poorly positioned computer terminal and involving reverse curvature of the spine. If prolonged, muscle fatigue and other more serious adverse health effects develop. The condition can often be corrected by use of a properly positioned backrest. (DEJ)

L

Labeled compound A chemical compound consisting, in part, of labeled molecules that have been tagged with radioactive isotopes. By observations of radioactivity or isotopic composition, it is possible to follow such a compound or its fragments through physical, chemical, or biological processes. *See also* Labeled molecule. (RMB)

Labeled molecule A molecule containing one or more atoms distinguished by nonnatural isotopic composition (with radioactive or stable isotopes). *See also* Labeled compound. (RMB)

Labile Refers to substance that is available to plants or readily undergoes transformation. (SWT)

Labile pool The total amount of an element in the soil solution and of that element solubilized or exchanged. (SWT)

Laboratory Any research, analytical, or clinical facility that performs health-care-related analysis or service. This includes medical, pathological, pharmaceutical, and other research, commercial, or industrial laboratories. (FSL)

Laboratory scale Refers to work with substances in which the container used for reactions, transfers, and other handling of substances is designed to be easily and safely manipulated by one person, as in a research project. Laboratory or small-scale procedures are in contrast to those of facilities or laboratories used to produce commercial quantities of materials. (FSL)

Lactogenic hormone Hormone that is produced by the pituitary gland of vertebrates, and that stimulates the secretion of progesterone from the corpus luteum and initiates milk production in female mammals. (FSL)

Lacustrine deposit Material that has been deposited in lake water and exposed by the rising of the land or lowering of the lake. (SWT)

Ladder cage A safety device used to reduce the possibility of accidental falls. The cage is fastened to the side rails of a fixed ladder greater than 20 feet in length or to the structure against which the ladder is placed. The cage must also enclose the climbing space of the ladder. (SAN)

Ladder pitch The angle between the horizontal plane and a ladder as measured on the side of the ladder opposite to the climbing side. (SAN)

Lagoon (1) A shallow pond where sunlight, microbial action, and oxygen decompose wastewater. (2) A place in which wastewaters or spent nuclear fuel rods are stored. (3) A shallow body of water, often separated from the sea by coral reefs or sandbars. *See also* Oxidation pond. (FSL)

Laminate To form or press into a thin sheet or layer, as may be done, for example, with several layers of plastic that are bound together by the partial melting of one or more of the layers, by an adhesive, or by impregnation. *See also* Delaminate. (FSL)

Land application Discharge of wastewater onto the ground for treatment or reuse. *See also* Irrigation. (FSL)

Land classification The arrangement of land units into categories based on the properties of the land or its suitability for a specific purpose—for example, as agriculture, industrial, or recreational uses. (SWT)

Land disposal Under the Resource Conservation and Recovery Act, includes, but is not limited to, any placement of specified hazardous waste in a landfill, surface impoundment, waste pile, injection well, land treatment facility, salt dome formation, salt bed formation, or underground mine or cave. (DEW)

Land farming A disposal process in which hazardous waste deposited on or in the soil is naturally degraded by microbes. This practice is now discouraged because of potential environmental impacts. (FSL)

Landfills (1) Land disposal sites for nonhazardous wastes at which the waste is spread in layers and compacted to the smallest practical volume, and to which cover material is applied at the end of each operating day. (2) Disposal sites for hazardous waste. They are limited to secure chemical landfills and are selected and designed to minimize the chance of release of hazardous substances to the environment. (FSL)

Landslide Movement of a mass of material downhill by gravity, often assisted by water in a material that has become saturated. (SWT)

Lanyard As a part of fall protection equipment, a flexible line of rope, wire rope, or strap used to secure a body belt or body harness to a deceleration device, lifeline, or substantial object. OSHA requires that all lanyards be capable of withstanding a tensile loading of 4000 pounds without cracking, breaking, or taking a permanent deformation. OSHA also requires that safety belt lanyards be a minimum of $\frac{1}{2}$-inch nylon or the equivalent, with a maximum length to provide for a fall of no greater than 6 feet. Lanyards should be used only for guarding workers. *See also* Fall protection; lifeline; safety belts. (SAN)

Larvicide An insecticide applied to destroy immature stages (larvae) of arthropods. *See also* Arthropods; insecticide. (FSL)

Laser Acronym for *light amplification by stimulated emission of radiation*. In one type, usually a continuous wave laser that has a continuous output, laser action takes place in a gas medium. Lasers are used in numerous applications, including surveying, construction, and surgery on humans and animals. The eyes are by far the most susceptible organ to damage by lasers and need to be protected appropriately. (FSL)

Latency period The time interval between exposure to toxic chemical agents and the onset of signs and symptoms of illness. (FSL)

Latent heat of fusion The amount of heat required to convert a unit mass of solid to liquid at the melting point. (FSL)

Latent heat of vaporization The amount of heat required to convert a unit mass of substance from a liquid to a gas at a certain temperature. (FSL)

Lateral exhaust hood A slot hood typically used to exhaust air contaminants from an open surface tank, and requiring full access to the top of the tank. The slots are narrow rectangular openings, usually located in a plenum at the rear of the tank opening. Also known simply as slot hoods. *See also* Biological safety cabinet; chemical hood. (DEJ)

Lateral sewers Pipes that run under city streets and receive the sewage from homes and businesses. *See also* Sewage; sewerage. (FSL)

Latex An aqueous suspension of a thick, white, milky-looking hydrocarbon polymer that occurs naturally in some species of plants or is made synthetically. Natural latex is used in the manufacture of surgical gloves and other medical equipment and as an adhesive for coating various products, such as tire cord. Synthetic latexes are used as binders in interior and exterior paints. (SMC)

Lattice structure The arrangement of atoms in a crystal structure. (SWT)

Lavoisier, Antoine Laurent (1743–1794) French chemist who applied the principles of chemistry to the respiratory process and described its functioning. He served on a committee that sought to establish uniform weights and measures in France, which led to the establishment of the metric system. (FSL)

LC_{50}/lethal concentration Median lethal concentration, a standard measure of acute toxicity. That concentration of a chemical that, when in the environment of test organisms, is estimated to kill 50% of the organisms in a specific time of observation and under the stated conditions of the test. *See also* LD_{50} (LD_{50} dose). (FSL)

LD_{50} (LD_{50} dose) Median lethal dose, a standard measure of acute toxicity. The dose of radiation or chemical, applied directly to experimental organisms, required to kill, within a specified period and under the stated conditions of the test, 50% of the population of the experimental group. (FSL)

LD LO The lowest concentration and dosage of a toxic substance that kills test organisms. (FSL)

Leachate A liquid that has passed through or emerged from solid waste and that contains soluble, suspended, or miscible materials removed from such wastes. (FSL)

Leachate collection system A system at a sanitary landfill for collecting the leachate that may percolate through wastes and enter the soil surrounding the landfill or the ground-water table. The collection system typically consists of a piped underdrain network and impermeable liners to prevent migration to the soil or ground water. *See also* Liner. (FSL)

Leaching The removal of soluble components of a rock, soil, or ore by the percolation of water. Leaching is a major process in the development of porosity in limestone. (FSL)

Leaching requirement The leaching fraction necessary to keep soil salinity, chloride, or sodium from exceeding the tolerance level for crops. (SWT)

Lead (Pb) An element with an atomic weight of 207.19, and an atomic number of 82. Lead is used in storage batteries, pottery, cable coverings, ammunition, solder, babbitt, and other bearing alloys. It is toxic to humans and other mammals and may accumulate in biological systems. Small doses may produce behavioral changes. Paralysis, blindness, and death may result from large doses. Lead is toxic by ingestion and by inhalation of fumes (TLV, fumes and dust for inorganic compounds, 0.15 mg/m^3 of air). The ambient-air EPA standard is 1.5 micrograms/m^3. FDA regulations require zero lead content in foods and less than 0.05% in house paints. In addition to the occupational concerns involving lead, there are also potential exposure hazards associated with individuals residing in close proximity to smelter operations and hazardous-waste disposal sites containing lead tailings and other wastes. Lead is also a particular concern involving children residing in deteriorating dwellings, particularly those constructed before 1960, which are apt to contain woodwork coated with multiple layers of lead-based paints. The evaluation of lead-exposed children involves the detection of blood lead levels. The following classification scheme and methodology has been developed.

Class	Blood lead concentration (micrograms/dL)	Comment
I	≤9	Child is not considered to be lead-poisoned.
IIA	10–14	The presence of many children (or a large proportion of children) with blood lead levels in this range should trigger community-wide childhood lead-poisoning prevention activities. A child in this range may need to be screened more frequently,
IIB	15–19	Child should receive nutritional and educational intervention and more frequent screening. If the blood level persists in this range, environmental investigation and intervention should be carried out.
III	20–44	Child should receive environmental evaluation and remediation and a medical evaluation. Such a child may need pharmacologic treatment for lead poisoning.
IV	45–69	Child will need both medical and environmental intervention, including chelation therapy.
V	≥70	Child requires emergency medical treatment. Medical and environmental management must begin immediately.

(FSL)

Leaded gasoline Gasoline to which either tetraethyl or tetramethyl lead has been added to raise the octane level, eliminate "pinging," and improve the efficiency of internal-combustion engines. Its uses for these purposes have been eliminated because of the potential for contamination of the air, soil, and water, and the known impact of lead on human health. (FSL)

Leaf The trailing digit in a stem-and-leaf diagram. (VWP)

Least-squares criterion A measure of how closely a curve approximates the pattern established by a set of data points. If there are n distinct data points (x_1,y_1), (x_2,y_2), . . . (x_n,y_n), and a curve, C, with points (x,y) is used to approximate the data for each data point, (x_i,y_i), where $i = 1, 2, \ldots, n$, there is a deviation or residual, $O_i = y - y_i$. Of all curves approximating a given set of data points, the curve having the property that $\sum_{i=1}^{n} O_i$ is a minimum is a best-fitting curve. (VWP)

Le Châtelier Mixture Rule A rule for determining the limits of flammability of mixtures.

$$L = \frac{100}{C_1/N_1 + C_2/N_2 + \ldots + C_n/N_n}$$

where L is the lower or upper limit of flammability of the mixture in air, C is the volumetric percentage of the combustible gas in the air-free and inert-gas-free mixture, and N is the respective lower or upper limit of flammability of the substance in air. *Also known* as Châtelier Mixture Rule. (DEJ)

Leeuwenhoek, Anton van (1632–1723) Dutch microscopist who was the first to observe protozoa and bacteria under the microscope. In 1674, he provided the first accurate description of red blood corpuscles. (FSL)

Leghemoglobin An iron-containing pigment that is similar to mammalian hemoglobin. It is produced in root nodules during the symbiotic association between the fungus *Rhizobium* and leguminous plants. (SWT)

Legionnaires' Disease The term used to describe a form of pneumonia that was first observed in Philadelphia, Pennsylvania, in the summer of 1976. The disease primarily affected persons attending an American Legion convention. About 5 months after the outbreak, investigators showed that the causative agent was a Gram-negative bacterium not previously shown to be a cause of disease in humans. Outbreaks of the disease have occurred in hospitals, hotels, and other buildings with or without central air conditioning. Soil, water from evaporative condensers, and cooling towers have been implicated as sources of the *Legionella pneumophilia* organism and cases of human illness. *See also* Pontiac fever. (FSL)

Lentic water Standing water, as in lakes, ponds, and marshes. (SMC)

Leptokurtic (statistics) A distribution having a relatively high peak. (VWP)

Lesion Any injury, wound, or local degeneration. (FSL)

Leucism The absence of pigment from fur or feathers; a type of albinism. (FSL)

Leukemia A disease characterized by an overproduction of leukocytes and their precursors, and enlargement of the spleen. The disease is variable, at times running a more chronic course in adults than in children. Although the causative agents are generally unknown, continued exposure to low intensities of ionizing radiation has been established as one possible cause. *See also* Leukocyte. (FSL)

Leukemogen Any substance or entity (e.g., benzene, ionizing radiation) considered to be a causal factor in the occurrence of leukemia. (FSL)

Leukocyte A white blood corpuscle in the blood, lymph, or tissues that plays a major role in the body's defense against disease. *See also* Lymphocyte. (FSL)

Leukoderma Loss of skin pigment, often caused by burns, injury to melanin cells, anti-oxidant chemicals, adhesives, cutting fluids, sanitizing agents, and rubber. (DEJ)

Levee An embankment, on a stream, creek, or river, specifically designed and constructed to prevent flooding. (FSL)

Level of concern (LOC) The concentration in air of an extremely hazardous substance above which there may be serious immediate health effects to anyone exposed to it for short periods of time. (FSL)

Level of confidence *See* Confidence coefficient. (VWP)

Level of significance (α) (statistics) *See* Probability of type-one error. (VWP)

Liable Legally bound or obligated to make good any loss or damage that occurs in a transaction; responsible. (FSL)

Liana Any luxuriantly growing woody tropical vine that roots in the ground and climbs around tree trunks. (SMC)

Lifeline A component of fall-protection equipment that consists of a flexible line for connection to an anchorage at one end, so as to hang vertically (vertical lifeline), or for connection to anchorages at both ends, so as to stretch horizontally (horizontal lifeline). It serves as a means for connecting other components of a personal fall arrest system to the anchorage. OSHA requires that lifelines be secured above the point of operation to an anchorage or structural member capable of supporting a minimum dead weight of 5400 pounds. Lifelines must be in dedicated service and can be used only for worker safeguarding. *See also* Fall protection; lanyard, safety belts. (SAN)

Life table A table describing the age-specific survival of cohorts of individuals in relation to their age. (VWP)

Lift In a sanitary landfill, a compacted layer of solid waste and the top layer of cover material. (FSL)

Lifting station *See* Pumping station. (FSL)

Lighting The luminous environment that allows humans to see. *See also* Candlepower; footcandle; footlambert; glare; illumination; lumen; luminance; reflectance; specular reflection; veiling reflection; visibility; visual angle; visual field; visual task. (DEJ)

Lignite A type of coal with a high moisture content and woody structure. It usually burns with little smoke and soot. (FSL)

Lime Calcium oxide (CaO), which is used extensively in agricultural operations to adjust the pH of soil and to provide calcium. Lime is also used in cement and mortar. (FSL)

Lime concretion Material that has formed an aggregate by precipitation and joined by calcium carbonate. (SWT)

Lime-pan A hardened soil layer cemented by calcium carbonate. (SWT)

Lime requirement The amount of lime needed to change soil to a specified pH. (SWT)

Limestone scrubbing Process in which sulfurous gases moving toward a smokestack are passed through a limestone and water solution to remove the sulfur compounds before they reach the atmosphere. (FSL)

Limiting factor A factor that limits or determines establishment, growth, reproduction, or population size. For example, in the case of desert plants, water is a limiting factor. A condition whose absence or excessive concentration is incompatible with the needs or tolerance of a species or population and that may have a negative influence on its ability to grow or survive. (FSL, SMC)

Limnology The study of the physical, chemical, meteorological, and biological aspects of fresh water. (FSL)

Lindane ($C_6H_6Cl_6$) A organochlorine hydrocarbon insecticide that was used in the United States for the control of insect soil pests and aphids. Also referred to as BHC (benzene hexachloride) or as the gamma isomer of BHC, this compound is no longer registered for use in the United States. Registration for products containing BHC has been canceled because of the potential for bioaccumulation and persistence in the soil. *See also* Organochlorine compound. (FSL)

Linear absorption coefficient A factor expressing the fraction of a beam of radiation absorbed in a unit thickness or material. In the expression $I = I_0e - \mu x$, I_0 is the initial intensity, I is the intensity after passage through a thickness of the material, μ (mu) is the linear absorption coefficient, and x is the distance traveled. (RMB)

Linear accelerator A device for accelerating particles and employing alternate electrodes and gaps that are arranged in a straight line and so proportioned that when their potentials are varied properly in amplitude and frequency, particles passing between them receive successive increments of energy. (RMB)

Linear amplifier A pulse amplifier in which the output pulse height is proportional to the input pulse height for a given pulse shape, up to a point at which the amplifier overloads. (RMB)

Linear correlation coefficient A number r, where $-1 \leq r \leq 1$, that is a measure of the strength of the linear relationship between two variables. It determines the precision with which predictions can be made using a regression line. *See also* Pearson's product moment. (VWP)

Linear energy transfer (LET) The linear rate of loss of energy locally absorbed by an ionizing particle traversing a material medium. (RMB)

Linear regression The estimation or prediction of the value of one variable from the given value of another variable, using a linear equation. (VWP)

Line breaking The intentional opening of a pipe, line, or duct that is or has been carrying flammable, corrosive, or toxic material, inert gas, or any fluid at a pressure or temperature capable of causing injury. *See also* Blinding. (SAN)

Liner A continuous layer of natural or synthetic materials, beneath or on the sides of a disposal site or disposal site cell, that restricts the downward or lateral escape of solid waste, solid waste constituents, or leachate. *See also* Leachate; leachate collection system. (FSL)

Lipase An enzyme that breaks fat into fatty acid and glycerol. (FSL)

Lipids A comprehensive term for fats and fat-derived materials that denotes substances extracted from animal or vegetable cells by nonpolar or "fat-soluble" solvents. Lipids are among the chief structural components of living cells and include fatty acids, glycerides, glycerol ethers, phospholipids, sphingolipids, alcohols, waxes, terpenes, steroids, and vitamins A, D, and E. (FSL)

Lipid solubility The maximum concentration of a chemical that will dissolve in fatty substances; lipid-soluble substances are insoluble in water. If a substance is lipid-soluble, it will very selectively disperse via living tissue. (FSL)

Lipolytic bacteria Bacteria that break down (hydrolyze) lipids. This hydrolysis results in the release of fatty acids and glycerol, and may cause rancidity. The bacteria proliferate in soil containing a high concentration of fats. They can become a problem in any food-processing operation involving fat-containing products—e.g., dairy products. Common lipolytic bacteria include *Pseudomonas* spp. and *Achromobacter* spp. Appropriate cleaners and sanitizers must be selected if these organisms are to be effectively removed. (HMB)

Liquefaction Changing of a solid into a liquid. (FSL)

Liquefied Petroleum Gas (LPG) A compressed or liquefied gas obtained as a by-product of petroleum refining or the manufacture of natural gasoline; so-called "bottled gas," which is composed of butane, propane, and pentane and that has been derived from crude oil. LPG is marketed in pressured containers and is widely used for heating and cooking purposes in recreational vehicles and by campers. The gas is highly flammable and a dangerous fire and explosion risk, and has a TLV of 1000 ppm in air. (FSL)

Liquid limit The minimum weight percentage of water that will cause soil to flow under standard treatment. (SWT)

Listed waste Waste listed as hazardous under RCRA but that has not been subjected to the toxic characteristics listing process because the dangers it presents are considered self-evident. (FSL)

Lister, Baron Joseph (1827–1912) English physician who is considered the founder of antiseptic surgery. His early work involved the use of carbolic acid (phenol) in the treatment of open wounds. (FSL)

Listeriosis An illness caused by the bacterium *Listeria monocytogenes*. The genus *Listeria* is classified among the "genera of uncertain affiliation" in Bergey's *Manual of Systematic Bacteriology*. *L. monocytogenes* is one of eight *Listeria* species not definitely identified and is a Gram-positive, non–spore-forming, non–acid-fast, pleomorphic rod-shaped bacterium. Although three of the eight are considered pathogenic, only *L. monocytogenes* is considered to be of major public health significance. *L. monocytogenes* is widely distributed in the environment, having been isolated from soil, vegetation, fecal material, sewage, and water. The primary habitat is probably soil and decaying vegetation, where the organism is a saprophyte.

The clinical manifestations of listeriosis vary widely and may be confused with those of other illnesses at times. Nine or more types of manifestations have been identified; however, human cases cannot always be classified in these categories, because combinations of two or more may occur simultaneously or in succession. The types of manifestations include: listeriosis during pregnancy; meningitis; meningoencephalitis and encephalitis; listeriosis of the newborn (granulomatosis infantiseptica); septicemia with pharyngitis and mononucleosis; cutaneous form; cervicoglandular form; oculoglandular form; granulomatosis septica and typhoid-pneumonic form; and possibly other forms. Pregnant women, the newborn, the elderly, and immunologically compromised individuals seem to be most susceptible to the disease.

Acquisition of human listeriosis through consumption of contaminated food was first suggested in 1926 following animal feeding studies. It wasn't until the early to mid-1980s that outbreaks of foodborne listeriosis associated with the consumption of contaminated coleslaw and Mexican-style cheese (in which deaths in

17 of 41 cases and 48 of 142 cases, respectively, occurred) that its place as a foodborne pathogen was firmly established.

Current reports indicate that *L. monocytogenes* is a very resistant organism. This lends even greater imperative to the effective application of sanitation measures in food processing and handling, especially with those products in which *Listeria* is thought to occur frequently in the unprocessed product. (HMB)

Liter In the metric system, a unit of measurement equivalent to 1.0567 quarts. (FSL)

Lithic contact A boundary between soil and underlying material that is too hard to be manipulated with a shovel or spade. *See also* Paralithic contact. (SWT)

Lithium (Li) A light, silvery-white element with an atomic weight of 6.939 and an atomic number of 3. It is the lightest solid known; it is used in alloys and as a fuel in fusion reactors. (FSL)

Lithosequence A group of related soils that differ in properties as a result of differences in parent rock as a soil-forming factor. (SWT)

Litter **(1)** Freshly fallen leaves, twigs, etc., on the surface of the forest floor. **(2)** Discarded scraps of human trash. (SMC, SWT)

Live load The total static weight of workers, tools, parts, and supplies that lifting equipment is designated to support. (SAN)

Loam A soil textural class containing a moderate amount of sand, silt, and clay. (SWT)

Loamy Refers to an intermediate in soil texture between fine and coarse. (SWT)

Local emergency planning committee Required by the Emergency Planning and Community Right-to-Know Act, a local committee appointed by the state emergency response commission to prepare an emergency response plan for hazardous-substance releases. 42 USC §§11003. *See also* Emergency Planning and Community Right-to-Know Act; comprehensive emergency response plan; Emergency Response Commission. (DEW)

Local exhaust ventilation system Air-handling system designed to capture and remove process emissions before they can escape into the workplace or the environment. A local exhaust ventilation system, as distinguished from a dilution ventilation system, consists of a hood, conveying ductwork, an air-cleaning device, a fan, and an exhaust stack. (DEJ)

Lockout device A device that utilizes a positive means such as a lock, key, or combination type, to hold an energy-isolating device in a safe position and prevent the energizing of machines or equipment. Included are blank flanges and bolted slip blinds. The placement of a lockout device on an energy-isolating device, in accordance with an established procedure, ensures that the device and the equipment being controlled cannot be operated until the lockout device is removed. *See also* Energy-isolating device; lockout/tagout; tagout device. (SAN)

Lockout/tagout Procedure for the control of hazardous energy, including specific steps for shutting down, isolating, blocking, and securing machines or equipment to control hazardous energy. The process includes specification of procedural steps for the placement, removal, and transfer of lockout devices or tagout devices and the responsibility for them; and specification of requirements for testing a machine or equipment to determine and verify the effectiveness of lockout devices, tagout devices, and other energy control measures. Activities in which lockout/tagout procedures are required include maintenance, inspection, cleaning, adjusting, and servicing of equipment (electrical, mechanical, or vessels) that requires entrance into or close contact with the equipment for which the main power disconnect switch or valve, or both, controls the source of power. *See also* Lockout device; tagout device. (SAN)

Locus The position on a chromosome occupied by a particular gene or its allele. (FSL)

Loess Soil material transported and deposited by wind. (SWC)

Log phase A growth phase during which all bacterial cells are dividing at a maximum rate that is characteristic of the microorganism and influenced by the conditions of growth. During this phase the number of dividing cells far exceeds the number of dying cells. The rate of reproduction is affected by factors such as pH, temperature, and nutrients. Also known as the exponential growth phase. (HMB)

Log rank test A test used to compare incidence rates for an event between two exposure groups, where incidence varies over the period of follow-up time. This test is used to analyze data for which the time of an event is important, rather than simply the fact of its occurrence. (VWP)

Long, Crawford Williamson A Georgia physician (1815–1878) who administered an anesthetic (ether) to a patient prior to the removal of a neck tumor in 1842, in one of the first recorded uses of ether as an anesthetic in surgery. (FSL)

Longitudinal study An investigation of data collected from the same group over a period of time. (VWP)

Lordosis A convex, forward curvature of the lower lumbar region of the spine. (DEJ)

Loss control Set of procedures developed to control conditions that may be responsible for (monetary) losses to an insurance company. Loss-control personnel attempt to reduce the morbidity and mortality associated with occupational accidents. (SAN)

Lotic water Flowing water, as in streams and rivers. (SMC)

Loudness An observer's impression of a sound's amplitude. (DEJ)

Low-hazard permit space A confined space in which there is an extremely low likelihood that an immediately dangerous to life or health (IDLH) or engulfment hazard could be present, and in which all other potentially serious hazards have been controlled. *See also* Permit-required confined space. (SAN)

Lower confidence limit (LCL) The minimum value expected for a parameter, using a given level of confidence. (VWP)

Lower flammable limit (LFL) **(1)** Often referred to as the lower explosive limit (LEL), the lowest concentration of gas or vapor in the air that will propagate a flame if a spark or source of heat is present. **(2)** The concentration of a substance in air, usually expressed as a volume percent, below which combustion cannot be supported at ordinary room temperature because of insufficient fuel. Combustion of a flammable material in air can occur only in concentrations between the lower and upper flammable limits in the presence of sufficient oxygen. Each substance has distinct lower and upper flammable limits. The lower flammable limit of most flammable solvents is well above their health hazard levels (TLV), and thus is not intended to protect against adverse health effects. For example, the LFL of an explosive gas can be as low as 1% in air, or 10,000 ppm, which is well above most occupational health exposure limits. Also known as lower explosive limit, or LEL. (DEJ)

Lowest achievable emission rate Under the Clean Air Act, the rate of emissions that reflects **(1)** the most stringent emission limitation contained in the implementation plan of any state for such source, unless the owner or operator of the proposed source demonstrates that such limitations are not achievable, or **(2)** the most stringent emissions limitation achieved in practice, whichever is more stringent. Application of this status does not permit a proposed new or modified source to emit pollutants in excess of existing new source standards. (FSL)

Low-level radioactive waste (LLRW) Radioactive material that **(1)** is not high-level radioactive waste, spent nuclear fuel, or by-product material (e.g., uranium or thorium mill tailings), as defined in section 11e(2) of the Atomic Energy Act and **(2)** has been classified by the Nuclear Regulatory Commission (NRC) as LLRW consistent with existing law and in accordance with (1). In addition to the NRC, the Department of Energy and EPA share responsibility for managing this category of waste. *See also* High-level radioactive waste. (FSL)

Low-Level Radioactive Waste Policy Act Sets out the responsibility of each state to dispose of low-level radioactive waste either alone or with the cooperation of other states through the formation of regional compacts. 42 USC §§2021b–d. (DEW)

Lumen Used to represent total light output; the unit of luminous flux emitted through a unit solid angle from a uniform point source of one candela. (DEJ)

Luminance **(1)** A measure of the light emitted by a luminous body or reflected from a surface. **(2)** The relationship, also called contrast, between an object and its background. (DEJ)

Lung function test A test, usually employing a spirometer, that measures an individual's breathing capacity and, indirectly, ability to wear a respirator. Such a test may include measurement of inspiratory reserve volume, expiratory reserve volume, residual volume, and tidal volume. (DEJ)

LUST Leaking underground storage tank. (SMC)

Luxury uptake Nutrients absorbed by plants in excess of their need for growth. (SWT)

Lye *See* Sodium hydroxide. (FSL)

Lymph A clear yellowish fluid that is found in tissues throughout the body and is eventually added to the venous blood circulation. It contains varying numbers of white blood cells (lymphocytes) and plays a major role in the body's defense against disease. *See also* Lymphocytes. (FSL)

Lymphocyte A white blood cell found in lymphatic tissues (e.g., lymph nodes, spleen, thymus) that is immunologically important and that attacks invading pathogens. *See also* Leukocyte. (FSL)

Lyse Destruction of red blood cells, bacteria, and other structures by a specific lysin, in a process usually referred to in terms of the structure destroyed—e.g., hemolysis, bacteriolysis. (FSL)

Lysergic acid diethylamide ($C_{20}H_{25}N_3O$) (LSD) A colorless, tasteless, odorless hallucinogenic drug, known by the street name "acid," that is manufactured from lysergic acid, a substance extracted from the ergot plant. (FSL)

Lysol™ Brand name for a phenolic compound used as a disinfectant and antiseptic. (FSL)

Lysozyme An enzyme that occurs in mammalian body fluids and destroys bacteria by dissolving their cell walls. (FSL)

M

Machine grounding *See* Ground.

Machine guard A barrier that prevents entry of an operator's hand or fingers into the point of operation or that prevents the operator from being struck by flying chips and sparks. The guard also protects clothing, hair, or body parts from being entrained into moving equipment (such as ingoing nip points and rotating parts such as chains and belts) as a result of close proximity. *See also* Point-of-operation guard. (SAN)

Macromolecule A molecule of colloidal size, usually organic, composed of an aggregation of hundreds or thousands of atoms—e.g., proteins, nucleic acids, and polysaccharides. Cellulose is the most common example found in nature. (FSL)

Macronutrient A plant nutrient (such as nitrogen, phosphorus, or potassium) that has a concentration greater than 500 mg/kg of plant tissue. (SWT)

Macrophage A large amoeboid phagocytic cell that arises from monocytic cells in bone marrow. Macrophages, which are widely distributed throughout the body, are typically large and long-lived and play a role in immunity by either engulfing microorganisms, through the production of antibodies, or by presenting antigens to lymphocytes for destruction. *See also* Lymphocyte. (FSL)

Made land An area filled with earth or with mixed earth and trash. (SWT)

Magazine A building or structure approved for the storage of explosive materials. (FSL)

Magistrate Generally, a court official vested with limited authority to administer judicial action. (DEW)

Maintenance application Application of fertilizer in the amount needed to maintain soil nutrients for plant growth. (SWT)

Maleic hydrazide [HC:CHC(O)NHNHC(O)] A herbicide of the growth-regulator type that is used for the control of grasses, as a post-harvest sprouting inhibitor, and as a systemic for the treatment of tobacco plants. This chemical, which inhibits growth by preventing cell division, is toxic for humans by ingestion. (FSL)

Malignant tumor A tumor capable of metastasizing, or spreading cancerous cells from one part of the body to another. (FSL)

Man lift A device consisting of a power-driven endless belt moving in one direction only, with steps or platforms and handholds attached to it and designed for the transportation of personnel from floor to floor. Man lifts are intended for the conveyance of persons only. All man-lift designs must meet the requirements of ANSI A90.1-1969. OSHA specifies the minimum size for floor openings, landing surface construction, guards for entrances, brake requirements, belt composition and size, and the location of handholds. It is not permitted to run man lifts in excess of 80 feet per minute. (SAN)

Management of change OSHA process safety standard 1910.119 requires employers to establish and implement written procedures to manage changes in the processing of chemicals, in technology, in equipment, and in facilities. The procedures must assure that the following are addressed prior to any change: the technical basis for the proposed change; impact of the change on safety and health; modifications to operating procedures; necessary time period for the change; and authorization requirements for the change. Employees involved in the process must be informed of, and trained in, the change in the process as early as practicable prior to its implementation. (SAN)

Maneb [(SSCNCH$_2$CH$_2$NHCSS)Mn] A carbamate fungicide used on food crops and ornamentals for the control of blights and molds. Maneb is toxic by inhalation and ingestion and is an irritant to the eyes and mucous membranes. (FSL)

Mangan A cutan containing enough manganese to effervesce with the application of hydrogen peroxide. (SWT)

Manganese (Mn) A red-white, hard element that occurs as the ore pyroluate. An essential nutrient for plants, manganese has an atomic weight of 54.938 and an atomic number of 25. It is used in the manufacture of steel, to improve corrosion resistance and hardness, and in the manufacture of aluminum. The dust or powder of manganese is flammable and toxic (TLV, fume, 1 mg/m^3 of air). (FSL)

Manifest As pertaining to hazardous wastes, a form or document used for identifying the quantity, composition, origin, routing, and destination of hazardous waste during transportation from the point of generation, through any intermediate points, to the point of disposal, treatment, or storage. *See also* Generator. (FSL)

Mann-Whitney V test *See* Wilcoxon rank sum test. (VWP)

Manometer Instrument used for measuring the pressure of any fluid or the difference in pressure between two fluids, whether gas or liquid. (FSL)

Mantel-Haenszel test A test that evaluates the statistical significance of the relationship between disease and exposure. (VWP)

Manufacturer's formulation A list of substances, chemicals, or component parts, as described by the maker, of a coating, pesticide or other product. (FSL)

Manure Animal excreta, with or without a mixture of bedding, which may be fresh or at various stages of decomposition. The term may denote fertilizer in some instances. (SWT)

Maple bark disease An infection in humans that is due to the inhalation of spores from the *Cryptostroma* *corticale* mold, which grows under the bark of maple logs. Exposure primarily occurs in mills during the debarking of trees. (FSL)

Marburg disease A viral hemorrhagic disease first reported in Marburg, Germany, among laboratory workers exposed to African green monkeys. (FSL)

Marine sanitation device Equipment installed on board a vessel to receive, retain, treat, or discharge sewage. (FSL)

Marl Soft, unconsolidated calcium carbonate that is usually mixed with clay or other material. (SWT)

Marsh An area that is periodically wet or continually flooded, but whose surface is not deeply submerged. Such an area is usually covered with hydrophytic (water-loving) plants, and constitutes a wetland ecosystem in which grasses, sedges, cattails, or rushes form the dominant vegetation. Marshes typically have largely mineral soil rather than the highly organic peat characteristic of bogs and fens. *See also* Swamp; wetland. (SMC, SWT)

Marsh gas *See* Methane. (SMC)

Mass The fundamental measure of the quantity of matter. Mass is different from weight in that it does not depend upon gravitational force. (RMB)

Mass absorption coefficient The linear absorption coefficient in centimeters raised to the -1 power and divided by the density of the absorber in grams per cubic centimeter. It is frequently expressed as μ/p, where μ is the linear absorption coefficient and p is the absorber density. (RMB)

Mass number The number of nucleons (neutrons and protons) in the nucleus of an atom. Abbreviation: A. (RMB)

Mass spectrometer An electronic instrument used for the separation of electrically charged particles by mass. This instrument is used in qualitative analysis chemicals, particularly organic compounds. (RMB)

Mass unit A unit of mass based upon one-twelfth the weight of a carbon atom, taken as 12.00000. Abbreviation: mu, or, for atomic mass unit, amu. (RMB)

Material misrepresentation In insurance law, a misrepresentation that would influence a prudent insurer in deciding whether to accept a risk or in determining the amount of premiums. (DEW)

Material Safety Data Sheet (MSDS) A synopsis of critical information about the potential hazards of specific chemicals, as required by the OSHA Hazard Communication Standard (29 CFR 1910.1200). Material safety data sheets are required for those chemicals to which an employee might be exposed in the workplace. MSDSs must be kept on file and must be accessible to employees. *See also* Hazard communication standard. (FSL)

Materials handling The lifting, moving, and placing of equipment, parts, and goods used in the work setting. This can be performed manually or by the use of equipment such as jacks, hoists, cranes, and forklifts. Manual lifting, if done incorrectly, can cause strains and joint injuries. There are many inherent hazards in the use of equipment for moving materials. *See also* Safe lifting techniques. (SAN)

Materials recovery facility A solid-waste-handling facility that provides for the extraction from solid waste of recoverable materials, materials suitable for use as fuels or in soil amendment, or any combination of such materials. *See also* Recycling. (FSL)

Mature soil A soil with deep, well-formed horizons produced during the natural process of soil development. (SWT)

Maximal voluntary ventilation The volume of air breathed with maximum voluntary effort by an individual for a given period of time, usually 10–15 seconds, corrected to one minute. The test is often used to investigate complaints of dyspnea. Also known as maximum breathing capacity. (DEJ)

Maximum breathing capacity *See* Maximal voluntary ventilation. (DEJ)

Maximum contaminant level Under the Safe Drinking Water Act, the maximum permissible level of a contaminant in water that is delivered to any user of a public water system. 42 USC §300f(3). *See also* Primary drinking water regulation; Safe Drinking Water Act. (DEW)

Maximum contaminant level goal Under the Safe Drinking Water Act, the level of a contaminant in drinking water at which no known or anticipated adverse effects on the health of persons occur and that allows an adequate margin of safety. *See also* Maximum contamination level; primary drinking water regulation; Safe Drinking Water Act. (DEW)

Maximum growth temperature The elevated temperature at which microbial growth ceases. The range of temperature within which an organism can survive is determined largely by whether essential enzymes can function. At some temperature above the optimum, one or more essential enzymes may cease to function and the organism may no longer be able to reproduce. Knowledge of the maximum growth temperature is an extremely important factor used to ensure food safety. Cooking foods at temperatures beyond the maximum growth temperature or maintaining cooked foods at adequately high temperatures is one measure taken to ensure the safety of the products. *See also* Minimum growth temperature; optimum growth temperature. (HMB)

Maximum permissible concentration (MPC) The amount of radioactive material that can be tolerated in the environment or in the body without producing a significant injury. (RMB)

Maximum permissible dose (MPD) For occupational exposures to radioactive substances, $MPD = 5(N - 18)$ where N is the age in years. (RMB)

McNemar's test A statistical test used to compare binomial proportions in paired samples of data. (VWP)

Mean **(1)** (Population, μ) The sum of the product of the data values and their respective probabilities:

$$\mu = \sum x \cdot P(x)$$

where the summation is taken over all x values in the population. **(2)** (Sample ($\bar{x}$)) *See* Arithmetic mean. (VWP)

Meander A curve or twist in a surface stream or river. (FSL)

Mean free path The average distance a particle or photon travels between collisions. (RMB)

Mean life The reciprocal of the decay constant λ (lambda). Abbreviation: χ. (RMB)

Means of egress A continuous and unobstructed way of exit travel from any point in a building or structure to a public way. Egress involves the means of exit access, the exit, and the means of exit discharge. A means of egress comprises the vertical and horizontal ways of travel and includes the intervening room spaces, doorways, hallways, corridors, and passageways. *See also* Evacuation plan. (SAN)

Mechanical turbulence Random irregularities of fluid motion in air, caused by mechanical, nonthermal processes. (FSL)

Media Specific environments, such as air, water, and soil, that are subjects of regulatory concern and activities. (FSL)

Median The middle value or the arithmetic mean of the two middle values of a set of numbers arranged in order of magnitude. (VWP)

Median lethal dose (MLD or LD_{50}) **(1)** (Radiation) The amount of radiation, received over the whole body, that would be fatal to about 50% of human beings, animals, or other organisms. It is usually accepted that a dose of 400 to 450 roentgens received over the whole body in the course of a few minutes represents the MLD for human beings. **(2)** (Toxicology) That dose of a particular substance that, when administered to all test animals under specified test conditions, is lethal to 50% of the animal test population. (FSL, RMB)

Medical consultation A discussion between an employee and a physician for the purpose of determining which medical examination, procedures, or laboratory tests are appropriate in cases in which exposure to a chemical, biological, or physical agent may have taken place. *See also* Medical monitoring; medical surveillance. (FSL)

Medical Literature Analysis and Retrieval System (MEDLARS) A computerized information system at the National Library of Medicine in Bethesda, Maryland. The *Index Medicus* is produced from this system. The on-line computerized bibliographic segment of the database is known as MEDLINE. *See also* TOXLINE. (FSL)

Medical monitoring The examination and surveillance by medical personnel of a worker's exposure to a contaminant after exposure to a potentially harmful agent. *See also* Medical consultation; medical surveillance. (DEJ)

Medical surveillance The process by which individuals exposed to hazardous agents are monitored for adverse health effects, usually by an occupational physician. Medical surveillance programs usually involve a medical history, an assessment of exposure potential, a physical examination, blood and/or urine sampling, pulmonary function tests, radiographic examinations, a written medical opinion regarding the individual's ability to wear certain types of respirators, and other specific medical tests and opinions. Medical surveillance programs are required under a number of OSHA and EPA standards (e.g., those relating to industrial workers exposed to lead, formaldehyde, or benzene, and those relating to hazardous-waste and emergency response workers). *See also* Medical consultation. (DEJ)

Medical waste As defined by the Medical Waste Tracking Act for regulation (40 CFR part 259) under the EPA demonstration program, any solid waste generated in the diagnosis, treatment (i.e., provision of medical services), or immunization of human beings or animals, in research pertaining thereto, or in the production or testing of biologicals. Specific forms of medical wastes include the following. **(1)** *Cultures and stocks:* Includes cultures and stocks of infectious agents and associated biologicals, such as: cultures from medical and pathological laboratories; cultures and stocks of infectious agents from research and industrial laboratories; wastes from the production of biologicals; discarded live and attenuated vaccines; and culture dishes and devices used to transfer, inoculate, and mix cultures. **(2)** *Pathological wastes:* Human pathological wastes, including tissues, organs, body parts, and body fluids that are removed during surgery, autopsy, or other medical procedures, and their containers. **(3)** *Human blood and blood products:* Liquid waste, human blood, products of blood, items saturated and/or wet with human blood, or items that were saturated and/or wet with human blood that are now caked with dried human blood. Includes serum, plasma, and other blood components, and their containers, used or intended for use in either patient care, testing, and laboratory analysis or the development of pharmaceuticals. Intravenous bags are also included in this category. **(4)** *Sharps:* Sharps that have been used in animal or human patient care or treatment or in medical, research, or industrial laboratories, including hypodermic needles, syringes (with or without the attached needle), Pasteur pipettes, scalpel blades, blood vials, needles with attached tubing, and culture dishes (regardless of presence of infectious agents). Also included are other types of broken or unbroken glassware that were in contact with infectious agents, such as used slides and cover slips. **(5)** *Animal waste:* Contaminated animal carcasses, body parts, and bedding of animals that were known to have been exposed to infectious agents during research (including research in veterinary hospitals), production of biologicals, or testing of pharmaceuticals. **(6)** *Isolation wastes:* Biological wastes and discarded materials contaminated with blood, excretions, exudates, or secretions from humans or animals that have been isolated to protect others from certain highly communicable diseases. **(7)** *Unused sharps:* Unused, discarded sharps such as hypodermic needles, suture needles, syringes, and scalpel blades.

Medical waste does not include EPA-specified hazardous waste, household waste, ash from incinerated medical waste, residues from treatment and destruction processes, and human corpses intended for interment or cremation. Includes any solid waste that is generated in the diagnosis, treatment (provision of medical services), or immunization of human beings or animals in research pertaining thereto, or in the production or testing of biologicals. *See also* Destroyed regulated medical waste; Medical Waste Tracking Act. (DEW)

Medical Waste Tracking Act of 1988 Amendment to the Solid Waste Disposal Act, establishing a demonstration program for tracking medical waste. Applies to the states of Connecticut, New Jersey, New York, and Rhode Island, and to Puerto Rico. The act also defined medical waste for purposes of the demonstration program. It sets standards for pretreatment of medical waste, including segregation, packaging, storage, decontamination, labeling of containers, and marking of containers for transport. Also establishes standards for specified generators, transporters, and treatment, destruction, and disposal facilities. Does not apply to the transport of etiologic agents regulated by the U.S. Department of Transportation and U.S. Department of Health and Human Services. 42 USC §6992–6992k. *See also* Etiologic agent; medical waste. (DEW)

Medicophobia Fear of physicians, often called the ''white coat'' phobia. (FSL)

Megacurie One million curies. *See also* Curie. (RMB)

Ménière's disease A disease of the inner ear, affecting both hearing and balance, characterized by episodic dizziness, nausea, vomiting, sensory hearing loss that is often progressive, tinnitus, and a sensation of fullness in the involved ear. (DEJ)

Mercaptans (C_2H_5SH) Also known as thiols, these are organic sulfur-containing compounds that have strong, repulsive odors and that are produced in oil refinery units and some paper mills. For pollution control purposes, these chemicals can be removed with scrubber units containing caustic soda. The mercaptans are toxic by inhalation and aliphatic thiols are flammable. They are used as warning agents in fuel gas lines and as chemical intermediates. (FSL)

Mercury (Hg) A liquid metal, also known as quicksilver or hydrargyrum, that is highly toxic to mammals via either inhalation of fumes or vapors or ingestion. Early occupational exposures to mercurial compounds among workers in hat manufacturing plants led to significant neurologic dysfunctions among these individuals and the designation ''mad hatters'' for those receiving large doses. In more recent times, mercury-contaminated fish have been responsible for ''Minamata disease'' among persons consuming such fish from Minamata Bay in Japan. The atomic weight of this metal is 200.59, and the atomic number is 80. (FSL)

Merokurtic Refers to peakedness of the normal (bell-shaped-curve) distribution of data. (VWP)

Mesic A soil temperature regime with mean annual temperatures of at least 8°C but less than 15°C, with less than a 5°C difference between summer and winter temperature. (SWT)

Mesofauna Nematodes, oligochaete worms, small insect larvae, and microarthropods. (SWT)

Meson A short-lived atomic particle, charged or uncharged, that has a mass that is a multiple of the electron mass. (RMB)

Mesophiles A physiological group of bacteria identified by their temperature range of growth. These organisms prefer moderate temperatures with an optimum temperature generally between 30°C and 45°C, a minimum temperature ranging from 5°C to 10°C, and a maximum growth temperature of about 45°C. These organisms are representative of many from human and animal origins and include most pathogens and many food spoilage agents. *See also* Psychrophiles; thermophiles. (HMB)

Mesothelioma A rare neoplasm that grows as a thick sheet in the pleura of the lungs and in the peritoneum. This condition has been demonstrated in workers who have had extensive exposures to asbestos. *See also* Asbestos. (FSL)

Metabolic heat load Metabolic heat is a by-product of the chemical and physical reactions occurring within cells, tissues, and organs. The net heat exchange between an organism and the environment is expressed as:

$$H = M +/-R +/-C - E +/-D$$

where H is body heat storage load, M is metabolic heat gain, R is radiant or infrared heat load, C is convective heat load, E is evaporative heat loss, and D is conductive heat load. (DEJ)

Metabolism The set of biochemical transformations that a chemical undergoes in the body. The sum of all the physical and chemical processes, both synthetic (anabolic) and degradative (catabolic), by which living organized substance is produced and maintained and by which energy is made available for the uses of the organism. (FSL)

Metabolite Any product (foodstuff, intermediate, waste product) of metabolism. *See also* Metabolism. (FSL)

Metal and Non-Metallic Mine Safety Act of 1966 A precursor to the Mine Safety and Health Act that regulated occupational safety and health matters in all mines other than coal mines. This law was combined with the Coal Mine Safety and Health Act of 1969 to form the basis for the Mine Safety and Health Act of 1977. (DEJ)

Metaldehyde (CH_3CHO) A pesticide (molluscide) used to kill slugs and snails in a variety of environments. The material is flammable, a dangerous fire risk, and a strong irritant to skin and mucous membranes. *See also* Molluscide. (FSL)

Metal-fume fever An acute condition, usually of short duration, caused by the inhalation of finely divided fumes of zinc, magnesium, copper, or their oxides, and possibly others produced during hot metal work (e.g., welding). Symptoms can appear in 4 to 12 hours after exposure, and consist of fever and chills. Most cases are the result of inhalation of zinc oxide from the welding of galvanized steel. (DEJ)

Metalizing An industrial process involving the coating of parts with molten metal, usually aluminum, by means of vacuum deposition. The process often presents occupational health hazards from metal fumes, dust, heat, and nonionizing radiation. Metalized plastic films are used for packaging yarns and labels. (DEJ)

Metamorphic rock Rock that originates from either igneous and sedimentary rocks through alteration of heat and pressure at great depths. This type of rock includes schist, gneiss, quartzite, marble, and slate. Rock that has been altered from its original condition by the combined action of heat and pressure—e.g., marble produced from limestone. (FSL, SWT)

Metaplasm The nonliving components of protoplasm. (FSL)

Metastable state An excited state of a nucleus, which then returns over a measurable half-life, to its ground state by the emission of a gamma ray or an internal conversion electron. (RMB)

Metastasis The transfer of malignant neoplastic cells from the original or parent site to a more distant one, with the resultant appearance of a neoplasm. *See also* Neoplasm. (FSL)

Methane (CH_4) The simplest hydrocarbon; the product of anaerobic decomposition of organic materials that is also known as marsh gas in its natural form. Methane can be produced by the anaerobic decomposition of manures and other agricultural wastes and within lines carrying sewage. Methane can be used as a fuel and, when captured from digestors at sewage treatment plants, is frequently used for heating pump rooms and other outbuildings associated with the treatment plants. Methane is a severe fire and explosion hazard and can form an explosive mixture with air (5–15% by volume). *See also* Marsh gas. (SMC, FSL)

Methanol (CH_3OH) A colorless liquid, also known as methyl alcohol or wood alcohol, that is widely used as a solvent and fuel. It is flammable, a dangerous fire risk, and can cause blindness if ingested (TLV 200 ppm in air). Methanol is used in the manufacture of formaldehyde and acetic acid, as a fuel for utility plants, and as a home-heating-oil extender. (FSL)

Methemoglobin A compound that does not combine with oxygen and that is formed by the oxidation of iron from the ferrous state to the ferric state. The presence of methemoglobin in circulating blood can cause methemoglobinemia that results in cyanosis. Excessive levels of nitrates in drinking water supplies have been accountable for cases of "blue babies," or methemoglobinemia among infants. (FSL)

Methylene-blue reduction test A test that indirectly measures bacterial densities in milk in terms of the time interval required, after starting incubation, for a dye–milk mixture with a characteristic blue color to become white. The method depends upon the ability of bacteria in milk, when incubation is started, to grow and to utilize oxygen dissolved in the mixture, which in turn lowers the oxidation-reduction potential of the mixture. To demonstrate visibly the rate of oxygen utilization, a solution of methylene blue thiocyanate, an oxidation-reduction indicator that produces a blue color initially in the mixture, is added. The mixture is incubated at 36°C. In general, reduction time is inversely related to the bacterial content of the sample when the incubation starts. Advantages of the method include simplicity, speed, economy, and the fact that only viable cells actively reduce the dye. However, not all organisms reduce the dyes equally, and, therefore, the test is not applicable to foods containing reductive enzymes. The test can facilitate decisions regarding the bacteriological quality of a raw product, which in turn can improve the quality of the final product. *See also* Resazurin reduction test. (HMB)

Metric ton (MT) 1000 kilograms; equal to 2204.6 lbs. avoirdupois or 2679.23 lbs. troy. (FSL)

Meuse Valley Incident An air-pollution episode that occurred in the Meuse Valley, France, in 1930, when a combination of cold weather, winds, and a temperature inversion caused smog to persist for several days. Hundreds of people became ill and about 60 died, and many cattle had to be destroyed. (FSL)

MeV (mega-electron volt) A unit of energy commonly used in nuclear physics. It is equivalent to 1.6×10^6 ergs. Approximately 200 MeV of energy are produced for every nucleus that undergoes fission. *See also* Electron volt. (RMB)

Mica pneumoconiosis A disease of the lung, caused by excessive inhalation of mica dust, usually over a number of years. Mica is a group of minerals composed of complex silicates that crystallize in thin, flexible, easily separated layers. Such silicates are used in vacuum tubes, incandescent lamps, roofing, rubber, wallpaper and wallboard joint cement, and electrical equipment. Also known as biotite, lepidolite, muscovite, lepidomelane, paragonite, and phlogopite. *See also* Pneumoconiosis. (DEJ)

Microaerophilic Refers to organisms that are aerobic but that require reduced concentrations of oxygen (pressures lower than 0.2 atmosphere) and elevated levels of carbon dioxide in order to grow. (HMB)

Microbes Microscopic organisms, such as algae, viruses, rickettsia, bacteria, fungi, and protozoa, some of which cause disease in human, plants, and/or animals. *See also* Microorganism. (FSL)

Microbial growth inhibition Control of growth of microorganisms by various methods that can also be used to destroy the microorganisms. Heat treatment is the most widely utilized and probably most widely studied method. Thermal inactivation studies have determined the appropriate times and temperatures required to eliminate pathogens or spoilage organisms in a product. The time and temperature of heat treatment (e.g., pasteurization) appropriate is dependent upon the characteristics of the organism(s) and the nature of the product. Dehydration is effective in microbial inhibition, because all organisms require moisture for survival. The amount required is contingent on the specific organism and the nature of the medium. Refrigeration does little to destroy organisms. A repeated freeze–thaw cycle will rupture some cells, but others will survive. Inhibition by low-temperature storage is based on slowed growth of the microorganisms because of slowed metabolic reaction rates. Chemicals may be used as inhibitors, the most common example of which is salt. Generally, any substance or method used to inhibit microorganisms in a product is seldom used by itself. Combinations (e.g., pasteurization and cold storage) are usually employed. *See also* D value; Water activity A_w; Z value. (HMB)

Microbial load The number of microorganisms in a substance (e.g., a food item) or on a surface. The so-called microbial load is often a reflection of the sanitation effectiveness in a particular facility. The term may imply the total number of detectable microorganisms, measured by means of nonselective microbiological media, or specific types, measured by means of specific selective and differential growth medium. (HMB)

Microbial pesticide A microorganism, such as *Bacillus thuringiensis*, that is used to control a particular pest species. Such microorganisms are of low toxicity to people. *See also* Pesticide. (FSL)

Microbiological safety index The reciprocal of the logarithm of the number of surviving microorganisms after the treatment of materials by sterilization. (JHR)

Microbiology The science concerned with the study of microscopic and ultramicroscopic organisms; protistology. (FSL)

Microcurie One-millionth of a curie. *See also* Curie. (RMB)

Micron A one-millionth part of a meter; i.e., 10^{-6} meter or 10^{-4} centimeter. It is roughly four one-hundred-thousandth (4×10^{-5}) of an inch. Abbreviation: μm. (FSL)

Micronutrient A plant nutrient (such as boron, iron, or zinc) with a concentration of less than 100 mg/kg in the plant tissue. (SWT)

Microorganism Any microscopic or submicroscopic organism, especially any of the viruses, rickettsia, bacteria, or protozoa. (FSL)

Microphobia Fear of microorganisms. (FSL)

Microrelief Local, small-scale differences in topography. (SWT)

Microscopy The use of a microscope; investigation by means of a microscope. (FSL)

Microsecond One-millionth of a second. Abbreviation: μs. (RMB)

Microwave Nonionizing radiation that has a wavelength of 3 meters to 3 millimeters (a frequency of 100 to 100,000 hertz), often used in cooking, radar, communications, and diathermy. Microwaves can be sufficiently intense to cause heating of tissue, with the effects generally being related to wavelength, power intensity, and length of exposure. The longer wavelengths produce greater penetration and temperature increase in deeper tissues. Threshold limit values have been developed for various types of microwave radiation in the occupational environment. (DEJ)

Miliaria A skin rash or prickly heat, of occasional occupational origin, often caused by high temperature and humidity. The condition results in waterlogging of the keratin layer of the skin and occlusion of the sweat ducts and skin openings. (DEJ)

Millirem One-thousandth of a rem. *See also* Rem. (RMB)

Milliroentgen One-thousandth of a roentgen. Abbreviation: mR. *See also* Roentgen. (RMB)

Millisecond One-thousandth of a second. (RMB)

Mine dumps Areas that have been covered with overburden and other waste materials from mining operations. These areas usually have little or no vegetative cover. (SWT)

Mineral A naturally occurring inorganic solid with a fixed chemical composition, such as a compound of copper, iron, phosphorus, calcium, or iodine. (FSL)

Mineralization The conversion, by microbial decomposition, of an organic form to an inorganic form. (SWT)

Mineralogical analysis The determination of the types and amounts of minerals in soil or rock. (SWT)

Mineral ore *See* Ore.

Mineral soil A soil that consists predominantly of, and has its properties determined by, mineral matter. (SWT)

Mine wash Water-deposited accumulations of materials recently eroded from mining operations. (SWT)

Minimum erythemal dose The amount of energy (usually ultraviolet) expressed in microwatt-seconds per square centimeter of skin, to which skin can be safely exposed. *See also* Threshold limit value. (DEJ)

Minimum growth temperature The lowest temperature at which a microbial population can grow, the range of temperatures within which microorganisms reproduce being determined by the functioning of cellular enzymes. *See also* Maximum growth temperature; optimum growth temperature. (HMB)

Miscellaneous land area A map unit used in soil surveys that shows landscape within which little or no vegetation occurs. This may be due to the fact that there is little or no soil, or that unfavorable soil conditions, active erosion, washing by water, or the activities of people prevent plant growth. Miscellaneous land areas, such as beaches, dumps, rock outcrops, and badlands, are named for the limiting condition. (SWT)

Misrepresentation In insurance law, a statement that is untrue, material to the risk, and made with knowledge that it is untrue and with intent to deceive. Can also be a statement made as if positively as true but that is not known to be true, and that has a tendency to mislead. (DEW)

Mist Liquid particles, measuring 40 to 500 microns, that are generated by condensation from the gaseous to the liquid state or by the breakup of a liquid into a dispersed state, such as by splashing, foaming, or atomizing. In contrast, fog particles are smaller than 40 microns in size. (FSL)

Mitigation Measures taken to reduce adverse impacts on the environment. *See also* National Environmental Policy Act. (FSL)

Mixed liquor A mixture of activated sludge and raw or settled wastewater contained in an aeration basin in the activated-sludge sewage treatment process. (HMB)

Mixed liquor suspended solids (MLSS) Suspended solids in activated sewage sludge mixed liquor, with concentration expressed in milligrams per liter. *See also* Mixed liquor volatile suspended solids. (HMB)

Mixed liquor volatile suspended solids (MLVSS) The fraction of suspended solids in activated sewage sludge mixed liquor that can be volatilized during heating at 55°C. It is indicative of the concentration of viable organisms available for biological oxidation. *See also* Mixed liquor; mixed liquor suspended solids. (HMB)

Mode The value that occurs with the greatest frequency in a set of numbers. A unique mode may not exist for a given set of numbers. (VWP)

Modeling An investigative technique using a mathematical or physical representation of a system that accounts for all or some of the system's known properties. Models are often used to test the effect of changes of system components on the overall performance of the system. (FSL)

Moder Forest humus that is transitional between mull and mor. *See also* Mull; mor. (SWT)

Moderator A material used to slow down neutrons by reducing the energy at which they are released to some lower energy (usually thermal). Neutrons lose energy through scattering collisions with the nuclei of the moderator. A good moderator has a high scattering cross-section, a large number of nuclei per unit volume (to increase collision probability), a low capture cross-section (to reduce neutron losses), and a low atomic weight (to increase the energy transferred per collision). Although no material has an ideal combination of nuclear and mechanical properties, hydrogen (in water, plastics, or paraffin), deuterium (in heavy water), beryllium, and graphite have all found extensive use as moderators. (RMB)

Moisture mass percentage Moisture content of soil expressed as a percentage of the oven-dry mass of soil. (SWT)

Moisture-retention curve A graph of soil moisture content versus applied tension. (SWT)

Moisture tension The pressure that must be applied to soil water to bring it to hydraulic equilibrium, through use of a porous permeable wall or membrane and a pool of water of the same composition. (SWT)

Moisture volume percentage The ratio of the volume of water in soil to the total bulk volume of the soil. (SWT)

Mold *See* Fungus. (FSL)

Mole In chemistry, that amount of pure substance containing the same number of chemical units as there are atoms in exactly 12 grams of carbon-12, 6.023×10^{23} (Avogadro's number). (FSL)

Molecular weight The relative weight of a molecule of any substance as compared to the weight of an atom of carbon-12 (12.00000). (FSL)

Molecule Ultimate unit quantity of a chemical compound that can exist by itself and retain all the properties of the original substance. (FSL)

Mollic epipedon A mineral surface horizon that is dark and relatively thick, contains at least 5.8 g/kg organic carbon, is not massive, hard, or very hard when dry, has a base saturation of >50% when measured at pH 7, has <110 mg P/kg soluble in 1% citric acid, and is dominantly saturated with bivalent cations. *See also* Mollisol. (SWT)

Mollisol A mineral soil that has a mollic epipedon overlying mineral material and that has a base saturation of 50% or more when measured at pH 7. *See also* Mollic epipedon. (SWT)

Molluscide A chemical substance, such as copper sulfate or metaldehyde, used for the destruction of snails and other molluscs. *See also* Copper sulfate; metaldehyde. (FSL)

Molybdenum (Mo) A hard white metal, with an atomic weight of 95.94 and an atomic number of 42, that occurs as molybdenite. Used as an alloy agent in steels and cast iron; in pigments for printing inks, paints, and ceramics; and in special batteries. Molybdenum dust or powder is flammable and has a TLV of 10 mg/m^3 of air (soluble); 5 mg/m^3 of air (insoluble). (FSL)

Momentum The product of the mass of a body and its velocity; units: g-cm/s. (RMB)

Monitoring (radiation) Periodic or continuous determination of the amount of ionizing radiation or radioactive contamination present in an occupied region, as a safety measure, for purposes of health protection. Area monitoring is routine monitoring of the level of radiation or of radioactive contamination of any particular area, building, room, or equipment. Usage in some laboratories or operations distinguishes between routine monitoring and survey activities. Personnel monitoring is monitoring any part of an individual, expired air, excretions, or any part of clothing. *See also* Radiological survey. (RMB)

Monitoring survey Periodic or continuous determination of the amount of biological, chemical, or radioactive contamination in an area, on equipment, or on personnel, as a safety measure for purposes of health protection. (RMB)

Monitoring wells Wells drilled at a hazardous-waste management facility or Superfund site to collect ground-water samples for the purpose of physical, chemical, or biological analysis to determine the amounts, types, and distribution of contaminants in the ground water beneath the site. (FSL)

Monochromatic radiation Electromagnetic radiation of a single wavelength, or in which all the photons have the same energy—for example, lasers. *See also* Lasers. (RMB)

Monoclonal antibodies Molecules in an organism that selectively find and attach to other molecules to which their structure conforms. This could also apply to equivalent activity by chemical molecules. Also called MABs and MCAs. (FSL)

Monoecious "One house": bisexual; having staminate (male) and pistillate (female) flowers on the same plant. (SMC)

Monoenergetic radiation Particulate radiation of a given type (alpha, beta, neutrons, etc.) in which all particles have the same energy. (FSL)

Monofill A method of solid waste handling that involves the burial of a single waste type or wastes having similar characteristics in a segregated trench or area that is physically separated from areas containing dissimilar or incompatible waste. *See also* Landfill. (FSL)

Monomer A molecule or compound usually containing carbon of relatively low molecular weight that, either alone or with another monomer, forms various types and lengths of molecular chains called polymers. Thus, for example, styrene is a monomer from which polystyrene resins are produced, and vinyl chloride is the monomer of polyvinyl chloride. (FSL)

Montmorillonite An aluminosilicate clay mineral with a 2:1 expandable lattice structure, in which two silica tetrahedron sheets enclose an alumina octahedron sheet. Considerable expansion may occur because of the movement of water into the sheets. (SWT)

Mor A forest humus in which there is little mixing of surface organic matter with mineral soil—i.e., in which the transition to the A horizon is abrupt. *See also* Moder. (SWT)

Morbidity *See* Illness; morbidity rate. (FSL)

Morbidity rate An incidence rate that includes all persons in a defined population who became clinically ill during a specified time period. The population may be defined according to race, age group, or sex, for example. Morbidity rates for a specific disease can be shown as follows

Incidence

$$= \frac{\text{Number of new cases reported during a given year}}{\text{Estimated population as of July 1 of the same year}} \times 10^{x}$$

Prevalence

$$= \frac{\text{Number of cases existing at a given time}}{\text{Estimated population at the same time}} \times 10^{x}$$

See also Illness. (FSL)

Morbidity Report An official report that is a compilation of the occurrence of reportable disease in a community. (FSL)

Morphology Form or shape; the branch of biology that deals with the form and structure of animals and plants. (SMC)

Mortality rate A rate calculated by use of a numerator equal to the number of deaths occurring in the population during a stated period of time, usually a year, and of a denominator equal to the estimated population at some specific point in time.

Crude mortality rate

$$= \frac{\text{Number of deaths reported during a given year}}{\text{Estimated population as of July 1 of the same year}} \times 1000$$

Mortality rate for a Specific Disease

$$= \frac{\text{Deaths assigned to a specific cause during a given year}}{\text{Estimated population as of July 1 of the given year}} \times 1000$$

Infant Mortality Rate

$$= \frac{\text{Deaths under one year of age reported during a given year}}{\text{Births reported during the same year}} \times 1000$$

(FSL)

Mortar A mixture of sand and cement that is used to bind bricks, stones, and other masonry materials. (FSL)

Mosaic (soil) An assemblage of overlapping aerial photographs or images whose edges have been matched to form a continuous pictorial representation of the earth's surface. (SWT)

Mosquito A winged insect of the genus *Aedes*, *Anopheles*, or *Culex*, capable of sucking blood from humans or animals, and transmitting diseases, such as dengue, yellow fever, and malaria. (FSL)

Mothproof The use or application of compounds such as paradichlorobenzene (PDB) or other repellents or insecticides in confined spaces for the specific purpose of eliminating clothes moths or other nuisance insect species. These compounds can also be applied to clothing in commercial dry cleaning establishments as a means of mothproofing certain types of clothing—e.g., wools, which have an affinity for clothes moths. (FSL)

Motion Generally, an application to the court for a particular rule of order. (DEW)

Mottled zone (soil) A soil layer that is marked with spots or blotches of different hues or shades of color. The pattern of mottling and the size, abundance, and color contrast of the mottles may vary considerably and should be specified in soil description. (SWT)

Mottling Spots or blotches of different hues or shades of color interspersed with the dominant color. (SWT)

Muck Organic matter that has been decomposed to a point at which plant parts that were originally present are not recognizable. Contains more mineral matter than peat does. (SWT)

MUG test A sensitive and specific assay method for the detection and enumeration of *Escherichia coli*, an important organism in the coliform group used as an indicator organism of possible fecal contamination. A nonfluorescent substrate, 4-methylumbelliferyl-b-d-glucuronide (MUG) is hydrolyzed by over 95% of varieties of *E. coli*. The compound 4-methylumbellifone, which fluoresces under long-wave UV light, is produced. Consequently, both a total coliform count and an *Escherichia coli* count can be determined on the same medium after 24 hours of incubation at 35°C. The assay is very specific for b-glucuronidase production by these organisms. Approximately 50% of *Shigella* spp., some salmonellae, and a few strains of *Yersinis enterocolitica* are the only other members of the families *Enterobacteriaceae* and *Vibrionaceae* that produce this enzyme. *See also* Coliform; fecal coliform. (HMB)

Mulch Any material, such as straw, sawdust, plastic film, and loose soil, that is placed on the surface of the soil to protect the soil and plants from erosion, freezing, and evaporation. (SWT)

Mulch farming Planting and tillage operations that result in maximum incorporation of plant residues or other mulch into the soil. (SWT)

Mull A type of forest humus in which the organic material is incorporated into the A horizon. *See also* Moder. (SWT)

Multiple antibiotic resistance The resistance of microorganisms to the antimicrobial effects of more than one antibiotic. Antibacterial drugs at one time were widely used in livestock and poultry production. This practice has now been largely eliminated, but the practice has resulted in the development and maintenance of populations of antimicrobial-resistant organisms. Many of these organisms owe their resistance to transferrable drug resistance factors (R factors). This has resulted in concerns about animal and human health implications. For example, many concerns have been raised regarding the spread of pathogenic, resistant organisms from animals to humans, as well as the colonization of the human intestinal tract by relatively nonpathogenic, antimicrobial-resistant organisms such as *E. coli*. Little is known about the role played by multiple antibiotic resistance in the comprehensive picture of microorganism persistence in animal foods and the food-processing environment. (HMB)

Multiple regression *See* Regression analysis. (VWP)

Multiple use Use of land for more than one purpose—e.g., grazing of livestock, wildlife protection, recreation, watershed, and timber production. Could also apply to the use of bodies of water for recreational purposes, fishing, and water supply. (FSL)

Municipality A city, town, borough, county, parish, district, or other public body created by or pursuant to state law, or any combination thereof acting cooperatively or jointly. (FSL)

Munsell color system A color-designation system that specifies the relative degrees of three simple variables of color: hue, value, and chroma. For example: 10YR 6/4 designates a soil color with a hue of 10YR, a value of 6, and a chroma of 4. (SWT)

Muscarine A highly toxic alkaloid derived from certain species of mushrooms—e.g., *Amanita muscaria*. *See also* Alkaloid. (FSL)

Mustard gas [$S(CH_2CH_2Cl)_2$] One of the first chemical warfare agents used during World War I; this compound, dichlorodiethylsulfide, was given its familiar name because of its pungent odor. It is a vesicant-type (blistering) war gas that causes conjunctivitis and blindness. The vapor is extremely poisonous and can be absorbed by the skin. (FSL)

Mutagen A chemical substance that has the ability to produce a change in the genetic composition of the DNA in a cell. The change is capable of being passed on to succeeding generations. Such changes can be brought about by ionizing radiation and chemicals such as nitrogen and sulfur mustards, epoxides, and methylsulfonate. *See also* Carcinogens; teratogens. (FSL)

Mutant An individual that has been altered as a result of mutation—i.e., from a change in the character of a gene that is perpetuated in subsequent division of the cell in which it occurs. (FSL)

Mutation A change in the characteristics of an organism, produced by an alteration of the DNA of living cells. (FSL)

Mutually exclusive events Outcomes defined in such a way that the occurrence of one precludes the occurrence of any of the others. (VWP)

Mycorrhiza The symbiotic association of a specific fungi with the roots of higher plants. (SWT)

Mycotoxin A toxin produced by a mold growing on a specific substrate. The term applies to a broad class of highly toxic secondary metabolites capable of causing acute mycotoxicosis (usually observed in animals, not humans). Examples of mycotoxins produced by various genera of fungi are aflatoxin, ochratoxin, patulin, rubratoxin, and citrinin. In humans, conditions ranging from severe inflammatory reaction and alimentary toxic aleukia (produced by trichothecenes) to ergotism (produced by the ergot alkaloids) have been recorded. Many of the mycotoxins are known to be potent carcinogens. It is difficult, however, to correlate the occurrence of different types of carcinoma with human consumption of mycotoxins. *See also* Aflatoxin. (HMB)

Myelin The white, fatty substance that forms a sheath around certain nerve fibers. (FSL)

Myiasis An infestation of humans or animals by the parasitic larvae of certain species of flies. (FSL)

Mylar A plastic material; as used in alpha radiation survey instruments, a plastic coated with aluminum. (RMB)

Myoglobin A protein, found in muscle, that promotes the short-term storage of oxygen. (FSL)

Myriapodology The study of myriapods (Class Myriapoda, Phylum Arthropoda), which include centipedes and millipedes. (SMC)

Myrmecology The study of ants (order Hymenoptera, Class Insecta, Phylum Arthropoda). (SMC)

Mysophobia Fear of dirt and filth. (FSL)

Myxomatosis A viral disease that affects rabbits and hares. It is endemic in South America, where it is transmitted by mosquitoes. The disease was intentionally introduced in Australia in the 1950s, to eradicate rabbits and hares, but the populations have since recovered. (FSL)

Myxovirus A virus, such as influenza and Newcastle virus, that may cause disease in chickens and other poultry flocks, and measles, mumps, or influenza in humans. (FSL)

N

Natality rate A measure of the frequency of births in a particular population during a specific period of time. *See also* Rate. (VWP)

National Advisory Committee on Occupational Safety and Health A body, established under Section 7(a) of the OSHA Act, whose basic function is to advise, consult with, and make recommendations to the Secretary of Labor (i.e., OSHA) and the Secretary of Health and Human Services (i.e., NIOSH). The group consists of 12 members representing management, labor, occupational safety, occupational health professionals, and the public. (DEJ)

National ambient air quality standards (NAAQS) Under the Clean Air Act, standards promulgated by the Environmental Protection Agency for specific air pollutants, consisting of levels of the pollutants allowed in the ambient air. *See also* Clean Air Act; national primary ambient air quality standards; national secondary ambient air quality standards. (DEW)

National Center of Toxicological Research (NCTR) A component of the Food and Drug Administration that conducts research in toxicology and that is located in Jeffersonville, Arkansas. (FSL)

National Conference of States on Building Codes and Standards (NCSBCS) An organization founded in 1967 as a result of congressional interest in building-code reform. NCSBCS is an executive branch of the National Governors Association and fosters interstate cooperation in the area of building codes and standards. (FSL)

National Contingency Plan (NCP) Also known as the National Oil and Hazardous Substances Pollution

Contingency Plan, the plan was promulgated by the Environmental Protection Agency in 1980, in order to implement the requirements of the Comprehensive Environmental Response Compensation and Liability Act (Superfund) and Section 311 of the Clean Water Act. Its purpose is to provide for efficient, coordinated, and effective response to discharges of oil, and releases of hazardous substances, pollutants, or contaminants. The plan sets out the national framework for such response activities and assigns responsibilities to the federal, state, and local governments involved. The bulk of the plan sets out detailed requirements for response actions under CERCLA. 40 CFR Part 300. *See also* Comprehensive Environmental Response, Compensation, and Liability Act; National Response Center; national response team; regional response team. (FSL)

National Electrical Code First issued in 1897, this model code provides for the practical safeguarding of persons and property from hazards arising from the use of electricity. (FSL)

National Emission Standard for Hazardous Air Pollutants (NESHAPS) Standards established by the Environmental Protection Agency for the emission of specified hazardous air pollutants under the Clean Air Act. 42 USC §7412. *See also* Clean Air Act; hazardous air pollutants. (DEW)

National Environmental Policy Act (NEPA) Federal law that sets out the national environmental policy, which is to use all practicable means and measures, including financial and technical assistance, in a manner calculated to foster and promote the general welfare, to create and maintain conditions under which humans and nature can exist in productive harmony, and to fulfill the social, economic, and other requirements of present and future generations of Americans. Specifically requires that the consideration of the environmental impact of major federal actions significantly affecting the quality of the human environment be incorporated into the planning and decision-making process through the preparation of environmental impact statements. Also established the Council on Environmental Quality to implement the provisions of the act and monitor its compliance by federal agencies. 42 USC §4321 *et seq.*, 40 CFR Parts 1500, 1510. *See also* Categorical exclusion; Council On Environmental Quality; environmental assessment; finding of no significant impact. (DEW)

National Fire Protection Association (NFPA) A private-sector organization that collects, analyzes, and prepares reports on fire experience data. It also develops guidance documents as to fire hazards associated with various classes of chemicals. The NFPA, which is located in Quincy, Massachusetts, frequently conducts fire investigations in conjunction with federal agencies, such as the U.S. Fire Administration and the National Institute of Science and Technology. Through a memorandum of understanding signed in 1983, NFPA conducts on-scene investigations of selected fires in conjunction with representatives of the model building-code organizations in the country. (FSL)

National Gas Pipeline Safety Act of 1968 Federal law authorizing the Secretary of Transportation to establish minimum safety standards for pipeline facilities and the transportation of gas in commerce. 49 USC 1671 *et seq.* 49 CFR Part 192. (DEJ)

National Highway Traffic Safety Administration (NHTSA) The federal agency that conducts programs to reduce the frequency of motor vehicle crashes, the severity of injuries, and the economic losses that result. NHTSA administers federal motor vehicle safety standards that establish requirements for motor vehicles. The agency has testing responsibility for vehicles to assess damage susceptibility, crashworthiness, ease of repair, and compliance with standards. Research and development projects to improve the safety of motor vehicles and related equipment are conducted and motor vehicle fuel economy standards are established. (SAN)

National Institute for Occupational Safety and Health (NIOSH) A major component of the Centers for Disease Control and Prevention, Public Health Service, and U.S. Department of Health and Human Services. NIOSH conducts technical assistance and training, research, and epidemiologic investigations of work-related disease and injuries. A nonregulatory agency, NIOSH provides data to OSHA in the Department of Labor, which in turn develops standards and regulations and enforces these requirements in the workplace. *See also* Occupational Safety and Health Administration. (FSL)

National Institute for Occupational Safety and Health (NIOSH) Lifting Guidelines Recommendations for improving lifting safety through job design that are based on selection of an optimum (lifting) weight, adjusted for various factors related to task variables. The guidelines, published by NIOSH, apply only to: smooth lifting; two-handed symmetric lifting directly in front of the body; compact load with handles; specification of an unobstructed, good-friction floor; and specification of a favorable visual and thermal environment. There are two limits: an action limit (AL), which represents a load acceptable to all workers, and either a maximum permissible limit (MPL) or a maximum weight to lift, both of which are based on certain key assumptions about the general work population. The AL can be performed by almost all workers, whereas the MPL can be performed by only a few. (SAN)

National Institute of Building Sciences (NIBS) Authorized by Congress in 1974 by PL 93–383, this is a nongovernmental, nonprofit corporation. The purpose of the organization is to develop, promulgate, and maintain nationally recognized standards for safety, health, and public welfare that are suitable for adoption by building-regulating jurisdictions and agencies. NIBS has established a consultative council with membership from private and public trade and professional and labor organizations to carry out its mission. (FSL)

National Institute of Standards and Technology (NIST) Formerly the National Bureau of Standards, which was a part of the Department of Commerce. NIST conducts research and develops codes and standards for fire protection and prevention, fire equipment, and safety of consumer and building products. (SAN)

National Oceanic and Atmospheric Administration (NOAA) The federal scientific support organization serving regulatory agencies that is charged with the enforcement of environmental laws affecting oceans and the atmosphere. (FSL)

National Pesticide Information Retrieval System (NPIRS) An automated database, operated by Purdue University, that contains information on EPA-registered pesticides, including reference-file material safety data sheets. *See also* Material safety data sheets. (FSL)

National Pollutant Discharge Elimination System (NPDES) Under the Clean Water Act, a program for issuing permits for the discharge of pollutants from any point source into the waters of the United States. 33 USC §1842. *See also* Clean Water Act; discharge of pollutants; point source. (DEW)

National primary ambient air quality standards Under the Clean Air Act, standards that define levels of air quality that the Administrator of EPA judges necessary, with an adequate margin of safety, to protect the public health. 40 CFR Part 50. *See also* Clean Air Act. (DEW)

National Priorities List (NPL) Under the Comprehensive Environmental Response, Compensation, and Liability Act (CERCLA), the list of priorities for long-term remedial evaluation and response, compiled by the EPA, for uncontrolled hazardous-substance releases in the United States. Sites are evaluated for inclusion according to the hazard ranking system; in addition, each state may designate its highest-priority site for inclusion. 42 USC §9605, 40 CFR §300.5. *See also* Comprehensive Environmental Response, Compensation, and Liability Act; hazard ranking system; hazardous substance release. (DEW)

National Response Center (NRC) A national communications center, located at U.S. Coast Guard headquarters, continuously staffed for handling activities related to response actions under the National Contingency Plan. Acts as the single point of contact for all pollution-incident reporting and as the national response team communications center. Discharges of reportable quantities of oil or hazardous substances are reported to the NRC pursuant to provisions of CERCLA and the CWA. 40 CFR §300.125. *See also* National response team; reportable quantities. (DEW)

National response team Team responsible for national response and preparedness planning, for coordinating regional planning, and for providing policy guidance and support to regional response teams, pursuant to the National Contingency Plan. Consists of representatives from various federal agencies that have responsibilities under the plan. 40 CFR §300.100. *See also* National Contingency Plan; regional response teams. (DEW)

National secondary ambient air quality standards Under the Clean Air Act, standards that define levels of air quality that the administrator of EPA judges necessary to protect the public welfare from any known or anticipated adverse effects of a pollutant. 40 CFR Part 50. *See also* Clean Air Act. (DEW)

National Toxicology Program (NTP) A program conducted by the Institute of Environmental Health Sciences within the Department of Health and Human Services as mandated by Section 301(b)(4) of the PHS Act, as amended by Section 262, PL 95–622. The program consists of an evaluation of carcinogenic substances, performed by scientists in the NTP and other federal health research and regulatory agencies. The NTP prepares an annual report on carcinogens. Listing a substance in the report is the first step in hazard identification, but is only qualitative in nature and represents only one step in the total risk-assessment process. Further evaluation of the degree of potential human risk from the substance listed in the NTP report is the purview of the federal, state, and local health regulatory and research agencies authorized to implement the laws relating to carcinogens. Agencies that represent the NTP Working group include: Agency for Toxic Substances and Disease Registry (ATSDR), Centers for Disease Control and Prevention/National Institute for Occupational Safety and Health (CDC/NIOSH), Consumer Product Safety Commission (CPSC), Food and Drug Administration (FDA), National Institutes of Health/National Cancer Institute (NIH/NCI), National Institutes of Health/National Institute of Environmental Health Sciences (NIH/NIEH), National Institutes of Health/National Library of Medicine (NIH/NLM), and U.S. Department of Labor/Occupational Safety and Health Administration (DOL/OSHA). (FSL)

National Transportation Safety Board (NTSB) Federal agency with the responsibility for investigating transportation accidents, and for developing recommendations for government agencies and transportation industries on safety, the transport of hazardous materials, and accident investigation methodology, regulation, and reporting. (SAN)

Natural gas A mixture of naturally occurring gases of hydrocarbon and nonhydrocarbon components, usually associated with petroleum deposits. The main component is methane, although some hydrogen sulfide and carbon dioxide may be present. Natural gas, an efficient heating fuel, is usually transported by pipeline or in a liquified state by tank trucks. (FSL)

Natural history of a disease A comprehensive description of the characteristics, sources, and distribution of a disease agent; the characteristics and ecology of the agent's reservoir; mechanisms of transmission, and the effect of a disease on people. *See also* Epidemiology. (FSL)

Naval stores A term that historically related to the pitch used on wooden ships. In modern usage, it refers to all products derived from pine wood and stumps, such as turpentine, pine oils, and their derivatives. (FSL)

Navigable waters Under the Clean Water Act, the waters of the United States, including the territorial seas. Includes: all waters that are currently used, were used in the past, or may be susceptible to use in interstate or foreign commerce, including all waters that are subject to the ebb and flow of the tide; interstate waters, including interstate wetlands; all other waters such as intrastate lakes, rivers, streams (including intermittent streams), mudflats, sandflats, and wetlands, the use, degradation, or destruction of which would affect or could affect interstate or foreign commerce. Navigable waters also include: any such waters that are or could be used by interstate or foreign travelers for recreational or other purposes; waters from which fish or shellfish are or could be taken and sold in interstate or foreign commerce; waters that are used or could be used for industrial purposes by industries in interstate commerce; all impoundments of waters otherwise defined as navigable waters under the CWA; tributaries of waters, including adjacent wetlands; and wetlands adjacent to waters identified above. Waste treatment systems are not waters of the United States. 40 CFR §110.1. (DEW)

Near miss A situation in which personnel injury or illness, or equipment or environmental damage is narrowly averted. *See also* Accident; incident. (SAN)

Nebulizers High-quality generators that produce aerosols of uniform composition that are used in many types of analytical equipment. (DEJ)

Necropsy An examination or autopsy of a dead animal or tissue from a dead animal. Refers to examination of animal bodies, organs, tissues, or other biological specimens following controlled exposures in toxicological studies. (DEJ)

Necrosis The death of one or more cells or a portion of a tissue or organ, resulting from irreversible damage. Necrosis can discolor areas on plants or kill them entirely. (FSL)

Negative air machine A device consisting of a fan and ductwork, and often a high-efficiency particulate air (HEPA) filter. The device moves air from one area to another to maintain a state of negative air pressure inside a contaminated area and thus prevent leakage from the contaminated area into a noncontaminated area. Frequently used in asbestos abatement work and other environmental mitigation activities. *See also* Negative air pressure. (DEJ)

Negative air pressure Air pressure, in a room or duct, that is less than the pressure in the adjacent area. For example, if a laboratory room is under negative air pressure with respect to an adjacent hallway, air will flow from the hallway into the room to balance the pressure. Leakage of air will therefore be in one direction only. Similarly, if an air duct is under negative pressure with respect to the room, then any leakage will occur into the duct, not out of it. This principle is often employed to control migration of building air from contaminated zones into noncontaminated zones. (DEJ)

Negative predictive value The probability that a person does not have a characteristic being tested for if the test being used to detect the characteristic has a negative result. (VWP)

Negligence Failure to exercise that degree of care that an ordinarily prudent person would exercise in the same or similar circumstances. (DEW)

Negligence per se An action that is considered negligent as a matter of law because it is either in violation of a law or standard or is so clearly beyond the bounds of common prudency that no careful person would have performed it. *See also* Nuisance. (DEW)

Neighborhood For a value, v, the interval $(v - \alpha, v + \alpha)$ where α is specified. (VWP)

Nematocide A pesticide, such as dibromochloropropane (DBCP) ($CH_2BrCHBrCH_2Cl$), used to control worms infesting various types of root crops, such as potatoes, beets, and turnips. DBCP is a carcinogen and is reported to cause sterility. Its use is regulated by EPA. *See also* Pesticide. (FSL)

Neoplasm A new growth of cells that is more or less unrestrained and not governed by the usual limitations of normal growth. The growth is benign if there is some degree of growth restraint and no spread to distant parts. It is malignant if it invades other tissues of the host, spreads to distant parts, or both. *See also* Metastasis; oncogenic. (FSL)

Neoprene A synthetic rubber produced by the polymerization of chloroprene. This material is highly resistant to oils, chemicals, heat, light, and oxidization, and is used in many types of personal protective equipment, such as gloves and protective clothing. *See also* Levels of protection. (FSL)

Nephrotoxin A toxin known to have a deleterious effect on kidney tissue. (FSL)

Neuron A nerve cell, the basic unit of the nervous system. It consists of a nerve cell body, dendrites, and an axon. (FSL)

Neutralization Decrease in the acidity or alkalinity of a substance by addition to it of alkaline or acidic materials, respectively. (FSL)

Neutral soil A soil in which the surface layer has a pH of approximately 7. (SWT)

Neutron A neutral particle (i.e., one without electrical charge) of approximately unit mass, present in all atomic nuclei, except those of ordinary (or light) hydrogen. Its rest mass is 1.00893 amu. Neutrons are used to initiate the fission process, and large numbers of neutrons are produced by both fission and fusion reactions in nuclear blasts. (RMB)

Neutron chain reaction A process in which some of the neutrons released in one fission event cause other fissions to occur. There are three types of chain reactions: **(1)** *Nonsustaining chain reaction:* Reaction in which an average of less than one fission is produced by each neutron released by previous fission. **(2)** *Sustaining chain reaction:* Reaction in which an average of exactly one fission is produced by each neutron released by previous fission. **(3)** *Multiplying chain reaction:* Reaction in which an average of more than one fission is produced by each neutron released by previous fission. (RMB)

Neutron flux The product of the neutron density (number of neutrons per cubic centimeter) multiplied by the velocity; the flux is expressed as neutrons per square centimeter per second. It is numerically equal to the total number of neutrons passing, in all directions, through a sphere of 1 square centimeter cross-sectional area per second. (RMB)

Neutron-induced gamma radiation Gamma radiation that is incident to nuclear explosions and that is generated by the interaction of neutrons with neutron-sensitive elements that, having absorbed (captured) a neutron, become radioactive and emit gamma radiation. Abbreviation: n, γ. (RMB)

New source Under the Clean Water Act, any source the construction of which is commenced after the publication of proposed regulations prescribing a standard of performance that will be applicable to the source. 33 USC §1316(a)(2). *See also* Clean Water Act; standard of performance. (DEW)

New source performance standards (NSPS) Uniform national EPA air emission and water effluent standards that limit the amount of pollution allowed from new sources or from existing sources that have been modified. (FSL)

Nickel Carbonyl [$Ni(CO)_4$] A toxic, explosive chemical that is also a respiratory irritant and a possible human carcinogen. The compound, which is flammable, is produced by passing carbon monoxide over nickel and has a TLV of 0.005 ppm in air. (FSL)

Nicotine ($C_5H_4NC_4H_7NCH_3$) An alkaloid abstract from tobacco leaves that is used as an insecticide for the control of aphids and other sucking insects. Usually applied as a nicotine sulfate formulation, the material is an excellent contact insecticide; however, it has few residual properties. Thus, it has been replaced by other pesticide chemicals. Nicotine is toxic by ingestion, inhalation, and skin absorption; its TLV is 0.5 mg/m^3 of air. Nicotine is considered to be one of the primary agents associated with lung cancers in humans as a result of tobacco smoking. (FSL)

NIFE The acronym for a mixture of nickel and iron that is thought to be the major component of the Earth's core. (FSL)

Night soil A euphemism for human feces. In some cultures, feces, or "night soil," are gathered, sometimes by special collectors in towns and cities, and spread on fields (or applied directly to fields at night) to improve the fertility of soils. (SMC)

NIMBY Acronym for "not in my back yard." A term frequently used to refer to individuals opposed to the siting of landfills, incinerators, sewage treatment plants, and related facilities in close proximity to their property. (DEW)

NIOSH *See* National Institute for Occupational Safety and Health. (DEW)

Nitrate A salt or ester of nitric acid, such as sodium nitrate ($NaNO_3$). Nitrates generally possess antimicrobial properties. They may be reduced to nitrites in foods through the action of microorganisms. They are toxic by ingestion and their use in cured meat, fish, and other food products is restricted. The EPA limit on nitrates in drinking water is 10 ppm. *See also* Nitrites. (FSL, HMB)

Nitrate reduction The reduction of nitrates by plants and microorganisms to ammonia for cell synthesis, or to lower oxidation states by bacteria that use the nitrates as the terminal electron acceptors in anaerobic respiration. (SWT)

Nitric acid (HNO_3) A highly corrosive, colorless acid that has a characteristic pungent odor. In addition to its uses as a laboratory reagent, it is frequently incorporated into glassware-rinsing solutions because of its ability to oxidize small bits of organic matter and thus impart a "sparkle" to glass. It is used in the manufacture of ammonium nitrate for fertilizers and explosives, for etching steel, and in reprocessing spent nuclear fuel. Nitric acid is the eleventh-highest-volume chemical produced in the United States. It is highly toxic by inhalation and caustic to skin and mucous membranes. (FSL)

Nitric oxide (NO) A poisonous gas formed by combustion under high temperature and high pressure in an internal combustion engine. It changes into nitrogen dioxide in the ambient air and contributes to chemical smog. (FSL)

Nitrification **(1)** The oxidation of ammonia, to nitrite and nitrate by an organism; a biologically induced increase in the oxidation state of nitrogen. **(2)** The process whereby ammonia in wastewater is oxidized to nitrite and then to nitrate by bacterial or chemical action. (FSL, SWT)

Nitrites Compounds that contain the $-NO_2$ group. Sodium nitrite ($NaNO_2$) and sodium nitrate ($NaNO_3$) are used for curing meat because they stabilize color and inhibit some pathogenic and spoilage organisms. The nitrite ion is far more important than the nitrate ion as far as food preservation and food safety are concerned. The microorganism of most importance relative to nitrite inhibition is *Clostridium botulinum*. The nitrites are shown to have antimicrobial properties toward other foodborne pathogens as well, e.g., *Staphylococcus aureus*. They are ineffective against organisms that can in some instances be spoilage organisms (e.g., the lactic acid bacteria), and ineffective against the Enterobacteriaceae, including the salmonellae. Nevertheless, the use of nitrites is an important adjunct to other methods employed to ensure food safety. Nitrites can cause the oxidation of hemoglobin to methemoglobin; some individuals show sensitivity to this effect. High concentrations of nitrites in ground water supplies have been linked to the methemoglobinemia in infants (the so-called "blue-baby syndrome"). *See also* Nitrate. (HMB)

Nitrogenase The specific enzyme required for biological fixation of nitrogen. (SWT)

Nitroglycerin ($CH_2NO_3CHNO_3CH_2NO_3$) A major component of explosive materials. When used in combination with nitrocellulose and stabilizers, it is the primary ingredient in powder and solid rocket propellants. (FSL)

Nitrosamine [$(C_6H_5)_2NNO$] Compounds formed by the reaction between an amine and nitrous oxides or nitrites. These compounds occur in food products, cosmetics, and industrial settings such as tanneries, rubber factories, and iron foundries. They are also formed within the body by reaction of amine-containing drugs with the nitrites resulting from bacterial conversion of nitrates. These compounds have been shown to be carcinogenic in experimental animals. (FSL)

Nitrous oxide (N_2O) A product formed during combustion by the oxidation of nitrogen at high temperatures, also known as nitrogen monoxide. Nitrous oxide can form explosive mixtures with air and is narcotic in high concentration. It is used as an anaesthetic in surgery and as a propellant gas in food aerosols. (FSL)

Noble gas A gas that is either completely unreactive or reacts only to a very limited extent with other elements. The noble gases are helium, argon, neon, krypton, xenon, and radon. (FSL)

Nodal point **(1)** (Optics) Either of two points on the axis of a lens such that a ray entering the lens in the direction of one leaves as if from the other and parallel to the original direction. **(2)** (Physics) The point, line, or surface of a vibrating object, such as a string, where there is comparatively very little or no vibration. (DEJ)

Node The place upon a plant stem from which other structures can arise. Leaves are the most common structures developed at nodes but branches, flowering stems, and roots can also arise there. (SMC)

Nodule bacteria Bacteria that are contained in the nodules on the roots or leaves of plants and that exist in organized structures that fix nitrogen. (SWT)

Noise Unwanted sound, usually expressed in decibels (dBA). Any pressure variation, in air, water, or other media, that the human ear can detect. Noise is characterized by both frequency (pitch) and pressure (intensity). *See also* Anechoic room; audiometer; decibel; free field; frequency; loudness; noise reduction coefficient; octave band; peak level; pitch; resonance; reverberation; sound; sound intensity; sound power; sound pressure; transmission loss; ultrasonic; wavelength; white noise. (DEJ)

Noise control Engineering principles involving the substitution of a noisy process with a quieter one, modification of the noise source, or modification of sound waves by confinement or absorption. Examples of engineering noise control methods include the use of mufflers, enclosures lined with sound-absorbing materials, vibration-damping systems, and industrial process modification. (DEJ)

Noise Control Act of 1972 Federal law designed to control environmental noise, requiring the EPA administrator to publish noise criteria, identify processes or products that are major noise sources in a community, publish noise emission regulations, and identify low-noise-emission products. (DEJ)

Noise exposure Exposure to any unwanted sound. Overexposure to noise in occupational settings in the United States is considered to be 90 dBA, taken as an 8-hour time-weighted average (29 CFR 1910.95). The threshold limit value (TLV) is 85 dBA as an 8-hour time-weighted average. (DEJ)

Noise reduction coefficient The average of the sound absorption coefficients of a material at levels of 250, 500, 1000, and 2000 Hz. (DEJ)

Noise reduction rating (NRR) A measure of the efficacy of a given hearing protector, usually expressed in decibels. Assuming a complete and perfect fit, the noise reduction rating is the difference between the sound pressure levels outside the ear and inside the ear (i.e., inside the hearing protector). (DEJ)

Nominal scale data Data that can be classified into categories that are not specifically ordered. (VWP)

Non-artesian well A type of water well that penetrates rock formations in which ground water is located. Pumping from the well lowers the water table in the vicinity of the well, and water flows toward the well under the pressure differences that are thus artificially created. (FSL)

Non-food contact surfaces Surfaces of equipment, in either a food-service or food-processing establishment, not intended for contact with food, but that are exposed to splash or food debris or that otherwise require frequent cleaning. Regulations generally require that such surfaces be designed and fabricated to be smooth, washable, free of unnecessary ledges, projections, or crevices, and readily accessible for cleaning and sanitizing. Such surfaces must be fabricated of such material and be in such repair that they can be maintained in a clean and sanitary condition. (HMB)

Non-food zone Any exposed surface other than food and splash zones. *See also* Non-food contact surfaces. (HMB)

Nonionic wetting agent Any of several compounds having properties of wetting agents and considered to be synthetic detergents. These agents have low sudsing properties and are incorporated in detergents. They are not adversely affected by water hardness. Common types include alkyl amine oxides and ethoxylated alcohols. *See also* Surfactant. (HMB)

Nonionizing radiation Electromagnetic radiation, such as ultraviolet laser, infrared, microwave, and radio-frequency radiation, that does not cause ionization. (FSL)

Nonparametric methods Statistical procedures that do not depend on the distribution of a population. (VWP)

Nonpoint source pollution Contamination of water, soil, or air from a variety of combined sources, such as air pollution from auto exhausts, and soil and water pollution from storm-water runoff near agricultural land. (FSL)

No observable adverse effect level (NOAEL) The highest concentration at which a chemical causes no demonstrable adverse effect in the animal test species under investigation. (FSL)

Normal distribution (bell-shaped curve) Curve that depicts the spread around the population mean of a random variable x. It is defined, for all real values x, by

$$f(x) = \frac{1}{\sigma\sqrt{2\pi}} e^{-\frac{1}{2}\left(\frac{x-\mu}{\sigma}\right)^2}$$

where μ is the population mean, σ^2 is the population variance, and $e \approx 2.71828$. Such a curve is illustrated below. (VWP)

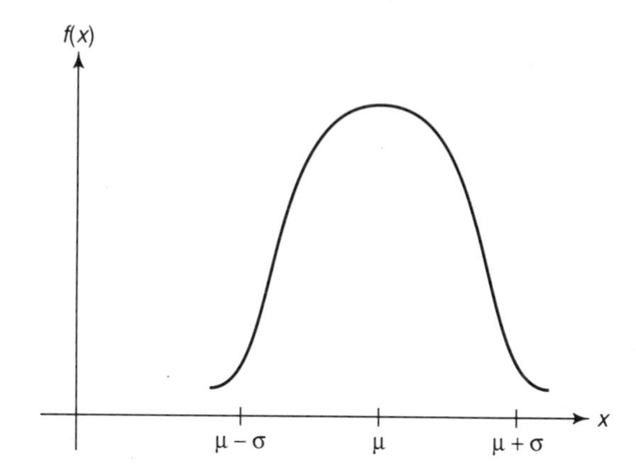

Nosocomial infection An infection that results from exposure to a source within a health-care facility and that was neither present nor incubating at the time of admission. The term is applied to such infections occurring among inpatients, visitors, and hospital personnel. The clinical manifestations may first become apparent after an individual is discharged from the hospital. The most common nosocomial infections are those associated with surgical wounds. (FSL)

Nuclear disintegration A process resulting in the change of a radioactive nucleus through the emission of alpha or beta particles. (RMB)

Nuclear fission A type of nuclear transformation characterized by the splitting of a nucleus into at least two other nuclei and the release of a relatively large amount of energy. Two or three neutrons are usually released during this type of transformation. (RMB)

Nuclear fusion The joining (coalescence) of two or more nuclei. (RMB)

Nuclear radiation spectrum The frequency distribution of nuclear or ionizing radiation with respect to energy. The gamma-ray spectrum of the initial radiation or the gamma and beta spectra of the residual radiations may be approximated in the laboratory by high-energy

X-ray or gamma-ray sources. Similarly, a neutron spectrum may be approximated by means of laboratory sources such as fission plates or a Godiva reactor. (RMB)

Nuclear reactor An apparatus in which nuclear fission may be sustained in a self-supporting chain reaction. The usual components of a nuclear reactor are as follows: fissionable material (fuel), such as uranium or plutonium; moderating material (except in the case of fast reactors); a reflector to conserve escaping neutrons; provision for heat removal; measuring and controlling elements. The terms ''pile'' and ''reactor'' are often used interchangeably. (RMB)

Nuclear Regulatory Commission (NRC) The federal agency that administers the laws and regulations relating to the operation of nuclear power plants and that controls the use of radioisotopes in biomedical research. The NRC is the regulatory body that enforces the Atomic Energy Act of 1954, through an independent executive commission created by the Energy Reorganization Act of 1974 (42 USC 5841). The NRC's standards for protection against radiation are found at 10 CFR Part 20. (DEJ)

Nuclear synthesis The fusion of light nuclei to form more complex nuclei, with an attendant release of energy resulting from conversion of matter. An example is a reaction that occurs in the sun whereby helium is synthesized by the fusion of hydrogen atoms with an overall reduction of mass that results in an equivalent release of energy. *See also* Nuclear fusion. (RMB)

Nuclear weapon A general name given to any weapon in which the explosion results from the energy released by reactions involving atomic nuclei, either fission, fusion, or both. Thus, the A-bomb (atomic bomb) and the H-bomb (or hydrogen bomb) are nuclear weapons. It would be equally correct to call all such bombs atomic weapons, because the energy of atomic nuclei is involved in each case. However, it has become more or less customary to refer only to weapons in which all the energy results from fission as atomic bombs. In order to make a distinction, those weapons in which at least part of the energy results from thermonuclear (fusion) reactions among the isotopes of hydrogen are called hydrogen bombs. (RMB)

Nucleon Common name for a constituent particle of the nucleus. It is applied to protons, neutrons, and any other particle found to exist in the nucleus. (RMB)

Nucleon number Total number of neutrons and protons in the nucleus. Also called the mass number. (RMB)

Nucleus **(1)** (Radiation, or atomic nucleus) The small, centrally located, positively charged region of an atom, that carries essentially all the mass. Except for the nucleus of ordinary (light) hydrogen, which is a single proton, all atomic nuclei contain both protons and neutrons. The number of protons determines the total positive charge, or atomic number; this number is the same for all the atomic nuclei of a given chemical element. The total number of neutrons and protons, called the mass number, is closely related to the mass (or weight) of the atom. The nuclei of isotopes of a given element contain the same number of protons, but different numbers of neutrons. They thus have the same atomic number, and so are the same element, despite their different mass numbers. **(2)** (Biology) The structure within cells that contains chromosomes and one or more nucleoli. (RMB)

Nuclide A general term referring to any nuclear species,either stable (of which there are about 270 in number) and unstable (of which there are about 500), of the chemical elements. (RMB)

Nuisance Generally, anything that is a danger to life or health, offends the senses, or interferes with the comfortable enjoyment of property. Nuisances can be litigated by common-law theory of tort liability or by specific statute. Nuisances can be of three types: those affecting a limited number of persons (private nuisances), those affecting the general public (public nuisances), or those that will pose an attraction to children, who will be unlikely to recognize their dangerous properties (attractive nuisances). *See also* Abatement. (DEW)

Nuisance particulates *See* Particulates not otherwise classified (PNOC). (DEJ)

N-unit The quantity of neutron radiation, measured in a Victoreen condenser R-meter, that will produce the same amount of ionization as one roentgen of X-radiation. (RMB)

Nutrient Any substance assimilated by living things that promotes growth. The term is generally applied to nitrogen and phosphorus in wastewater, but is also applied to other essential and trace elements. (FSL)

Nutrient antagonism A situation in which plant nutrients may reduce the uptake and availability of one another. (SWT)

Nutrient diffusion Movement of nutrients in soil as a result of a concentration gradient. (SWT)

Nutrient-efficient plant A plant that is able to absorb and utilize nutrients even under conditions of low availability. (SWT)

Nutrient stress A condition of plant growth that results when an inadequate nutrient supply restricts growth. (SWT)

Nyctophobia Fear of the dark. (FSL)

Nymph The immature phase of an insect that undergoes incomplete metamorphosis—i.e., the development of the insect (e.g., locusts) proceeds to the adult form through successive instars without any intermediate or pupal stage. In contrast, some insects (e.g., houseflies) undergo complete metamorphosis, in that development proceeds from the egg stage to the larval, pupa, and mature adult stage. (FSL)

Nystagmus Involuntary movement of the eyeballs, often experienced by workers who continuously subject their eyes to abnormal or unaccustomed movements. The condition is often accompanied by headaches, dizziness, and fatigue. The most prevalent form of occupational nystagmus occurs in miners. (DEJ)

Obligate anaerobes Microorganisms that are strictly intolerant of oxygen in their environment. These organisms are inhibited by the presence of oxygen probably because of direct oxygen toxicity. *See also* Anaerobes. (HMB)

Obstruction detector A safety control that will stop a suspended or supported unit from moving in a given direction if an obstruction is encountered, and will allow the unit to move only in a direction away from the obstruction. This type of control is typically installed on powered industrial platforms. (SAN)

Obturate To prevent the escape of a gas or vapor by obstructing openings, such as cracks or crevices in rock formations, through which it might pass. (FSL)

Occam's razor Refers to the principle that the simplest explanation of available data is often the best. (DEJ)

Occupational exposure limits Various exposure limits found in literature or listed on a material safety data sheet that are based primarily on time-weighted average limits, ceiling values, or other parameters. The values indicated in the following table can be used as a reference for determining relative toxicity and can assist in the selection of appropriate personal protective equipment.

Value	Abbreviation	Definition
A. Threshold limit value	TLV	Airborne concentrations of substances and represents conditions under which it is believed that nearly all workers may be repeatedly exposed day after day without adverse effect

Value	Abbreviation	Definition
(1) Threshold limit value–time-weighted average (ACGIH)	TLV-TWA	The time-weighted average concentration, for a normal 8-hr workday and a 40-hr work week, to which nearly all workers may be repeatedly exposed, day after day, without adverse effect.
(2) Threshold limit value–short-term exposure limit (ACGIH)	TLV-STEL	The concentration to which workers can be exposed continuously for a short period of time (provided that the daily TLV-TWA is not exceeded) without suffering from irritation, chronic or irreversible tissue damage, or narcosis of sufficient degree to increase the likelihood of accidental injury, impair self-rescue, or materially reduce work efficiency
(3) Threshold limit value–ceiling (ACGIH)	TLV-C	The concentration that should not be exceeded during any part of the work exposure
B. Permissible exposure limit (OSHA)	PEL	Same as TLV-TWA
C. Immediately dangerous life and health (OSHA)	IDLH	The maximum concentration (in air) from which one could escape within 30 minutes without any escape-impairing symptoms or any irreversible health effects
D. Recommended exposure limit (NIOSH)	REL	Highest allowable airborne concentration that is not expected to injure a worker, expressed as a ceiling limit or time-weighted average for an 8- or 10-hr work day.

Occupational illness Any abnormal condition or disorder, other than one resulting from an occupational injury, caused by exposure to environmental factors associated with employment. The definition includes acute and chronic illnesses or diseases that may be caused by inhalation, absorption, ingestion, or direct contact. The following categories are used in the OSHA 200 Log to classify recordable occupational illnesses: skin diseases or disorders, dust diseases of the lungs (e.g., pneumoconiosis), respiratory conditions due to toxic agents, poisoning (systemic effects of toxic materials), disorders due to physical agents (heat, cold, noise, radiation), and illnesses such as anthrax, brucellosis, and food poisoning. *See also* Occupational injury; OSHA 200 Log of Occupational Injuries and Illnesses. (SAN)

Occupational injury Any injury, such as a cut, fracture, sprain, or amputation, that results from a work accident or from a single instantaneous exposure in the work environment. This includes conditions resulting from the bites of animals, insects, or snakes, and from one-time (acute) exposures to chemicals. *See also* Occupational illness; OSHA 200 Log of Occupational Injuries and Illnesses. (SAN)

Occupational medicine A branch of medicine devoted to the appraisal, maintenance, restoration, and improvement of the health of workers through the scientific application of preventive medicine, emergency medical care, rehabilitation, epidemiology, and environmental medicine. (DEJ)

Occupational Safety and Health Act of 1970 This law (PL 91–596), the William-Steiger Act, was signed on December 29, 1970, and took effect on August 28, 1971. It provided the stimulus for the first uniform standards for occupational safety and health across the country, regulating safety and health in the workplace. Generally, the law provides for the promulgation of safety standards or guidelines for general industry and specific occupational settings. With regard to chemical substances, it requires that material safety data sheets (MSDS) on chemicals in use and other information be made available to the workers. It also sets limits and conditions on the use of chemicals and levels of exposure, and sets out monitoring, education, training, and medical surveillance requirements. The law is administered by the Occupational Safety and Health Administration (OSHA) of the Department of Labor and approved state programs. Occupational research, health-based recommendations, and consultation are provided by the National Institute for Occupational Safety and Health of the Public Health Service, Department of Health and Human Services. 29 USC §651 *et seq.*, 29 CFR Parts 1910, 1915, 1918, 1926. (DEW)

Occupational Safety and Health Review Commission An independent agency, established under Section 12(a) of the Occupational Safety and Health Act, charged with adjudication of enforcement actions brought by the Secretary of Labor against employers. The commission is composed of three Presidential appointees who serve 6-year terms. (DEJ)

Ochric epipedon A surface horizon of soil that is light in color and is both hard and dense when dry. (SWT)

Octane number (or rating) A parameter used to measure the quality of gasoline. A number representing the antiknock properties of a fuel mixture determined by the percentage of isoctane, in a mixture with normal heptane, that produces the knocking quality of the fuel being tested. The higher the number, the greater the antiknock properties. Higher octane numbers are characteristic of more highly refined fuels to which chemicals such as platinum have been added to prevent or eliminate engine knocking. (FSL)

Octave band As applied to noise, a bandwidth that has an upper band frequency that is twice its lower band frequency. (DEJ)

Octave band analyzer A device that measures sound pressure levels at specific frequency ranges, using a series of noise filters. Typically, such devices measure sound pressure levels in decibels at 63 Hz, 125 Hz, 250 Hz, 500 Hz, 1000 Hz, 2000 Hz, 4000 Hz, 8000 Hz, and 16000 Hz. (DEJ)

Odds If p is the probability that an event will occur, the odds that an event will occur are $p/1 - p$ and the odds against the occurrence of that event are $1 - p/p$. (VWP)

Odds Ratio (OR) If p_1 and p_2 are the probabilities for success for two populations, then

$$OR = \frac{p_1/(1 - p_1)}{p_2/(1 - p_2)} = \frac{p_1(1 - p_2)}{p_2(1 - p_1)}$$

(VWP)

Odor The characteristic of a substance that makes it perceptible to the sense of smell, whether pleasantly (fragrance) or unpleasantly (stench). (FSL)

Odor threshold The minimum concentration of a substance that can be detected and identified by a majority of the individuals exposed to the material. The sensitivity of smell varies from individual to individual and will diminish with continued exposure to an agent. (FSL, MAG)

Offal Those portions of butchered animals that are considered inedible by humans; carrion. (FSL)

Office of Federal Agency Programs (OFAP) The organizational unit of OSHA that provides technical assistance to federal agencies and guidance in development and implementation of occupational safety and health programs for federal employees. (FSL)

Office of Technology Assessment (OTA) A federal organization, responsible to Congress, that conducts research and identifies policy alternatives on technology-related programs. (FSL)

Oil spill An accidental or intentional discharge of oil that reaches bodies of water. Such spills can be controlled by chemical dispersion, combustion, mechanical containment, and/or adsorption. *See also* Clean Water Act. (FSL)

Oil wasteland Land on which oily wastes have accumulated and adjacent areas that have been affected. (SWT)

Olefins A class of unsaturated hydrocarbons, such as ethylene, butene, and propylene, derived from petroleum and natural gas. (FSL)

Oleic acid [$CH_3(CH_2)_7CH{:}CH(CH_2)_7COOH$] A long-chain unsaturated fatty acid that is found in many animal fats and vegetable oils. (FSL)

Omnivore An organism that feeds on a variety of food types, including plant and animal tissue. (SMC)

Onchocerciasis A disease, also known as "river blindness," that is transmitted by blackflies of the genus *Simulium*. The microfilaria, which are transmitted to a susceptible host during the process in which the fly feeds on the host's blood, frequently reach the eyes, causing visual disturbances and ultimately blindness. (FSL)

Oncogene A viral gene, found in some retroviruses, that may transform the host cell from normal to neoplastic. More than 30 oncogenes have been identified in humans. *See also* Neoplasm. (FSL)

Oncogenic A substance that causes tumors, whether benign or malignant. (FSL)

One-tailed test A statistical test in which the values of the parameter being investigated under an alternative hypothesis are allowed to be either greater than or less than the values of the parameter under the null hypothesis. (VWP)

On-scene incident commander As defined by the OSHA HAZWOPER standard, the individual who will assume control of an incident scene beyond the first-responder awareness level and who has training above and beyond that of the first-responder operations-level personnel. (SAN)

Open dump A disposal site at which solid waste from one or more sources is consolidated and left to decompose, burn, or otherwise create a threat to human health or the environment. *See also* Landfill. (FSL)

Open-face filter cassette A cassette holding a filter that collects airborne particulates, usually fibers, on removal of the entire lid, not only the small inlet plug. (DEJ)

Operating and support hazard analysis (OSHA) A hazard analysis technique used in process safety procedures that identifies and documents hazards and risk-reduction alternatives during a system's operation. It includes a review of standard operating procedures, along with an identification of changes needed in system design, support equipment, facilities, and related elements. The process also identifies requirements for safety devices, personal protective equipment, warnings, hazardous materials, training, personnel certification, and emergency procedures. This analysis technique evolved from the U.S. Military Standard System Safety Program, which incorporates different analytical steps, including preliminary hazard analysis, subsystem hazard analysis, and system hazard analysis. (SAN)

Operating voltage As applied to radiation-detection instruments, the voltage across the electrodes in the detecting chamber required for proper detection of an ionizing event. (RMB)

Opportunistic infection An infection caused by a microorganism that does not ordinarily cause disease but can become pathogenic under some circumstances. *See also* Acquired Immunodeficiency Syndrome. (FSL)

Opportunity assessment A procedure that identifies practices that can be implemented to reduce the generation of hazardous wastes (source reduction) or to reduce the quantity that must subsequently be treated, stored, disposed, or recycled. (FSL)

Optical density A logarithmic expression of the degree of attenuation provided by a filter. (FSL)

Optimum growth temperature The temperature at which a population grows at its maximum rate. As temperature increases from a minimum growth temperature, metabolic activity increases accordingly, and a microbial population will reproduce at an increasing rate until it reaches this optimal temperature. *See also* Maximum growth temperature; minimum growth temperature. (HMB)

Oralloy Uranium enriched in the isotope U-235. This material is an excellent fission fuel and is capable of sustaining a chain reaction. Abbreviation: Oy. (RMB)

Order **(1)** (Taxonomy) The taxon in the classification hierarchy between class and family. **(2)** (Soil classification) The highest category in soil classification. The soil orders are aridsol, entisol, histisol, inceptisol, oxisol, ultisol, and vertisol. *See also* Taxonomy. (SWT)

Ordinal data Data that can be ordered but that do not have numerical values that can be arithmetically manipulated. (VWP)

Ordinate In mathematics, the vertical Cartesian coordinate on a plane, measured from the *x*-axis, along a line parallel with the *y*-axis. *See also* Abscissa. (VWP)

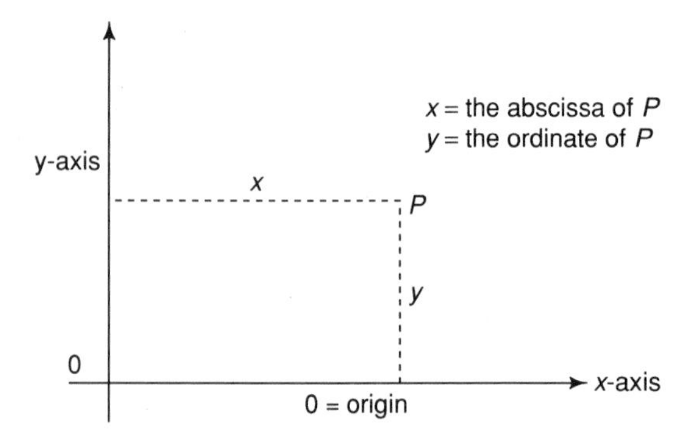

Ore An aggregate of valuable minerals and gangue (e.g., quartz, calcite, feldspar) from which one or more metals can be extracted, usually at a profit. (FSL)

Organ Organized group of tissues that perform one or more definite functions in an organism. (FSL)

Organan A cutan composed of a concentration of organic matter. *See also* Cutan. (SWT)

Organic **(1)** Of, pertaining to, or derived from living organisms. **(2)** Referring to substances containing carbon compounds. **(3)** Relating to the branch of chemistry that deals with carbon compounds. **(4)** Referring to farming practices, crops, or animal products that do not involve or are free from synthetic or chemical fertilizers, pesticides, hormone herbicides, and additives. (SMC)

Organic farming The production of crops without the use of artificial fertilizers or pesticides. (FSL)

Organic matter Carbonaceous waste contained in plant or animal matter and originating from domestic or industrial sources. (FSL)

Organic peroxide An organic compound that contains a bivalent —O—O— structure and that may be considered a structural derivative of hydrogen peroxide in which one or both of the hydrogen atoms has been replaced by an organic radical. Compounds, such as dioxane and diethyl ether, that form peroxides in high concentrations are considered potentially explosive and must be handled with caution. *See also* Shock-sensitive material; unstable material. (FSL)

Organic soil A soil that contains at least 150 but not more than 200 grams per kilogram of organic matter throughout the solum. (SWT)

Organic vapor meter *See* Halide meter. (FSL)

Organism Any living biological entity composed of one or more cells. (SMC)

Organochlorides *See* Chlorinated hydrocarbon insecticides; polychlorinated biphenyl (PCB). (FSL)

Organophosphates A group of nonpersistent insecticides that inhibit the action of cholinesterase, an enzyme in nerve cells that is responsible for the transmission of impulses. These compounds, such as parathion and guthion, are highly toxic to mammals by skin contact, inhalation, or ingestion, and must be used with caution and applied in accordance with instructions on the container label. The organophosphates have little residual action and do not accumulate in the ecosystem. *See also* Pesticides. (FSL)

Organotins Chemical compounds used in anti-fouling paints to protect the hulls of boats and ships, buoys, and dock pilings from marine organisms such as barnacles. They include butyl tin trichloride and dibutylin oxide. All the organotin compounds are highly toxic and accumulate in sediments. They have a TLV of 0.1 mg/m^3 of air. (FSL)

Origin In mathematics, the point in a Cartesian coordinate system at which the axes intersect. *See also* Abscissa; ordinate. (FSL)

Ornithology The study of birds (Class Aves, Phylum Chordata). (SMC)

Ornithosis An acute human disease, also known as psittacosis or parrot fever, caused by the agent *Chlamydia psittaci*. Outbreaks have been reported in people who have worked with or near aviaries and pigeon lofts. Infection is usually acquired by inhalation of the agent from dried droppings and secretions of infected birds. (FSL)

Orthotolidine ($C_{14}H_{16}N_2$) A reagent used for the detection of chlorine in water. Orthotolidine will turn yellow if sufficient chlorine has been applied to overcome the "chlorine demand" of the water, and an excess or residual chlorine content of between 0.2 and 0.3 mg/l exists. A deep-orange color indicates a larger excess of chlorine in the water. *See also* Chlorine; disinfectant; orthotolidine arsenate test. (FSL)

Orthotolidine arsenite test (OTA) A test, developed in 1939, that was used to detect the amount of "free" chlorine or chlorine residual in water. When orthotolidine arsenite reagent is added to water containing chlorine, a greenish-yellow color develops, the intensity of which is proportional to the amount of residual chlorine present. The color of the solution can be compared to that of a permanent standard and the concentration of chlorine noted. This test has been replaced by other techniques for the detection of chlorine, primarily because of the toxic properties of both orthotolidine and arsenic compounds. *See also* Chlorine. (FSL)

OSHA Refers to either the Occupational Safety and Health Act or the Occupational Safety and Health Administration, which is responsible for enforcing the act. The act states that U.S. employers are obliged to provide workplaces that are free of recognized health and safety hazards, to the fullest extent feasible. OSHA is part of the Department of Labor, and has issued a number of health and safety regulations and standards for general industry, construction, and maritime industries. Regulations for general industry are found at 29 CFR Part 1910; for the maritime industry, at 29 CFR 1915, and for the construction industry, at 29 CFR 1926. (DEJ)

OSHA frequency rates *See* Incidence rate; severity rate. (SAN)

OSHA Log of Occupational Injuries and Illnesses The OSHA No. 200 form that is used for recording and classifying occupational injuries and illnesses, and for noting the extent and outcome of each case. The log indicates when the occupational injury or illness occurred, to whom, what the injured or ill person's regular job was at the time of the injury or illness exposure, the department in which the person was employed, the kind of injury or illness, how much time was lost, and whether the case resulted in a fatality. An annual summary must be prepared for each establishment, must be posted conspicuously in the workplace by February 1 of the year following the report, and must remain in place until March 1 of the year of posting. *See also* Lost work-day case; occupational injury; occupational illness. (SAN)

OSHA recordable injuries and illnesses Disabling injuries or illnesses that occur in the workplace. *See also* Lost work-day case; occupational injury; occupational illness; OSHA Log of Occupational Injuries and Illnesses; restriction of work or motion. (SAN)

Osmosis Diffusion that proceeds through a semipermeable membrane separating two miscible solutions and that tends to equalize their concentrations. (FSL)

Osmotic Pertaining to osmosis. *See* Osmosis. (FSL)

Osmotic pressure The pressure developed when a solvent is separated from a solution by a membrane that allows the passage of solvent molecules only. (FSL)

Otolaryngologist A physician specializing in diseases or abnormalities of the ear, larynx, and, often, the upper respiratory tract, head, neck, tracheobronchial tree, and esophagus. This specialist performs hearing assessments for possible damage in industrial and other occupational settings. (FSL)

Outbreak The occurrence of two or more cases of disease that are epidemiologically related. *See also* Foodborne outbreak. (FSL)

Outcome A particular result of an experiment. (VWP)

Outer Continental Shelf Lands Act Federal law authorizing the head of the U.S. Coast Guard to develop, promulgate, and enforce regulations concerning lights and other warning devices and safety equipment on lands and structures or in adjacent waters. The act is codified at 43 USC 1331 *et seq*. (DEJ)

Outfall The place where an effluent from a sewer or individual source is discharged into receiving waters. *See also* Sewage. (FSL)

Outfall pipe A pipe or conduit that conveys either raw or treated municipal waste to a terminal point of discharge. *See also* Sewage. (FSL)

Outriggers Devices used to maintain or increase the stability of work platforms and lifting equipment such as cranes. Transportable outriggers are designed to be moved from one work location to another and are often attached to the equipment. *See also* Davit. (SAN)

Outwash Stratified glacial drift deposited by meltwater streams beyond active glacier ice. (SWT)

Oven-dry soil Soil that is dried at 105°C until it reaches a constant weight. (SWT)

Overburden Material recently deposited on the surface of a soil by erosion or the activities of people. Material that overlies a deposit of other materials, such as coal or ores. (SWT)

Overconsolidated soil deposit A soil deposit that has been subjected to an effective pressure greater than the present overburden pressure. (SWT)

Overcurrent Any current in excess of the rated electrical current of equipment or the capacity of a conductor. The conduction may result from overload, short circuit, or ground fault. A current in excess of rating may be accommodated by certain equipment and conductors for a given set of conditions. OSHA has rules for overcurrent protection that are specific for particular situations. *See also* Ground fault circuit interrupter; overload. (SAN)

Overfire air Air forced into the top of an incinerator or boiler to fan the flames. (FSL)

Overland flow A land application technique that cleanses wastewater by allowing it to flow over a sloped surface. As the water flows over the surface, the contaminants are removed and the water is collected at the bottom of the slope for reuse. (FSL)

Overload The operation of equipment in excess of the normal, full electrical load rating, or of a conductor in excess of rated ampacity. If overload persists for a sufficient length of time, damage or dangerous overheating will result. Faults such as a short circuit or ground fault are not considered overloads. *See also* Overcurrent. (SAN)

Overturn The mixing of water and nutrients in a lake when the thermal strata become mixed during spring and fall. It results in a uniformity of chemical and physical properties of the water at all depths. (FSL)

Oxalic Acid (HOOCCOOH · 2HOH) A toxic dibasic acid found naturally in fruits and in vegetables such as rhubarb and spinach. Oxalic acid is manufactured by the reaction of carbon monoxide and sodium hydroxide. It is toxic by inhalation and ingestation and is a strong irritant. Poisoning by this chemical is known as oxalism. The acid is used as a radiator cleanser, in leather tanning, as a laboratory reagent, and in printing and dyeing processes. (FSL)

Oxic horizon A subsurface horizon from which weathering has removed most of the combined silica, leaving a mixture composed mainly of hydrous oxide clays with some 1 : 1-type silicate minerals and quartz present. (SWT)

Oxidant A substance that contains oxygen and that may react chemically in air to produce a new substance. Some of the primary ingredients of photochemical smog are oxidants. (FSL)

Oxidase An enzyme (a dehydrogenase) that catalyzes oxidative reactions in which hydrogen is removed from a substrate and combines with free oxygen. (FSL)

Oxidation (Chemistry) The combination of oxygen with a substance or the removal of hydrogen from a substance, in which process an atom loses electrons. An increase of positive charge on an atom or the loss of negative charge. Univalent oxidation refers to the loss of one electron; divalent oxidation, the loss of two. (FSL)

Oxidation ditch An open channel in which liquid organic wastes are circulated for partial digestion and in which the circulation and aeration are carried out by means of a mechanical device. (SWT)

Oxidation pond An artificial lake or body of water in which waste is consumed by anaerobic bacteria. Such ponds are used in conjunction with other water-treatment processes. *See also* Lagoon. (FSL)

Oxidation-reduction potential Some chemical reactions in the cell or in an organism's environment involve the transfer of electrons from one molecule to another. Oxidation is the loss of electrons from a molecule and reduction is the gain of electrons. These reactions occur in pairs: for every oxidation, there is a corresponding reduction. The oxidation-reduction potential (also referred to as the redox potential) is a measure of the relative oxidizing and reducing power of an organism's substrate. It is one of the intrinsic factors affecting the growth of organisms. Some organisms require reduced conditions for optimal growth, whereas others require oxidized conditions. A low oxidation-reduction potential favors the growth of anaerobes, whereas aerobes favor high oxidation-reduction potential. Facultative anaerobes can function under both conditions. Oxidation-reduction potential figures significantly in terms of the types and likely numbers of organisms that survive in a given environment. (HMB)

Oxidizer A chemical, other than a blasting agent or explosive, that initiates or promotes combustion in other materials, thereby causing fire either in and of itself or through the release of oxygen or other gases. (FSL)

Oxisols Soils that occur in tropical and subtropical regions and are characterized by the presence of an oxic horizon. *See also* Oxic horizon. (SWT)

Oxygen (O) Nonmetallic, colorless, odorless, and tasteless gaseous element with an atomic number of 8 and atomic weight of 15.9994. Oxygen constitutes roughly 20% by volume of air at sea level. It is a moderate fire risk as an oxidizing agent. Therapeutic overdoses can cause convulsions. Liquid oxygen may explode on contact with heat or oxidizable materials. Oxygen ranks fifth among chemicals produced in the United States. (FSL)

Oxygenated solvent An organic solvent containing oxygen as part of its molecular structure. Alcohols and ketones are oxygenated compounds often used as paint solvents. (FSL)

Oxygen deficiency A condition in which the concentration of oxygen by volume is insufficient to maintain normal respiration, and atmosphere-supplying respiratory protection must be provided. It exists in atmospheres in which the percentage of oxygen by volume is less than 19.5%. (FSL)

Oxytocin A hormone, secreted by the pituitary gland, that stimulates contractions of the uterus and the production of milk. (FSL)

Ozonator A device that adds ozone to water. (FSL)

Ozone (O_3) A reactive chemical that is an irritant to the eyes and respiratory tissues. The ozone molecule consists of three atoms, rather than two, as in the case of ordinary oxygen. Ozone occurs at a concentration of about 0.01 part per million (ppm) in the atmosphere.

Levels above 0.1 ppm are considered toxic in the workplace. Ozone is formed by the recombination of oxygen atoms after ionization by electrical discharges or radiation such as ultraviolet light. Ozone is an unstable blue gas that has a pungent odor and is formed in the air by lightning and in the stratosphere by ultraviolet radiation. It is a dangerous fire and explosive risk when in contact with organic materials. Ozone is used in the disinfection of drinking water and for the purification of various industrial wastes. It can seriously affect the human respiratory system and is one of the most prevalent and widespread of all the criteria pollutants for which the Clean Air Act required EPA to set standards. *See also* Clean Air Act. (FSL)

Ozone depletion Destruction of the stratospheric ozone layer, which shields the earth from ultraviolet radiation harmful to life. This destruction of ozone is caused by the breakdown of certain chlorine-, fluorine-, and/or bromine-containing compounds (chlorofluorocarbons or other halons) that break down when they reach the stratosphere and catalytically destroy ozone molecules. *See also* Polybrominated biphenyl; polychlorinated biphenyl; ozone layer. (FSL)

Ozone layer Segments of the stratosphere in which ozone is found. In the stratosphere (the layer beginning 7 to 10 miles above the earth's surface), ozone is a form of oxygen that is found naturally and that provides a protective layer shielding the earth from the harmful effects of ultraviolet radiation. Ozone may occur in the troposphere (the layer extending up 7 to 10 miles from the earth's surface) in photochemical smog. There, the ozone is produced through complex chemical reactions of nitrogen oxides (which are among the primary pollutants emitted by combustion sources) and by hydrocarbons released into the atmosphere through the combustion, handling, and processing of petroleum products. The action of sunlight is typically involved. *See also* Polybrominated biphenyl; polychloronated biphenyl; ozone. (FSL)

P

Packed tower A pollution control device that forces dirty air through a tower packed with crushed rock or wood chips while liquid is sprayed over the packing material. The pollutants in the air stream either dissolve or chemically react with the liquid. (FSL)

Paired data A data set in which each data element from a sample has a corresponding element associated with it. (VWP)

Paired difference The value found by subtracting the data value of a first sample from the data value of a second sample from the same source. (VWP)

Pair Production The process whereby a gamma-ray (or X-ray) photon, with energy in excess of 1.02 MeV, in passing near the nucleus of an atom is converted into a positive electron and a negative electron. As a result, the photon ceases to exist. *See also* Photon. (RMB)

***Paleo-* (also spelled *palaeo-*)** **(1)** Prefix meaning *ancient*. **(2)** Prefix used in naming or referring to extinct or fossil organisms. (SMC)

Paleontology The study of fossil records or extinct organisms. (SMC)

Paleosol **(1)** (Buried) A soil that was originally from an earlier geologic period but was later buried by sedimentation. **(2)** (Exhumed) A formerly buried soil that has been exposed by erosion of the mantle of sediments. (SWT)

Palmitic acid [$CH_3(CH_2)_{14}COOH$] A combustible saturated fatty acid, also known as hexadecanoic acid, derived from animal and vegetable fats. It is used in the manufacture of soaps, lubrication oils, and food-grade additives. (FSL)

Palynology The study of ancient or fossilized pollen and other microfossils. Pollen deposits are often preserved beneath lakes, bogs, and other bodies of still water, and are used to determine the age of strata and the kind of plant life existing during former periods. (SMC)

Panacea A treatment or cure for all diseases. (FSL)

Pancytopenia The condition in which there is an abnormal depression of all the cellular components of the blood. (FSL)

Pandemic A disease affecting the population of a widespread region, a nation, or the world. *See also* Epidemic. (FSL)

Pan, genetic A naturally occurring subsurface soil layer of low or very low permeability, with a high concentration of small particles that differ in certain physical and chemical properties from the soil immediately above or below the pan. (SWT)

Pan, pressure or induced A subsurface horizon or soil layer that has a higher bulk density and a lower total porosity than soil directly above or below it, due to pressure applied during tillage or other artificial means. (SWT)

Pans Horizons or layers in soils that are compacted, indurated, or very high in clay content—e.g., claypan, fragipan, or hardpan. (SWT)

Panmixis The interbreeding of individuals in a population in which each individual has the potential capacity of mating with any other individual. (SMC)

Pan wells Depressions or recessed areas for holding pans of food as in a food cart in a food-service operation. It is essential that these structures be designed for easy cleaning and sanitizing. (HMB)

Paraldehyde ($C_6H_{12}O_3$) A chemical derived from the action of hydrochloric acid and sulfuric acid upon acetaldehyde. A sedative, previously used in the treatment of alcoholics, that has an unpleasant taste and imparts a foul odor to the breath. Paraldehyde is flammable; it is used in dyestuffs and as a solvent for fats, oils, and resins. (FSL)

Paralithic contact Similar to a lithic contact, except that the soil is softer and can be dug with a spade. *See also* Lithic contact. (SWT)

Paralytic Shellfish Poisoning (PSP) A condition in humans caused by naturally occurring nerve toxins that can accumulate in bivalve shellfish, such as clams, oysters, mussels, and scallops, during blooms of the dinoflagellate *Alexandrium catenella* (previously known as *Goynaulax catenella*). Although the PSP toxin does not harm the shellfish, in humans it causes symptoms such as tingling and numbness of the lips, tongue, and fingers, loss of balance and muscle coordination, and difficulty in swallowing. Death can result from suffocation. Cooking does not destroy the toxin. Cases have been reported along both the Atlantic and Pacific coasts. *See also* Red tide. (FSL)

Parameter A numerical feature of an entire population. (VWP)

Parametric Methods Statistical procedures in which the assumption is made that the parent population is at least approximately normally distributed or that the control limit theorem is applied to establish a normal approximation. (VWP)

Paraquat ($[CH_3(C_5H_4N)_2CH_3]\cdot 2CH_3SO_4$) An organic phosphate herbicide used for the control of broad-leaved weeds and grasses. A contact chemical, the material is highly toxic to humans and many aquatic organisms. It is highly toxic by ingestion, inhalation, and skin absorption and is now classified by EPA as a restricted-use pesticide. The TLV is 0.1 mg/m^3 of air. Excellent control of most aquatic weeds treated with paraquat can be expected. (FSL)

Parasite An organism (often microbial) that lives at the expense of the host in or on which the organism resides. Parasites are not necessarily harmful to their host. (FSL)

Parasitology The study of parasites and parasitism. (SMC)

Parathion [$(C_2H_5O)_2P(S)OC_6H_4NO_2$] An organic phosphate insecticide used for the control of a wide range of insects, including aphids, mites, and boll weevils. A highly toxic chemical, parathion does not accumulate in the environment and breaks down rapidly following application. Repeated exposure may, without symptoms, be increasingly hazardous. Ethyl parathion is more toxic than methyl parathion and, for this reason, the latter material is often applied by aircraft for the control of boll weevils. All uses of parathion are currently under review by the EPA. *See also* Special review. (FSL)

Parent material The material from which soil is formed. Parent material is unconsolidated, chemically mineral or organic matter from which the soil is created. (SWT)

Parorexia The tendency to ingest nonfood items, such as starch and paints. *See also* Pica. (FSL)

Part B permit Under the provisions of the Resource Conservation and Recovery Act (RCRA), a narrative section submitted by hazardous-waste generators in the permitting process. It is, in effect, a plan that details the procedures to be followed at a hazardous-waste generator's facility in order to protect human health and the environment. *See also* Generator; Resource Conservation and Recovery Act. (FSL)

Partially incinerated chemical compounds (PICC) The residue from the incineration of chemicals. It may include products such as dioxins and furans. (FSL)

Particle accelerator *See* Accelerator. (RMB)

Particle density The mass per unit volume of soil particles. (SWT)

Particles Small, distinct masses of solid or liquid matter, such as dust, fume, mist, or smoke. (FSL)

Particle size The effective diameter of a particle, measured by sedimentation, sieving, or micrometric methods. (SWT)

Particle-size analysis (soils) The procedure for determining the amounts of particle separates in a sample, usually by sedimentation, sieving, micrometry, or a combination of these methods. (SWT)

Particle-size distribution (soils) The quantification of various soil particle separates in a soil sample, usually expressed in percentages. (SWT)

Particulate-filter respirator *See* Respirator. (DEJ)

Particulates Fine liquid or solid particles, such as dust, smoke, mist, fumes, or smog, found in air or emissions. (FSL)

Particulates not otherwise classified A designation developed by the American Conference of Governmental Industrial Hygienists to describe airborne particulates that have the following effects when inhaled: **(1)** the architecture of the air spaces remains intact; **(2)** collagen (scar tissue) is not formed to a significant extent; and **(3)** the tissue reaction is potentially reversible. Inhalation of high concentrations of such dust (above 10 mg/m^3 as an 8-hour time-weighted average) in the workplace can cause injury to the skin or mucous membranes by chemical or mechanical action, produce excessive deposits in the eyes, ears, and nasal passages, or seriously reduce visibility. Sometimes called inert dusts. (DEJ)

Parts per million/parts per billion (PPM/PPB) A common way of expressing extremely low concentrations of pollutants in air, water, soil, human tissue, food, or other products. (FSL)

Party wall A wall that is situated on a common lot line of two buildings and that is common to both buildings. (FSL)

Passive humoral immunity The type of immunity attained either naturally by transplacental transfer from the mother, or artificially by inoculation of specific protective antibodies. (FSL)

Pasteur, Louis (1822–1895) French chemist and microbiologist who developed the method of vaccination, in his work with the cowpox virus vaccine, which provided immunity against smallpox. Pasteur was also noted for his work in fermentation study, particularly as applied to the wine industry, and for the heat sterilization (pasteurization) process that is now applied to dairy products and many perishable foods. Pasteur also developed an effective vaccine against rabies and his fame stimulated the formation of the Pasteur Institute, which stands today in France as a major scientific facility. *See also* Pasteurization. (FSL)

Pasteurization The process of heating to temperatures below the boiling point to destroy human pathogens. Almost all milk sold in the United States today is pasteurized. However, pasteurization can be applied to any food product that would sustain this type of heat treatment. The heating does not sterilize (kill all microorganisms) in the food, but is sufficient to kill all human pathogens. The time and temperature of pasteurization varies in the several acceptable methods of pasteurization that are applied to milk and other dairy products. They are as follows. **(1)** *Batch pasteurization, or low-temperature long-time (LTLT) treatment:* This is a method in which the milk is heated at approximately 63°C for 30 minutes. **(2)** *High-temperature short-time (HTST) treatment:* This is a method used in a continuous-flow system that pasteurizes at 72°C for 15 seconds. This process is the most desirable because it best kills microorganisms, produces fewer changes (e.g., in the flavor of milk), and allows large volumes to be processed easily. HTST may also be accomplished by heating at higher temperatures for shorter time periods—e.g., at 88°C for 1.0 second or at 90°C for 0.5 second.

Because a pasteurized product is not sterilized, sanitation practices surrounding production of such a food product are critical. With products of animal origin—e.g., milk—human pathogens may potentially be present. Postpasteurization contamination is always a probability. Therefore, a quality sanitation program is a crucial element in producing a safe product. (HMB)

Pathogen Any viral, rickettsial, bacterial, fungal, or parasitic microorganism capable of causing disease in humans, animals, or plants. (FSL)

Pathogenicity The capacity of an agent to cause disease in a susceptible host. *See also* Virulence. (FSL)

Pathology The branch of medicine concerned with all aspects of disease, but having special reference to the causes and development of abnormal conditions as well as structural and functional changes as the result of disease. *See also* Histopathology. (FSL)

Peak sound pressure level The maximum instantaneous level that occurs over any specified time period, and is usually measured in decibels. (DEJ)

Pearson's product moment (r) A number, r, such that $-1 \leq r \leq 1$, that is a measure of the strength of the linear relationship between two variables.

$$r = \frac{\Sigma(x - \bar{x})(y - \bar{y})}{(n - 1)s_x s_y}$$

where (x, y) is a data point, $(\bar{x}, \bar{y})$ is the centroid, n is the number of data values and s_x and s_y are the standard deviations of the x and y variables. *See also* Linear correlation coefficient. (VWP)

Peat Soil material that consists mainly of undecomposed organic matter that has accumulated under conditions of excessive moisture. (SWT)

Peat soil An organic soil containing >500 g/kg of organic matter. (SWT)

Ped A soil structure unit that has been formed by a natural process. Usually described as an aggregate, crumb, prism, block, or granule. (SWT)

Pediment On an eroded slope, a footslope that normally has a concave upward profile. (SWT)

Pedological features Recognizable units within a soil material that are distinguishable from the enclosing material because of their origin or because of differences in arrangements of the constituents. (SWT)

Pedon A three-dimensional section of soil that is large enough in area (from 1 to 10 square meters) to permit the study of horizon shapes and relations. (SWT)

Pelagic Living or feeding in the open ocean. (SMC)

Pellicle A layer of film or scum that forms on the surface of liquids. (FSL)

Peneplain An area that has been reduced by erosion from a high, rugged area to a low, gently rolling surface resembling a plain. (SWT)

Penetrability, soil The amount of force required to push a probe into the soil. (SWT)

Penetration A mechanism by which a substance enters through a barrier via an imperfection such as a rip or tear. *See also* Permeation. (MAG)

Peptidase An enzyme, such as pepsin, secreted in the stomach of vertebrates that breaks down peptides and, often, protein. (FSL)

Peptide A compound of two or more amino acids in which the alpha carboxyl group of one is united with the alpha amino group of another. This reaction results in the elimination of a molecule of water and the formation of a peptide bond. (FSL)

Peptize To increase the dispersion of a colloidal solution. To produce a colloidal solution by the action of alkaline material on soil containing proteins. (HMB)

Peptones Products that result from the breakdown of proteins, which consist of polypeptides. (FSL)

Peracetic Acid (CH_3COOOH) A strong oxidizing agent formed from equilibrium mixtures of acetic acid, hydrogen peroxide, and sulfuric acid as a catalyst. Like chlorine, this chemical has a broad spectrum of biocidal activity, and is effective against Gram-positive and Gram-negative bacteria, yeasts, molds, and fungal spores. It has application in diverse food-processing operations as a sanitizer. It is nonfoaming and noncorrosive, is effective in hard water, and produces safe degradation products. However, it is irritating to skin and eyes, has an unpleasant odor, and is highly reactive with metal contaminants, organic soils, and alkaline residues, thereby losing biocidal activity. (HMB)

Percentage depth dose Amount of radiation delivered at a specified depth in tissue, expressed as a percentage of the amount delivered at the skin. (RMB)

Percentiles Numbers that divide a set of ranked data into 100 equal parts. (VWP)

Percolation, soil The movement of water downward and radially through subsurface soil layers, usually continuing downward to the ground water. The results of soil percolation tests are used as the design criteria for the development of subsurface sewage disposal systems. (SWT)

Percolation test A field test designed to determine the acceptability of a site for a subsurface disposal system and to establish the design criteria for that system. The test involves placing six or more test holes to a depth that approximates that of the absorption trench. The time, in minutes, required for a given volume of water to fall one inch in the hole is defined as the percolation rate. This information, when applied to the required absorption area in square feet per bedroom, or other design criteria, can be used to determine the actual square footage of the trench area required. *See also* Soil absorption system. (FSL)

Perfect Botanical term that means hermaphroditic; it refers to plants that have stamens (male reproductive organs) and pistils (female reproductive organs) in the same flower. (SMC)

Perennial A plant that normally persists for more than two years. (SMC)

Pergelic Refers to a soil temperature regime that has a mean annual temperature of $<0°C$. (SWT)

Periodic table An arrangement of chemical elements in order of increasing atomic number. Elements of similar properties are placed one under the other, yielding groups and families of elements. Within each group there is a gradation of chemical and physical properties but in general a similarity of chemical behavior. However, there is a progressive shift of chemical behavior from group to group, from one end of the table to the other. (FSL)

Permafrost Permanently frozen soil horizon. (SWT)

Permanent gas Term used to describe a gas that cannot be liquefied at normal ambient temperatures. (FSL)

Permanent wilting percentage For a soil, the water content, by percentage, at which indicator plants wilt and do not recover. (SWT)

Permeability, soil The rate at which gases, liquids, or plant roots penetrate or pass through a bulk mass of soil or a layer of soil. (SWT)

Permeable Refers to a substance or barrier that affords passage or penetration. *See also* Penetration. (FSL)

Permeation Movement, on a molecular level, of a substance through a barrier, such as a chemically resistant glove. *See also* Penetration. (MAG)

Permissible dose (radiation) The amount of radiation that may be received by an individual within a specified period with expectation of no harmful result. For long-continued X-ray or gamma-ray exposure of the whole body, the dose permissible is 0.3 R per week, measured in air. (RMB)

Permissible exposure limits (PELs) The specific levels established by OSHA for airborne human exposure to industrial chemicals and also for exposure to noise. They were originally established when OSHA adopted the 1968 threshold limit value (TLV) list in an attempt to provide legal control of the work environment. They were then renamed permissible exposure limits, or PELs. For laboratory uses of OSHA-regulated substances, the employer must assure that laboratory employees' exposures to such substances do not exceed the permissible exposure limits specified in 29 CFR Part 19910, Subpart 2. PELs exist for several hundred chemicals and for noise. *See also* Ceiling exposure limit; short-term exposure limit; threshold limit value; time-weighted average. (FSL, DEJ)

Permit A written authorization that entitles the permittee to commence and/or continue an activity, such as operation of a facility, as long as both procedural and performance standards are met. It may include a functional equivalent, such as a registration or license. *See also* Part B permit. (DEW)

Permit-required confined space An enclosed space that is large enough and so configured that an employee can enter it and perform assigned work. This type of confined space has one or more of the following characteristics: **(1)** it contains or has a known potential to contain a hazardous atmosphere; **(2)** it contains a material that has the potential for engulfment of an entrant; **(3)** it has an internal configuration that might trap or asphyxiate an entrant because of inwardly converging walls or a floor that slopes downward and tapers to a smaller cross-section; **(4)** it contains any other recognized serious safety or health hazard. *See also* Confined space; low-hazard permit space; permitting systems. (SAN)

Permitting systems An employer's written procedures for preparing and issuing permits for certain activities, such as hot work, excavations, and confined space entry. The term can also apply to governmental agencies and their duties associated with the issuance of permits for food-service operations, water well construction, subsurface sewage disposal system installation, building permits, etc. (SAN)

Persistence The capacity of a chemical substance to remain stable. Most of the chlorinated hydrocarbon insecticides meet this criterion. Residues of these materials remain in the soil at detectable levels for extended periods of time. For this reason, compounds such as

DDT, endrin, and aldrin are no longer used as pesticides. (FSL)

Persistent pesticides Pesticides that do not break down chemically or that break down very slowly and that remain in the environment after a growing season. (FSL)

Person An individual, trust, firm, joint stock company, corporation (including a government corporation), partnership, association, state, interstate body, or department, agency, or instrumentality of the United States. (FSL)

Personal decontamination The removal of biological, chemical, or radioactive material from the body or clothing by appropriate mechanical and/or chemical means. (FSL)

Personal decontamination center **(1)** (Radiation) In a building or area, the space in which an individual who has been exposed to radioactive contamination may remove his or her clothing and be monitored for radioactivity, remove radioactivity by bathing and/or chemical procedures, and have clean clothing and dosimetry devices issued before resuming work. **(2)** (Chemicals) The facility established for removal of chemical contaminants and subsequent decontamination of individuals who may have been exposed to chemicals at an emergency response scene or hazardous-waste disposal site, or as the result of an industrial accident involving exposure to chemical compounds. **(3)** (Asbestos) A shower/wash station designed specifically for the removal of asbestos fibers from the personal protective clothing of workers who have been engaged in asbestos abatement. (RMB, MAG)

Personal hygiene **(1)** Those protective measures, primarily an individual responsibility, that promote health and limit the spread of infectious diseases, principally those spread by direct contact. **(2)** Applies to employees who are involved in handling foods during processing or in the preparation and service to a consumer, and who therefore represent possible sources of contamination of the foods and are a critical part of the food sanitation equation. Ill employees should not be allowed to work around foods, equipment, and utensils used in the preparation and serving of foods. Food handlers' hands and hair, as well as the materials they release during coughing and sneezing, are sources of organisms and can also contaminate foods. Proper health and hygiene precautions are usually identified in foodservice sanitation manuals. Statements like the following are included:

> No person, while infected with a disease in a communicable form that can be transmitted by foods or who is a carrier of organisms that cause such a disease or while afflicted with a boil, an infected wound, or an acute respiratory infection, shall work in a food service establishment in any capacity in which there is a likelihood of such person contaminating food or food-contact surfaces with pathogenic organisms or transmitting disease to other persons.
>
> Employees shall thoroughly wash their hands and the exposed portions of their arms with soap and warm water before starting work, during work as often as is necessary to keep them clean, and after smoking, eating, drinking, or using the toilet. Employees shall keep their fingernails clean and trimmed.

(HMB, FSL)

Personal Monitoring **(1)** (Radiation) The determination, using standard survey meters, of the degree of radioactive contamination on individuals, and the determination, by means of dosimetry devices, of dosage received. **(2)** (Industrial hygiene) The practice of having an individual wear monitoring or sampling equipment to measure exposure to various hazardous substances or agents—for example, attaching an air-collecting medium, often housed in a filter cassette or glass tube, which can then be attached to an air-sampling pump worn on the belt (or to no pump if a diffusion sampler is used). Personal noise and radiation monitoring are conducted by having an individual wear a noise or radiation dosimeter, usually for the duration of a workday. (DEJ, RMB)

Personal protective equipment (PPE) Any of a number of devices or types of equipment worn by an individual to provide protection against various hazards. It includes respirators, hardhats, safety shoes, gloves, face shields, chemical splash goggles, and protective clothing, such as aprons, laboratory coats, acid-resistant body suits, and other impermeables. (DEJ)

Pertussis An infectious disease, also known as whooping cough, that is due to an infection with the bacterial organism *Bordetella pertussis*. The infection is characterized by an explosive cough accompanied by a crowing or whooping sound. (FSL)

Pest An insect, rodent, nematode, fungus, weed, or other form of terrestrial or aquatic animal or plant life, or a virus, bacteria, or other microorganism that is injurious to health or to the environment. (FSL)

Pesticide For purposes of the Federal Insecticide, Fungicide, and Rodenticide Act, any substance or mixture of substances intended for preventing, destroying, repelling, or mitigating any kind of pest, and any substance or mixture of substances intended for use as a plant regulator, defoliant, or desiccant, with the general exception of specified animal drugs or feed. 7 USC §136(u). *See also* Contact pesticide; Federal Insecticide, Fungicide, and Rodenticide Act; microbial pesticide. (FSL)

Pesticide tolerance The amount of pesticide residue allowed by law to remain in or on a harvested crop. By using various safety factors, EPA sets these levels well below the point at which the chemicals might be harmful to consumers. (FSL)

Petition Generally, a formal written request that a specified action be taken. (DEW)

Petri plate Shallow flat-bottomed, covered dish with flat, overhanging covers, used for culturing microorganisms. Such glass or plastic plates are used in essentially all microbiology laboratories today. Solidified growth media are used to provide nutrients to grow microorganisms. The nutrient agar in plates provides a large area over which microorganisms can be spread, and the lids help to reduce contamination. *See also* Petri film. (HMB)

Petri film method A dry film method of microbiological assay consisting of the use of two plastic films attached together on one side and coated with plate-count agar ingredients and a cold-soluble jelling agent. This method, which simplifies the standard plate-count procedure, was recently developed. Following incubation, growth appears on the film as small colonies colored red by a dye incorporated in the agar. Selective medium films are also being developed. (HMB)

Petrocalcic soil horizon A continuous, indurated calcic horizon that is cemented with calcium carbonate or magnesium carbonate. It is usually impenetrable by plant roots. (SWT)

Petrochemical An organic compound, such as gasoline, kerosene, or petroleum, that has been obtained from petroleum or natural gas. (FSL)

Petrogypsic soil horizon A continuous, strongly adhered horizon that is cemented with calcium sulfate. (SWT)

pH The negative logarithm of the hydrogen ion concentration $[H^+]$. A measure of the degree to which a substance is acidic or basic. pH values are indicated on a scale of 1–14, with those above the neutral point (7) being characteristic of bases and those below being characteristic of acids. (FSL)

Phage typing The characterization of a bacterium by the identification of those bacteriophages to which the bacterium is susceptible; a means of strain differentiation. (FSL)

Phagocyte A cell that engulfs and destroys foreign particles or microorganisms by digestion. *See also* Inflammation; leukocyte; lymphocyte. (FSL)

Phagocytosis The process by which a cell ingests particles from its surroundings. (FSL)

Phalloidin A toxin, contained in the mushroom *Amanita phalloides*, that, following ingestion, can cause illness and in some cases death in humans. (FSL)

Pharmaceuticals A collective term that includes all types of drugs and medication and also related products such as vitamins, tonics, and dietary supplements. *See also* Pharmacology. (FSL)

Pharmacology The study of the preparation of drugs, their sources, appearance, chemistry, actions, and uses. *See also* Toxicology. (FSL)

pH-dependent charge In ion-exchange separation procedures, the portion of the cation or anion exchange capacity that varies with pH. (SWT)

Phenol (C_6H_5OH) Aromatic organic compound that is a by-product of petroleum refining, tanning, and textile, dye, and resin manufacturing. Phenol is also known as carbolic acid, phenylic acid, and hydroxybenzene. Phenol is a strong irritant to tissue and is toxic by ingestion, inhalation, and skin adsorption. Even low concentrations cause taste and odor problems in water. Higher concentrations can be harmful to aquatic life, humans, and animals. Phenol is widely used as a disinfectant in laboratories, in phenolic resins, and as a component of rubber chemicals. (FSL)

Phenol coefficient A measure of the effectiveness of a germicide. It is expressed as the ratio of the effectiveness, against a specific organism, of the test germicide to the effectiveness of phenol. For example, if a 1 : 128 dilution of a test germicide kills a standard population of *Salmonella typhimurium*, and the highest dilution of phenol that shows the same effectiveness is 1 : 32, the phenol coefficient of the test germicide is 128/32 or 4.0. This phenol coefficient indicates that the test germicide is four times as effective as phenol in killing *S. typhimurium*. The phenol coefficient is an important measurement because it is essential in the development of new antimicrobial agents to have a known standard to which their effectiveness can be compared. (HMB)

Pheromone A chemical, released by an animal (often an insect) into its environment, that influences the behavior or development of others of the same species. Both natural and artificial pheromones are used as attractants to lure certain insects into traps, where they can be eliminated. (FSL)

Phloem A specialized tissue in the vascular system of plants that conducts food in solution. The fluids carried by the phloem and xylem are commonly called sap. *See also* Xylem; sap. (SMC)

Phoresy The transport of one organism by another of a different species—e.g., the carrying of mites by insects. (SMC)

Phosgene ($COCl_2$) A toxic gas, known also as carbonyl chloride, produced by the reaction of chlorine compounds (e.g., trichloroethylene) and carbon monoxide. Used as a chemical warfare agent during World War I, phosgene interferes with the nervous system, causes inflammation of the lungs, and is a very strong irritant to the eyes. It is used in the synthesis of polyurethane and polycarbonate resins, carbamates, and pesticides, and in the manufacture of dyes. (FSL)

Phosphatase Test A test applied to dairy products to determine whether raw milk has been added to pasteurized milk or if the product has been properly pasteurized. Phosphatases are a large and complex group of enzymes, some of which are highly specific. Generally, phosphatases hydrolyze esters of phosphoric acid. In milk the enzyme is highly heat sensitive, making the test an excellent one for determining pasteurization effectiveness or the possibility of the presence of added raw milk.

The test is based on the reduction of phosphatase to a certain level of activity and it cannot be interpreted as an absolute test of proper pasteurization. It is possible for milk with a minute amount of raw milk present or for blends of over- and underpasteurized milk to yield a negative phosphatase test result. The principle of the test is that the alkaline phosphatase enzyme in raw milk liberates phenol from a disodium phenyl phosphate substrate or phenolphthalein from a phenolphthalein monophosphate substrate when the test is done at an appropriate temperature and pH. (HMB)

Phosphor A substance, either organic or inorganic, liquid or crystal, that has the ability to absorb energy from incident radiation and emit a portion of this energy as visible or ultraviolet light. (RMB)

Phosphorescence The emission of radiation by a substance as a result of previous absorption of radiation of shorter wavelength. In contrast to fluorescence, the emission may continue for a considerable time after cessation of the exciting radiation. (RMB)

Phosphoric acid (H_3PO_4) An acid formed by the action of sulfuric acid or hydrochloric acid on phosphate. Phosphoric acid is toxic by ingestion and inhalation and is an irritant to the skin and eyes. The TLV is 1.0 mg/m^3 in air. Among the highest-volume chemicals produced in the United States, phosphoric acid is used in fertilizers, soaps, sugar refining, and water treatment, and as a laboratory reagent. (RMB)

Phosphorus (P) An element that occurs in several forms, the most common of which is white and very toxic. Phosphorus can ignite spontaneously in air at 86°F and must be stored under water and away from heat. It is toxic by ingestion and inhalation, and contact with the skin can cause burns. The TLV is 0.1 mg/m^3 of air. White phosphorus is used as a rodenticide, in obscuring smoke screens, and as an analytical reagent. Red phosphorus is used in the manufacture of phosphoric acid, fertilizers, matches, and detergents. (FSL)

Photochemical oxidants Air pollutants, such as aldehydes, acids, and nitrates, formed by the action of sunlight on oxides of nitrogen and hydrocarbons. (FSL)

Photochemical smog Air pollution caused by the action of sunlight on various pollutants emitted from numerous sources. For example, olefins (open-chained hydrocarbons containing one or more double bonds) are highly photochemically reactive, as are many aromatic hydrocarbons (although benzene is an exception). (FSL)

Photochemistry The branch of chemistry that relates to the effects of light or other radiant energy in producing chemical action, as in photography. (FSL)

Photodosimetry *See* Photographic dosimetry. (RMB)

Photoelectric effect An effect observed in process by which a photon ejects an electron from an atom. All the energy of the photon is absorbed in ejecting the electron and in imparting kinetic energy to it. This effect as an absorption process is large for low-energy photons impinging on high-Z elements. (RMB)

Photofluorography *See* Fluorography. (RMB)

Photographic dosimetry The determination of the cumulative dose of ionizing radiation by use of photographic film. *See also* Film badge. (RMB)

Photomap A mosaic map composed of aerial photographs that have physical and cultural features similar to those shown on a planimetric map. (SWT)

Photometry The measurement of the intensity of light. (FSL)

Photon The quantum of energy emitted or absorbed in the form of electromagnetic radiation whose energy value is the product of its frequency and Planck's constant ($E = h\nu$). (RMB)

Photoneutron A neutron released from a nucleus in a photonuclear reaction. (RMB)

Photoneutron source A source of neutrons in which the nuclear reaction is induced by impingement of a gamma ray on a light element, such as beryllium or deuterium. (RMB)

Photosynthesis Synthesis by means of energy from light; specifically, the synthesis of carbohydrates by green plants in the presence of sunlight. The absorption of light by chlorophyll in green plants stimulates photochemical reactions in which oxygen is released from water and the light energy is converted into chemical energy by the formation of adenosine triphosphate (ATP). (FSL)

Phototropic Refers to organisms that obtain energy from sunlight. (FSL)

Phototropism The response in which an organism turns toward light. Plants are phototropic and either the leaves or the entire plant rotates toward the light. (FSL)

Phreatic water Water found below the water table, in the zone of saturation (ground water). (FSL)

Phyllo-, phyll- Prefix that indicates a leaf or a structure or something that pertains to leaves. (SMC)

Phyllosphere The surface of above-ground living plant parts. (SWT)

Phylogenic Refers to the lines of origin or lines of descent or evolutionary development of any species. (FSL)

Phylum The taxon in the classification hierarchy between kingdom and class. (SMC)

Physical and chemical treatment Processes generally used in large-scale wastewater treatment facilities. Physical processes may involve air-stripping or filtration. Chemical treatment includes coagulation, chlorination, and ozone addition. The term can also refer to treatment of toxic materials in surface waters and ground waters, treatment of oil spills, and some methods of dealing with hazardous materials on or in the ground. (FSL)

Physical hazard Term used by OSHA to classify those chemicals that are combustible liquids, compressed gases, explosives, flammable materials, organic peroxides, oxidizers, or pyrophoric, unstable (reactive), or water-reactive substances. *See also* Hazardous material; hazardous waste. (FSL)

Physical properties (soil) Those characteristics, processes, or reactions of a soil that are influenced by physical forces and that are described by or expressed in physical terms or equations. Examples are bulk density, water-holding capacity, hydraulic conductivity, and porosity. (SWT)

Physical weathering The breakdown of rocks and minerals into smaller particles by physical forces such as frost. (SWT)

Physiognomy In ecology, the shape or appearance of a community as determined by the growth form of the dominant vegetation (e.g., grassland, pine forest, beech–maple forest). (SMC)

Physiology The science concerned with the normal vital processes of organisms, especially as to their normal functioning rather than to their anatomical structure. *See also* Histology. (FSL)

Phyto-, -phyte Prefix that indicates a plant or something that pertains to plants. (SMC)

Phytolith A small opaline rock consisting primarily of fossil plant remains. (SWT)

Phytometer A plant used as a measure or indicator of the physical factors of the habitat in terms of physiological activities. (SWT)

Phytophagous Plant-eating. (SMC)

Phytoplankton Autotrophic plankton; plankton that can photosynthesize; ''plantlike'' plankton. *See also* Plankton. (SMC)

Phytotoxic Refers to a chemical known to be toxic to green plants. (FSL)

Pica A craving, brought on by nutritional deficiencies, for unnatural articles of food. This condition, frequently associated with children living in low socioeconomic conditions and also noted among mentally retarded children, is manifested by their chewing on crib rails, window sills, furniture, and similar articles. Because the surfaces of these items may be covered with lead-based paint, this activity represents a significant source of exposure to the toxic properties of lead. *See also* Lead poisoning; plumbism. (FSL)

Picocurie (pCi) A measurement of radioactivity. A picocurie is one million-millionth, or a trillionth, of a curie, and represents about 2.2 radioactive particle disintegrations per minute. (RMB)

Picocuries per liter (pCi/L) A unit of measure generally used for expressing levels of radon gas in air. *See also* Picocurie. (FSL)

Picric Acid [$C_6H_2(NO_2)_3OH$] Chemical, also known as trinitrophenol, that is used as a dye, tissue fixative, stain, and antiseptic. It is shock-sensitive and will detonate if allowed to dehydrate to the crystalline form. For this reason, the chemical is marketed in the hydrated state and must be periodically checked to make certain the crystals are kept moist. Picric acid is also used in matches, explosives, and in textile dyeing processes. *See also* Shock sensitive. (FSL)

Pig A container (usually lead) used to ship or store radioactive materials. The thick walls protect the person handling the container from radiation. (RMB)

Pile A nuclear reactor, so called because the earliest reactors were "piles" of graphite blocks and uranium slugs. (RMB)

Pinch point A location, other than the point of operation, at which it is possible for a part of the body to be caught between the moving or stationary parts of a machine or its auxiliary equipment. Common sources of pinch points, also known as nip points, are rotating gears and chains. OSHA requires that pinch points be guarded. *See also* Machine guarding. (SAN)

Pioneer species A plant species that invades, colonizes, or germinates in areas soon after a destructive disturbance to plants in an area. Pioneer species are typically early successional species that are easily and widely dispersed. Pioneer plant species are usually shade-intolerant, weedy species and often comprise the first community in secondary growth succession. (SMC)

Pitch An attribute of auditory sensation on which sounds may be ordered on a scale from low to high. Pitch is dependent upon the frequency of the sound stimulus, but also upon the sound pressure and wave form of the stimulus. (DEJ)

Pitchblende **(1)** A mixture of uranium oxides that occurs naturally in large masses. **(2)** A black mineral that contains uranium oxides and from which are obtained radium, uranium, and polonium. (RMB)

Placenta A vascular organ developed within the uterus of mammals that serves as the structure through which nourishment for the fetus is received from the circulatory system of the mother, and wastes of the fetus are eliminated. (FSL)

Placic horizon A thin, black to dark-reddish mineral soil horizon that is commonly cemented with iron and is slowly permeable. (SWT)

Plaggen epipedon A soil surface horizon that is formed by long-term manuring and mixing. (SWT)

Plague An infectious, usually fatal disease caused by the bacillus *Yersinia pestis*. The disease is spread from rats and other rodents by the bite of infected fleas. Plague can occur in three forms: bubonic, pneumonic, and septicemic. Bubonic plague is characterized by buboes, or inflammation of the lymph nodes. Pneumonic plague involves the lungs. Septicemic plague involves mainly the blood. Prevention of the disease can be brought about by effective rodent control and the elimination of their fleas or other ectoparasites. (FSL)

Plaintiff The party who initiates a lawsuit. (DEW)

Planck's constant A (ν) natural constant of proportionality (h) relating the frequency of a quantum of energy to the total energy of the quantum:

$$h = E/\nu = 6.6 \times 10^{-27} \text{ erg-s}$$

Plankton Free-floating aquatic organisms, most of which are microscopic. (SMC)

Plan review When a new food-service establishment is to be built, or an existing establishment is renovated, food-service regulations usually require that construction or renovation plans be reviewed and approved by public health officials. This entails a complete review of all elements of the facility to ensure that cleaning and sanitation are facilitated and food safety is optimized. Items reviewed parallel those covered in a sanitary survey. These include physical facilities such as floors, walls, ceilings, lighting, handwashing and toilet facilities, dressing rooms, and lockers. Also covered would be all aspects of plumbing, water supplies (with specification of appropriate temperatures for all types of usage), and sewage disposal. All kitchen facilities are reviewed to ensure that installation, design, and construction aspects facilitate cleaning and sanitation. These facilities include storage areas, coolers/freezers, cooking equipment, dishwashing machines, and kitchen ventilation devices. Also considered are insect and rodent control aspects and garbage disposal facilities. The plan review process should be thorough so that food sanitation problems can be avoided and are not inherent in the new or remodeled establishment. (HMB)

Plant geography The study of the distribution of plants and plant community types. (SMC)

Plant growth regulator A chemical, such as atrazine, that, when used as a herbicide, controls or inhibits the growth of undesirable plants in noncrop areas or industrial sites. Such materials are reported to inhibit the photosynthetic process in algae. (FSL)

Plant nutrient An element absorbed by plants and necessary for growth. (SWT)

Plasma **(1)** (Human) The fluid (noncellular portion) of the circulating blood, as distinguished from serum obtained after coagulation. **(2)** (Soil) That part of the soil material that is composed of colloidal-sized materials (mineral and organic). (SWT)

Plasmid A circular piece of DNA that exists apart from chromosomes and replicates independently of them. Bacterial plasmids generally carry information that renders the bacteria resistant to antibiotics. Plasmids are often used in genetic engineering to carry desired genes into organisms. (FSL)

Plastic A high polymer, usually synthetic, that has been combined with other ingredients, such as fillers, so as to allow the mixture to be formed or molded under heat. There are two major types of plastic: thermoplastic materials, which can be resoftened to their original condition by the application of heat, and thermosetting materials, which cannot be resoftened. Plastics are used in flooring, pipes, piping insulation, packaging, automobiles, and numerous other applications. Some types of plastics are combustible when exposed to flames, with the resultant generation of toxic fumes, particularly in the case of polyurethane. (FSL)

Plasticity **(1)** The capacity of an organism to adapt itself to various environmental conditions. **(2)** The expression of various phenotypes from the same genotype (e.g., all dandelions are very similar in genotype but vary widely in size and height, depending on moisture, mowing, and other factors. **(3)** The capacity of the soil to be changed in shape under applied stress and to retain the impressed shape after removal of the stress. (SMC)

Plastic limit The minimum amount of moisture percentage by weight at which a small sample of soil material can be deformed without rupture. (SWT)

Plastic soil A soil that is capable of being molded or deformed by moderate pressure. (SWT)

Plateau As applied to radiation detector chambers, the level portion of the counting-rate–voltage curve, at which changes in operating voltage introduce minimum changes in the counting rate. (RMB)

Platykurtic Refers to a flat-topped statistical distribution. (VWP)

Playa A flooded, vegetatively barren area on a basin floor that is veneered with fine-textured sediment that acts as a final barrier for drainage water. (SWT)

Plinthite A mixture of iron and aluminum oxides, clay, quartz, and other materials that commonly occurs as red soil that is usually mottled in platy, polygonal, or reticulate patterns. Upon repeated wetting and drying, plinthite changes irreversibly to ironstone hardpans or irregular aggregates. (SWT)

Plow layer The top layer of soil (15 to 20 cm) that is inverted during tillage operations. (SWT)

Plow pan *See* Pan; pressure. (SWT)

Plugging **(1)** The act or process of stopping the flow of water, oil, or gas into or out of a borehole or well penetrating a formation. **(2)** The act of stopping a leak or sealing off a pipe or hose. (FSL)

Plumbism A chronic poisoning of humans, caused by the absorption of lead or lead salts. *See also* Lead poisoning. (FSL)

Plume **(1)** A visible or measurable discharge of a contaminant from a given point of origin, as, for example, a plume of smoke. **(2)** The area of measurable and potentially harmful radiation leaking from a damaged reactor. **(3)** The area around a toxic release considered dangerous for those exposed to leaking fumes. (FSL)

Plutonium (Pu) A synthetic radioactive metallic element originally prepared in 1941 and chemically similar to uranium. Plutonium is the most radiotoxic of the elements and among the most toxic substances known; it is also a powerful carcinogen. Plutonium must be handled by remote control and with proper shielding. This element forms low-melting alloys with a number of metals (iron, tin, cobalt, and nickel); these alloys are often used as liquid reactor fuels. (FSL)

Pneumatic placing The insertion of blasting agents or explosives into a borehole, using compressed air as the loading force. (FSL)

Pneumoconiosis Inflammation often leading to fibrosis of the lungs and caused by the inhalation of dust associated with various occupations—e.g., mining. The disease is characterized by chest pains, cough, cyanosis, and fatigue. Bauxite pneumoconiosis (Shaver's disease) is caused by the inhalation of bauxite fumes emitted during the manufacture of alumina abrasives. *See also* Anthracosis; black lung disease; mica pneumoconiosis; Shaver's disease. (FSL)

Pocket chamber A small condenser ionization chamber used for determining radiation exposure. An auxiliary charging and reading device is usually necessary. (RMB)

Pocket dosimeter A direct-reading portable unit shaped like a pen with a pocket clip, generally used to measure X- and gamma radiation. The advantage of this type of dosimeter is that it allows an individual to determine radiation dose while still working with radioactive materials, without having to wait for the results of film badge or thermoluminescent dosimeter results. (RMB)

Pocosin A swamp that is partly or completely enclosed by a sandy rim and that contains an organic soil. (SWT)

Podzolization A soil formation process that results in the formation of podzols—that is, soils formed in temperate climates under coniferous or deciduous trees. (SWT)

Point estimate For a parameter, the approximation that is the value of a corresponding sample statistic. (VWP)

Point of operation guard A guarding device used in the area of a machine where material is actually positioned and work is being performed during any process, such as shearing, punching, forming, or assembling. Any point of operation that exposes an employee to injury must be guarded according to OSHA requirements. The guarding device must be in conformity with appropriate standards or, in the absence of applicable specific standards, must be designed and constructed so as to prevent any part of the operator's body from being in the danger zone during the operating cycle. Special hand tools for placing and removing material can be used to supplement a guard but cannot be used in lieu of one. Machines that usually require point-of-operation guarding include guillotine cutters, shears, alligator shears, power presses, milling machines, power saws, jointers, portable power tools, forming rolls, and calendars. *See also* Machine guarding. (SAN)

Point source Under the Clean Water Act, any discernible, confined, and discrete conveyance, including, but not limited to, any pipe, ditch, channel, tunnel, conduit, well, discrete fissure, container, rolling stock, concentrated animal feeding operation, landfill leachate collection system, vessel, or other floating craft from which pollutants are or may be discharged. 40 CFR §122.2. *See also* Clean Air Act; Clean Water Act. (DEW)

Poison Any substance that, when administered to a living organism, causes a harmful effect. Most substances are harmful at some dose and are harmless at very low doses. *See also* Hazardous material, toxic. (FSL)

Poisson distribution The spread of probabilities of K (which can be indefinitely large) events with a parameter λ, which is the expected number of events over a time period, is given by the equation

$$P(x) = \frac{\lambda^x e^{-\lambda}}{x!} \quad \text{where } x = 0, 1, 2, \ldots \quad \text{and } e \approx 2.71828.$$

The equation is used to approximate the binomial distribution for $n \geq 100$ and $P \leq 0.01$. (VWP)

Pollen Microspores of seed plants that are generally carried by the wind or insects prior to fertilization. Pollen is a background air pollutant important in the etiology of hay fever. (FSL)

Pollutant or contaminant **(1)** For purposes of the Clean Water Act, dredged spoil, solid waste, incinerator residue, sewage, garbage, sewage sludge, munitions, chemical wastes, biological materials, radioactive materials, wrecked or discarded equipment, rock, sand, cellar dirt, and industrial, municipal, and agricultural waste discharged into water. Does not include: sewage from vessels (as covered by 33 USC §1322); water, gas, or other material that is injected into a well or that facilitates production of oil or gas; or water derived in association with oil or gas production and disposed of in a well, if the well is used either to facilitate production or for disposal purposes and is approved as such by authority of the state in which the well is located. If such state determines that injection or disposal will not result in the degradation of ground or surface water resources, the activity will be permitted. 33 USC §1362(6). **(2)** For purposes of CERCLA, includes but is not limited to, any element, substance, compound, or mixture, including disease-causing agents, that, after release into the environment and upon exposure, ingestion, inhalation, or assimilation into any organism, either directly from the environment or indirectly by ingestion through food chains, will or may reasonably be anticipated to cause death, disease, behavioral abnormalities, cancer, genetic mutation, physiological malfunctions, or physical deformations in such organisms or their offspring. Excludes petroleum and petroleum fractions, and natural gas. *See also* Clean Air Act; Comprehensive Environmental Response, Compensation, and Liability Act (CERCLA); Clean Water Act; Release. (DEW)

Pollution *See* Pollutant or contaminant; environmental pollution. (FSL)

Polybrominated biphenyl (PBB) Heavy brominated compounds such as hexabromobiphenyl ($Br_3C_6H_2$–$C_6H_2Br_3$), which are incorporated into thermoplastics to increase the heat stability of such plastics, which are used in calculators, microfilm readers, and business machine housings. One PBB, a fire-retardant material called Firemaster BP-6, was accidently added to animal feed in Michigan in 1973 and resulted in the widespread poisoning of farm animals. This incident led to the removal of BP-6 from the market. PBBs are considered to be carcinogenic. *See* Polychlorinated biphenyl (PCB). (FSL)

Polycarbonates **[$(COOC_6H_5C(CH_3)_2C_6H_5O)_n$]** Thermoplastic resins derived from bis-phenyl A and phosgene. Polycarbonates are used in structural parts, tubes and piping, street-light globes, and household appliances. They are considered noncorrosive, resistant to weather and ozone, and nontoxic. (FSL)

Polychlorinated biphenyl (PCB) An aromatic compound containing two benzene rings with two or more substituted chlorine atoms. PCBs were once industrial chemicals widely used in electric capacitors, transformers, vacuum pumps, and liquid insulators, and as fire retardants, adhesives, and plasticizers. The use of PCBs was discontinued in the United States in 1976 because of their toxicity, persistence in the environment, and the degree of environmental contamination. Toxic effects in humans include chloracne, abnormal pigmentation of skin and nails, and gastrointestinal disturbances. Accidental contamination of rice bran oil in Japan with Kaneschlor 400, a PCB, led to an outbreak of what was called Yurbo disease. *See also* Chloracne; polybrominated biphenyl (PBB). (FSL)

Polycyclic Refers to organic compounds having three or more ring structures that may be similar or different, such as in anthracene or naphthalene, respectively. (FSL)

Polycyclic aromatic hydrocarbon (PAH) A chemical such as benzopyrene ($C_{20}H_{12}$), a polynuclear (5-ring) aromatic hydrocarbon that is found in cigarette smoke, coal tar, and in the atmosphere, as a result of incomplete combustion. Benzopyrene is highly toxic and is a carcinogen by inhalation. (FSL)

Polycythemia A disease characterized by overproduction of red blood cells. *See also* Erythrocytes. (FSL)

Polymer A substance formed by the chemical union of five or more identical units called monomers. Polymers include organic materials, such as cellulose, pectin, and vegetable gums, synthetics, such as polyvinyl chloride, polystyrene, and polyurethane, and materials, such as rayon and cellulose acetate. (FSL)

Polymerization A chemical reaction in which a high-molecular-weight material is produced by the addition to or condensation of a simpler compound—for example, the production of polystyrene from styrene. (FSL)

Polymerization gases Gases, such as vinyl chloride and butadiene, that are used in the process of forming polymers. (FSL)

Polyvinyl chloride [PVC, $(-H_2CCHCl-)_x$] A synthetic thermoplastic polymer that releases hydrochloric acid when burned. PVC is used in piping, conduits, plumbing, and other construction applications. The use of PVC in food containers such as bottles and boxes is under restriction by the FDA. (FSL)

Pontiac fever A flu-like illness, with little or no pulmonary involvement, that is caused by the organism *Legionella pneumophila*. The disease, which is named for the city of Pontiac, Michigan, is not as life-threatening as the pulmonary form, which is known as Legionnaires' disease. *See also* Legionnaires' Disease. (FSL)

Population **(1)** A collection of individuals or objects whose characteristics are to be investigated. **(2)** a group of interacting and (in the case of plants and animals) interbreeding organisms of the same species. (SMC, VWP)

Pore ice Frozen water in the interstitial spaces of a porous medium such as soil. (SWT)

Pore-size distribution The quantification of the various pore sizes in a soil, usually expressed as a percentage of bulk volume (soil plus pore space). (SWT)

Pore space In a bulk volume of soil, the space not occupied by soil particles. (SWT)

Porosity The percentage of the total bulk not occupied by solid particles. (SWT)

Portland cement A mixture of calcium carbonate, silica, and alumina that is the most common type of cement used in construction. The cement can be modified by the addition of various plastic latexes to improve adhesion strength and curing properties. (FSL)

Positive-displacement pump In an individual water distribution system, a pump that forces or displaces water through a pumping mechanism. Such pumps may include reciprocating, helical, and regenerative turbine types. (FSL)

Positive predictive value The probability that an organism really has a specific characteristic being tested for if the test being used to detect the characteristic has a positive result. (VWP)

Positron Particle equal in mass but opposite in charge to the electron—i.e., a positive electron. (RMB)

Post-closure period The time period following the shutdown of a waste management or manufacturing facility. For monitoring purposes, this is often considered to be 30 years. *See also* Comprehensive Environmental Response Compensation and Liability Act (CERCLA). (FSL)

Post-emergence herbicide *See* Atrazine.

Potable water Water that is fit to drink and free of physical, chemical, microbial, and radiologic contaminants in amounts that would be sufficient to cause disease or harmful physiological effects. Water from any source that has been treated chemically or by physical means to remove undesirable chemicals, microorganisms, suspended solids, or color, and is suitable as a drinking water supply. A safe, potable water supply is mandatory at all food-handling facilities. (HMB)

Potassium (K) A white, soft, reactive material with an atomic weight of 39.102 and an atomic number of 19. It is an essential plant nutrient and a major component of fertilizers. Potassium is a dangerous fire risk and reacts with moisture to form potassium hydroxide and hydrogen. It can ignite spontaneously in moist air and is difficult to extinguish. When stored under mineral oil, potassium metal may form peroxides that may explode violently when the metal is handled or cut. (FSL)

Potassium-supplying power of soil The ability of the soil to supply potassium as a plant nutrient from exchangeable and moderately available forms. (SWT)

Potential energy *See* Energy. (RMB)

Potentially responsible party (PRP) A business, individual, or organization most likely responsible for a pollution incident. The EPA seeks reimbursements from PRPs for costs of cleaning up waste disposal sites on the National Priority List and in other situations in which there has been a release of a hazardous material to the environment. *See also* Comprehensive Environmental Response, Compensation, and Liability Act (CERCLA). (FSL)

Potentiation An increased toxicologic effect by one agent on another, resulting in a combined effect that is greater than the simple sum of those of the individual agents. *See also* Synergism. (FSL)

Potentiometer An instrument for measuring, comparing, or controlling electric potential—e.g., the relative voltage at a point in an electrical circuit or field with respect to a reference point in the same circuit or field. (FSL)

Pother A dense cloud of smoke, dust, or similar material. (FSL)

Potomology The study of streams or stream sanitation. (FSL)

Power **(1)** A source of physical or mechanical force or energy; force or energy that is at work or can be put to work—for example, water or electric power. **(2)** The rate at which work is done. (RMB)

Powered air-purifying respirator (PAPR) *See* Air-purifying respirator. (DEJ)

Poza Rica A city in Mexico that was the location of an air pollution incident that killed 22 people and made 320 ill. The incident occurred in 1950 as a result of the discharge of large quantities of hydrogen sulfide into the atmosphere from an oil refinery during a thermal inversion. (FSL)

Precipitate A collection or deposit of solid particles that have settled out of solution. (FSL)

Precipitation **(1)** (Chemistry) The formation and settling out of solid particles from a solution.

(2) (Meteorological) The presence of rain, snow, sleet, etc. (FSL)

Precipitation interception The interception of and temporary holding of precipitation in any form by a vegetative canopy such as bushes, trees, shrubs, or vegetation residue. (SWT)

Precipitators Air pollution control devices that collect particles from an emission. In a precipitator, particles pass through an electrostatic field and become ionized (electrically charged). These particles are then attracted to collector plates that have the opposite charge. These precipitators have the advantage of negligible pressure drop. When used in dusty environments, they are usually preceded by primary collectors, which work under a different principle. Although they can collect mists, electrostatic precipitators are unable to collect nonparticulate gases or vapors. (FSL)

Precursor **(1)** (Photochemistry) A compound, such as a volatile organic compound (VOC), from which an oxidant is formed. Precursors react in sunlight to form ozone or other photochemical oxidants. **(2)** (Physiology) A physiologically inactive substance that is converted to an active enzyme, vitamin, hormone, etc., or to a chemical substance that is built into a larger structure in the course of synthesis of the latter. (FSL)

Predator Free-living organism that consumes other organisms (prey). If the organisms being eaten are plants, the predator is usually referred to instead as an herbivore. If the consuming organism is not free-living, it is usually called a parasite. However, the term "predator-prey relations" is often used to refer to all true predator-prey, herbivore-plant, and parasite-host relationships. (SMC)

Pre-discharge employee alarm An alarm that will annunciate at an established time prior to actual discharge of an extinguishing system, so that employees may evacuate the area before the system releases. These types of alarms are required for areas where gaseous-agent fire-extinguishing systems are installed. *See also* Suppression system. (SAN)

Preliminary assessment For purposes of an action under the Comprehensive Environmental Response, Compensation, and Liability Act, a review of existing information and an off-site reconnaissance, if appropriate, to determine whether a release may require additional investigation or action. May include an on-site reconnaissance, if appropriate. CFR §300.5 *See also* Comprehensive Environmental Response, Compensation, and Liability Act (CERCLA). (FSL)

Preliminary hazard analysis (PHA) A technique used in system safety that is performed in the early design stage of a project. It is a general, qualitative analysis that provides an estimate of potential hazards. PHA includes a review of the design and an identification of any additional safety criteria needed. Equipment and material specifications are also reviewed. There is no specific format for reporting the results of the study. PHA evolved from the U.S. Military Standard System Safety Program, which incorporates many different analytical steps. These methods can be performed in conjunction with a single project or may be applied independently. (SAN)

Premise A platted lot or part thereof or an unplatted lot or parcel of land, that is either occupied or unoccupied by any dwelling or other structure and that may include any building located thereon. (FSL)

Presbycusis Hearing loss due to normal aging, as opposed to hearing loss due to environmental or occupational exposure to noise. (DEJ)

Prescriptive zone (PZ) The range of environmental conditions in which deep body temperature is determined primarily by work intensity. Used to determine health stress exposure limits. *See also* Environmental-driven zone (EDZ). (DEJ)

Preservation The protection of natural resources or habitats. Generally, preservation implies no human changes or lasting impacts on the resource or habitat. (SMC)

Preservatives Chemicals, such as ascorbic acid, ammonium sulfate, and calcium propionate, that prevent the spoilage or decomposition of foods and other products. Any materials applied during procedures such as canning, pickling, or salting to prevent or minimize growth of microorganisms in foods. *See also* Disinfectants; sanitizer; sterilization. (FSL)

Pressure Force applied or distributed over a surface and measured as force per unit area. (FSL)

Pressure-demand respirator A type of respirator that provides a positive pressure during both inhalation and exhalation and delivers an air flow of about 115 L/min. (4 cfm) before a negative pressure is measurable at the face piece. These respirators are used in situations in which an inward leakage at the face seal may be unacceptable and in which it is not practical to have the relatively high air consumption characteristic of a constant-flow unit. *See also* Air-purifying respirator. (FSL)

Pressure membrane A membrane that is permeable to water and only slightly permeable to gas when wet, through which water can escape, in response to a pressure gradient from a soil sample. (SWT)

Pressure sewer A system of piping in which water, wastewater, or other liquid is transported to a higher elevation by use of pumping force. (FSL)

Pressure-type connection A type of cross-connection. May also be described as a direct-type connection. This type is one in which the water supply is connected to another line or a pressurized vessel. A pressure-type vacuum breaker is used to avoid a cross-connection of this type, usually in nonpotable systems. A pressure-type connection is used typically when a toxic liquid (i.e., any liquid that, when introduced into the water supply, creates, or may create, a danger to the health and well-being of the consumer) is a part of the system. *See also* Backflow; back-siphonage; cross-connection; vacuum breakers. (HMB)

Pressure vessels Tanks that can tolerate internal or external pressure. Operating pressure must be factored into the design. The design pressure is usually higher than the operating pressure. The American Society of Mechanical Engineers (ASME) has developed a boiler and pressure vessel code that details the minimum requirements for materials, design, fabrication, and inspection of these vessels. All vessels with an internal or external design pressure of 15 psi or more are considered pressure vessels by ASME. Division-1 pressure vessels have pressures ≤ 3000 pounds; Division-2 pressure vessels have pressures above that. Vessels that meet the requirements of this standard qualify for the ASME cloverleaf stamp. ASME requires that pressure vessels be hydrostatically tested. ASME also requires that safety relief valves be incorporated into the vessel and be set to release if the pressure exceeds the maximum allowable pressure by more than 10%. *See also* Hydrostatic testing; pneumatic testing; safety relief valve. (SAN)

Pretreatment Processes used to reduce, eliminate, or alter the nature of wastewater pollutants from non-domestic sources before discharge into publicly owned treatment works. (FSL)

Prevalence The probability of currently having a disease. It is calculated by dividing the number of people who have a disease by the number of people in the population being studied. *See also* Incidence; morbidity rate; mortality rate. (VWP)

Prevalence rate *See* Rate. (VMP)

Prevention of significant deterioration Program under the Clean Air Act to prevent degradation of air quality in areas of the country in which the existing air quality is better than the levels established by national standards. *See also* Clean Air Act; national ambient air quality standards. (FSL)

Preventive maintenance A system that involves monitoring, inspection, maintenance, and repair of equipment, engineering controls, and personal protective equipment on a regular schedule. Preventive maintenance programs are designed to anticipate potential hazards and prevent their occurrence. Critical equipment to be included in the program are pressure vessels, piping, emergency relief systems, alarms and interlocks, bellows-type expansion joints, flame arresters, emergency vents, destruct systems such as scrubbers and flares, and rotating machinery. Shutdown systems receive special treatment. This may involve inclusion of a backup system, assignment of a person to monitor the system, and immediate repair of critical safety equipment. *See also* Process safety. (SAN)

Preventive medicine The branch of medicine that deals with prevention of physical and emotional disease and injury, in contrast to treatment of the sick and injured. There are three levels of prevention: primary, secondary, and tertiary. Primary prevention is prevention of the occurrence of disease or injury by immunization against infectious disease, by chlorination of a community water supply, etc. Secondary prevention involves early detection and intervention to prevent advancement of a particular disease or condition—e.g., screening for diabetes. Tertiary prevention is intended to minimize the effects of disease and disability to prevent further deterioration—e.g., the care of pressure points and bladder function in paraplegics. (FSL)

Prey Organisms that are consumed by other organisms (predators). If the predator is an herbivore or a parasite, then the prey is usually called a food plant or a host, respectively. *See also* Predator. (SMC)

Primary blast An initial blast that loosens rock from its natural location. This is in contrast to a secondary blast, which is used after the primary blast to reduce rocks to a smaller size to facilitate handling. *See also* Explosion. (FSL)

Primary clarifiers Settling basins that receive raw wastewater prior to biological treatment. Such basins are a part of primary treatment. Their configuration may be either rectangular or circular. *See also* Clarifier; clarification; primary wastewater treatment; secondary clarifier. (HMB)

Primary drinking water regulation Under the Safe Drinking Water Act, a regulation that sets a maximum contaminant level for a specific contaminant in a public water system or that sets out known treatment techniques to reduce the level of the contaminant. 42 USC §300f(1). *See also* Safe Drinking Water Act; maximum contaminant level. (DEW)

Primary electron The electron ejected from an atom by an initial ionizing event, as caused by a photon or beta particle. (RMB)

Primary food contamination A source of food contamination produced directly by an infected food animal or its discharges. For example, an animal may be sent to market while it is infected with a pathogen or contaminated with a chemical or other residues, e.g., antibiotics. Animal excrement containing pathogens may cross-contaminate other animals or the food-processing environment. Care must be taken to avoid fecal contamination of carcasses when viscera are removed. (HMB)

Primary protective barriers Barriers sufficient to reduce the useful radiation beam to the permissible dose rate. (RMB)

Primary radiation Radiation arising directly from the target of an X-ray tube or from a radioactive source. (RMB)

Primary settling basin The initial settling tank that is used for the removal of settleable solids and through which wastewater passes in a treatment plant. *See also* Primary wastewater treatment. (HMB)

Primary wastewater treatment Unit processes consisting of one or more of the following chemical or physical operations: screening comminution and grinding, grit removal, retention, preaeration, flocculation, sedimentation, skimming, flotation, and sludge removal. It is the first major process in a wastewater treatment facility and does not involve biological oxidation. The treatment method consists of clarification (sedimentation) and sludge removal. *See also* Advanced wastewater treatment; secondary wastewater treatment; clarifier; clarification. (HMB)

Primate Members of the Order Primates, which includes humans, monkeys, and lemurs. (FSL)

Primatology The study of the nonhuman primates (Order Primates), including apes, monkeys, and lemurs. (SMC)

Prismatic soil structure A soil structure with prism-like aggregates that have a vertical axis much longer than the horizontal axis. (SWT)

Privacy Act Federal law that provides individuals with access to federal executive-branch agency records that pertain to those individuals. Also sets out restrictions on the release of those records to others without the prior written consent of the individuals, except for specified uses. 5 USC §552a. (DEW)

Privately owned/government-operated facility (POGO) A facility (buildings or space) that the government leases for its operations. (DEW)

Privy An outhouse; an outdoor toilet, without water, used in rural, recreational, and other unsewered areas for the disposal of human excreta. There are several types of privies, including pit, vault, and box-and-can. *See also* Cesspool; excreta; septic tank; sewage. (FSL)

Probability The number of times an event will probably occur over the range of possible occurrences, expressed as a ratio. The following principles apply to probability. **(1)** *Addition law:* If A and B are mutually exclusive events, $P(A \text{ or } B) = P(A) + P(B)$. **(2)** *Distribution:* A spread of the probabilities associated with each of the values of a random variable. **(3)** *Multiplication law:* $P(A \text{ and } B) = P(A) \cdot P(B|A)$, where $P(B|A)$ is the conditional probability for B, given A. **(4)** *Probability of type-one error (α):* The probability of rejecting the null hypothesis, given that H_0 (the null hypothesis) is true. **(5)** *Probability of type-two error (β):* The probability of accepting the null hypothesis given that H_1 (the alternative hypothesis) is true. *See also* Level of significance. (VWP)

Process hazard analysis The procedure that addresses: the hazards of a process, engineering and administrative controls applicable to the hazards, and their interrelationships; consequences of failure of these controls; and a consequence analysis of the effects on all workplace employees. The process hazard analysis should be performed by a team that has expertise in engineering and process operations. The team should include at least one employee who has experience and knowledge specific to the process being evaluated. A system must be in place for the following purposes: to address the team's findings and recommendations; to document actions taken; to communicate them to operating, maintenance, and other employees whose work assignments are in the facility, and who are affected by the recommendations or actions; and to assure that the recommendations are implemented in a timely manner. *See also* Checklist; event-tree analysis; failure mode and effects analysis; fault-tree analysis; hazard and operability study; what-if. (SAN)

Process safety management The proactive identification, evaluation, and mitigation or prevention of chemical releases that could occur as a result of failures in process, procedures, or equipment in the industrial setting. The major objective of process safety management of highly hazardous chemicals is to prevent releases of such chemicals that could expose employees and/or the community to serious hazards. Using this approach, the process design, process technology, operational and maintenance activities and procedures, nonroutine activities and procedures, emergency preparedness plans and procedures, training programs, and other elements that have an impact on the process are considered in the evaluation. *See also* Checklist; event-tree analysis; failure mode and effects analysis; fault-tree analysis; HAZOPS. (SAN)

Process weight Total weight of all materials, including fuel, used in a manufacturing process. It is used to calculate the allowable particulate emission rate in the process. (FSL)

Prodromal period Period of time between the initial symptom of disease and the first sign or symptom upon which a diagnosis can be based. *See also* Sign; symptom. (FSL)

Prodrome A symptom indicating the onset of disease. (FSL)

Producer *See* Autotroph. (SMC)

Productive soil A soil that has physical, biological, chemical properties that are suitable for crop production in a specific location. (SWT)

Product liability The legal doctrine that governs a manufacturer's liability for injuries and illnesses resulting from the use (and misuse) of their products. The basic legal theories involved relate to negligence and breach of warranty (which deal with the actions of the manufacturers, including their duty to warn consumers of potential hazards), and liability (which deals with the qualities of the product). *See also* Product safety; negligence. (SAN)

Product safety management The branch of safety management that deals with the testing and design of products to minimize the likelihood of injury resulting from their use. However, because misuse also can result in injury, product safety management includes communication of the risks of products by means of labels and technical data sheets. *See also* Negligence; product liability. (SAN)

Profile, soil A vertical section of the soil that extends through all its horizons into the parent material. (SWT)

Prognosis A forecast of the probable cause and/or outcome of a disease or injury. *See also* Epidemiology. (FSL)

Prompt gamma Gamma radiation emitted at the time of fission of a nucleus. (RMB)

Prompt neutrons Neutrons emitted during the fission process. (RMB)

Propergols A generic term that refers to all chemical propellants used in rocket technology. (FSL)

Prophylaxis Preventive measures against disease—e.g., immunization against hepatitis B virus. (JHR)

Proportional counter An instrument in which a gas-filled radiation detection tube or chamber receives pulses that are proportional to the number of ions formed in the gas by the primary ionizing particles. (RMB)

Proportional region In radiation detection, the voltage range in which the gas amplification is greater than 1 and in which the charge collected is proportional to the charge produced by the initial ionizing event. (RMB)

Propylene Oxide (C_3H_6O) Chemical that is produced by the chlorohydration of propylene followed by saponification with lime, and is among the highest-volume chemicals produced in the United States. Propylene oxide is highly flammable, a dangerous fire risk, and an irritant, and has a TLV of 20 ppm in air. It is used in surfactants and detergents and synthetic lubricants and as a fumigant. *See also* Pesticides. (HMB)

Pro rata Proportionately; as divided by a certain rate, percentage, or share. (DEW)

Prospective study A method of identifying causal associations between host or environmental characteristics and disease occurrence. Such associations are identified by comparing the future disease experience of two defined population groups (cohorts), only one of which has the characteristics under study. *See also* Epidemiology. (FSL)

Protection factors Numbers assigned to various types of respirators that indicate, with 95% confidence, the degree of leakage. Defined as the ratio of the concentration of the contaminant outside the mask divided by the concentration of the contaminant inside the mask. Depending on the type of cartridge used and the contaminant under consideration, protection factors for the following face pieces and respirators have been assigned by the National Institute for Occupational Safety and Health:

Device	Protection Factor
Half-mask air-purifying respirator	10
Full-facepiece air-purifying respirator	50
Powered air-purifying loose-fitting helmet or hood respirator	25
Powered air-purifying full-facepiece respirator	50
Supplied air, continuous-flow, loose-fitting helmet or hood respirator	25
Supplied air, continuous flow, full facepiece	50
Supplied air, full facepiece, pressure demand plus escape canister	2000
Self-contained breathing apparatus, positive-pressure mode	10,000

(DEJ)

Protective barriers Barriers of radiation-absorbing material, such as lead, concrete, and plaster, used to reduce radiation hazards. (RMB)

Protective creams *See* Barrier creams. (FSL)

Protective laboratory practices Those laboratory procedures, practices, and associated equipment accepted by laboratory health and safety experts as effective (or that employees can show to be effective) in minimizing the potential for exposure to hazardous chemicals or biological or physical agents. (JHR)

Proteins Complex nitrogenous organic compounds of high molecular weight that contain amino acids as their basic units and that are essential for growth and repair of animal tissue. Many proteins are enzymes. (FSL)

Proteolytic bacteria Bacteria that are capable of breaking down proteins. These organisms produce various proteases that degrade proteins and form products such as peptides, proteoses, peptones, and amino acids. These organisms are very important to control in food processing. For example, in dairy processing, various *Staphylococci*, *Pseudomonas*, *Micrococci*, and other microorganisms may be present, causing defects in product quality or making the product wholly unacceptable. Effective cleaning and sanitation programs should be used to control the presence of these and other organisms in milk and milk products, as well as in other high-protein or protein-containing foods. (HMB)

Proton An elementary nuclear particle with a positive electric charge equal numerically to the charge of the electron, and having a rest mass of 1.007575 atomic mass units. (FSL)

Protozoans Microscopic, single functional cell units or aggregations of nondifferentiated cells loosely held together and not forming tissues. Some protozoans are pathogenic for humans—e.g., *Balantidium coli* infects the colon and causes diarrhea and vomiting. *Entamoeba histolytica* may act as a commensal (may establish a symbiotic relationship) or may invade tissues and give rise to intestinal or extra-intestinal disease. (FSL)

Proximate cause A cause that, in a natural and continuous sequence, unbroken by any efficient intervening causes, produces injury, and without which the result would not have occurred. (DEW)

Psamments Entisol (mineral soils) that have a texture at least as coarse as that of loamy fine sand in all parts, that have $<35\%$ coarse fragments, and that are not saturated with water for periods long enough to limit their use for most crops. (SWT)

Psammophytes A group of plants that tolerate or prefer sand. (SWT)

Pseudomonas A genus of Gram-negative, aerobic, motile, rod-shaped bacteria (e.g., *Pseudomonas cepacia* and *Pseudomonas fluorescens*) that can be opportunistic pathogens in humans. *Pseudomonas mallei* is the causative agent of glanders in horses. (FSL)

Psychrometer An instrument used for the measurement of dry- and wet-bulb temperature. (FSL)

Psychrophiles A physiological group of bacteria identified by their temperature range of growth. These organisms are capable of growth at 0°C, with a growth range varying from about 0° to 20°C. Some scientists define the group on the basis of optimal growth temperature, as organisms that have an optimum temperature for growth of 15°C or lower, and a minimum growth temperature of 0°C or lower. It is important to control these organisms in any food sanitation program. They may account for a significant percentage of the spoilage noted in refrigerated foods. *See also* Mesophiles; thermophiles. (HMB)

Psychrotrophic Refers to microorganisms that are cold-temperature-tolerant. Many organisms whose optimum temperature is higher than 20°C can tolerate and thrive at temperatures below 20°C. These organisms are not to be confused with true psychrophilic organisms, which prefer cold temperatures. The psychrotrophic microorganisms are a serious concern for the food-processing industry. Despite the fact that refrigeration effectively inhibits the growth of other "mesophilic" organisms, those psychrotrophs capable of rapid growth at these temperatures are sustained. Examples of psychrotrophs include species of *Pseudomonas* and *Micrococcus. See also* Psychrophiles; mesophiles; minimum growth temperature; optimum growth temperature. (HMB)

Public health survey An analysis of the prevailing conditions, actual or potential, and the forces producing those conditions that exert or may exert a favorable or unfavorable influence on the health of the inhabitants in a given area. *See also* Sanitary survey. (FSL)

Public law *See* Statute. (DEW)

Publicly Owned Treatment Works (POTW) Any device or system that is owned by a state or municipality and is used in the treatment (including recycling and reclamation) of municipal sewage or industrial wastes of a liquid nature. Includes sewers, pipes, or other conveyances only if they convey wastewater to a POTW providing treatment. 40 CFR Part 122.2c. *See also* Clean Water Act; primary sewage treatment; secondary sewage treatment. (DEW)

Pull-out device In the industrial setting, a mechanism attached to an operator's hands and connected to the upper die or slide of a mechanical power press or other equipment. The device is designed, when properly adjusted, to withdraw the operator's hands as the dies close, if the hands are within the point of operation. *See also* Machine guarding. (SAN)

Pumice A mildly abrasive substance containing aluminum, potassium, and sodium and used for finishing and polishing various materials. (FSL)

Pumping station Set of mechanical devices installed in sewers, water systems, or other liquid-carrying pipelines that convey the liquids to a higher elevation. (FSL)

Pure tone Sound characterized by a single frequency. (DEJ)

Purging The initial step in adjusting to acceptable standards the atmosphere in a confined space. It is accomplished either by displacing the atmosphere in the space with fluid or vapor (inert gas, water, steam, and/or cleaning solution), or by carrying out forced-air ventilation. *See also* Confined space; permit-required confined space. (SAN)

Purpura Large hemorrhagic spots in or under the skin or mucous tissues. The spots are initially red, gradually darken to purple, fade to brownish yellow, and usually disappear in two to three weeks. (FSL)

Push-pull ventilation system An exhaust ventilation system consisting of a nozzle that pushes air and an exhaust hood that receives and removes the push jet. The system intercepts contaminated air and carries it over relatively long distances to the exhaust hood opening. Often used for open surface tanks in production or processing operations. (DEJ)

Putrefaction The anaerobic decomposition of organic matter, with incomplete oxidation and the production of gases such as oxides of nitrogen and hydrogen and odorous compounds such as mercaptans, ammonia, and hydrogen sulfide. (FSL)

Putrescible Able to decay quickly enough to cause odors and attract flies and other insects. (FSL)

Putrescible wastes Wastes that are capable of being quickly decomposed by microorganisms. Examples include kitchen wastes, animal manure, offal, hatchery and poultry processing-plant wastes, and garbage. *See also* Garbage; offal. (FSL)

Putrescine A diamine that was initially isolated from decaying meat and that is now known to occur in small quantities in most cells. (FSL)

p Value The smallest level of significance for which observed sample information becomes significant, provided the null hypothesis is true. (VWP)

Pyocyanin A blue-green pigment, produced by the organism *Pseudomonas aeruginosa*, that accounts for the color in "blue pus." (FSL)

Pyrethrum A natural insecticide made from the flowers of a kind of chrysanthemum plant native to Kenya, Ecuador, and Japan. This material has little residual, breaks down rapidly, and has a minimal effect on non-target species. It is used for the control of flies, mosquitoes, and certain garden insects, and is approved for use in food-service preparation areas. Pyrethrum is toxic by ingestion and inhalation, and there have been reports of allergic problems in humans who have been exposed to high concentrations of this insecticide. (FSL)

Pyridine [$N(CH)_5$] A pale yellow or colorless, foul-smelling derivative of coal tar, tobacco, and many types of organic matter. It is also derived during coal carbonization and from coke oven gases. It is flammable and toxic by ingestion and inhalation, and has a TLV of 5 ppm in air. It is used in rubber chemicals, antifreeze mixtures, solvent waterproofing materials, and fungicides. (FSL)

Pyrite (FeS_2) A lustrous, yellow mineral compound of iron sulfide, known as iron pyrite or fool's gold. It is used in the manufacture of sulfur, sulfuric acid, and sulfur dioxide. (FSL)

Pyroclastics A term applied to volcanic materials that have been explosively or aerially ejected from a volcanic vent. (SWT)

Pyrogen An agent that is capable of causing a rise in temperature. Pyrogens are produced by bacteria, molds, viruses, and yeasts, and commonly occur in distilled water. (FSL)

Pyrolosis A process, such as incineration or distillation, that results in the complete destruction of organic material. The chemical decomposition of a material by heat in the absence of oxygen. (FSL)

Pyromania A preoccupation with fires, or an obsessive compulsion to set fires. (FSL)

Pyromet powder A mixture of diammonium phosphate, protein, sodium chloride, and a waterproofing and flow-promoting agent. It has been shown to be useful in extinguishing fires involving calcium, sodium, and zirconium. (FSL)

Pyrophoric Refers to materials that ignite spontaneously in the presence of sufficient oxygen. *See also* Autoignition; pyrophoric chemical; spontaneous combustion. (FSL)

Pyrophoric chemical A chemical that ignites spontaneously in dry or moist air at a temperature of 130°F (54.5°C) or less. *See also* Autoignition; pyrophoric; spontaneous combustion. (FSL)

Q fever An infection with the rickettsial organisms *Coxiella burnetti*, often seen among meat and livestock handlers. Humans usually contract the disease by inhalating the organism, which is carried by dust particles in the air. The organisms are found on the hides of sheep and cattle. Human infections occur as a result of contact not only with such animals, but also with infected humans, air, dust, wild reservoir hosts, and other sources. Symptoms include sudden high fever, chills, headache, muscle pains, and coughing. The disease can be treated with antibiotics. (FSL)

Quality (Radiology) The characteristic spectral energy distribution of X-radiation. It is usually expressed in terms of effective wavelengths or half-value layers of a suitable material—for example: up to 20 kV (peak), cellophane; 20 to 120 kV (peak), aluminum; 120 to 400 kV (peak), copper; and over 400 kV, tin. (RMB)

Quality assurance/quality control A system of procedures, checks, audits, and corrective actions intended to ensure that laboratory research design and performance, environmental monitoring and sampling, and other technical and reporting activities are of the highest achievable quality. (FSL)

Quantile The value of a random variable that divides data into equal subdivisions. (VWP)

Quantum The smallest quantity of energy, responding to the energy of electromagnetic radiation, that can be associated with a given phenomenon. (RMB)

Quantum theory Theory based on the concept that energy is radiated intermittently in units of definite magnitude called quanta, and absorbed in like manner. (RMB)

Quarantine The application of measures to prevent contact between uninfected persons and persons suspected of being infected. The limitation of freedom of movement of well persons or domestic animals that may have been exposed to a communicable disease, enforced for a period of time equal to the longest usual incubation period of the disease, in a manner to prevent contact with persons not so exposed. (FSL)

Quaternary ammonium compounds (QACs) Typically, chlorine or bromine salts of ammonia in which the hydrogens are substituted with *n*-alkyl, benzyl, methyl, or ethyl groups. Quaternary ammonium compounds are used as chemical sanitizers in food sanitation and food processing. They are cationic wetting agents with bactericidal activity, are active over a wide pH range, and can be used at high temperatures. They are also nonirritating and noncorrosive. QACs are generally less bactericidal than either iodophors or chlorine sanitizers. *See also* Disinfectants. (HMB)

Quartiles Numbers that divide a set of ranked data into four equal parts. (VWP)

Quench To put out, extinguish, or suppress fires. (FSL)

Quenching The process of inhibiting continuous or multiple discharge in a radiation counter tube that uses gas amplification. (RMB)

Quenching gas Polyatomic gas used in Geiger-Mueller radiation counters to extinguish avalanche ionization. (RMB)

Quench tank A water-filled tank used to cool incinerator residues and materials during industrial processes. (FSL)

Quicklime *See* Calcium Oxide. (FSL)

Quicksilver *See* Mercury. (FSL)

Quintiles Numbers that divide a set of ranked data into five equal parts. (VWP)

R

Race gas A gas such as argon, helium, or xenon. (FSL)

Radappertization Radiation sterilization, or achievement of ''commercial sterility'' as it is defined in the food-canning industry. Usually, radiation treatment at this level is in the 3- to 4-megarad dose range. (HMB)

Radiac An acronym derived from ''radioactivity detection indication and computation.'' A generic term applying to the various types of radiological instruments or equipment, normally used as an adjective modifying ''instruments,'' but also used as a noun to designate the equipment itself. (RMB)

Radiant energy The energy of electromagnetic waves, from sources such as radio waves, visible light, infrared rays, X-rays, and gamma rays. *See also* Radiant heat load. (RMB)

Radiant exposure The total exposure to thermal energy. It is usually expressed in terms of calories per square centimeter (cal/cm^2), where the calorie is defined as the heat required to raise the temperature of 1 g of water from 15°C to 16°C at 760 mm mercury (Hg) pressure. (RMB)

Radiant heat load Energy that is transformed into heat when it strikes an object. The human body can emit and receive radiant energy. The term is often used synonymously with *infrared heat load*. (DEJ)

Radiation The emission and propagation of energy, in the form of waves, through space or through a material medium—for instance, the emission and propagation of electromagnetic waves or of sound and elastic waves. **(1)** The term *radiation* or *radiant energy*, when unqualified, usually refers to electromagnetic radiation; such radiation is commonly classified according to frequency, as Hertzian, infrared, visible (light), ultraviolet, X-ray, and gamma ray. **(2)** By extension, corpuscular emissions, such as alpha and beta radiation, or rays of mixed or unknown type, as cosmic radiation. *See also* Photon. (RMB)

Radiation absorbed dose (rad) The unit of absorbed dose, which is 100 ergs/g. The rad is a measure of the energy imparted to matter by ionizing particles per unit mass of irradiated material at the place of interest. It is a unit that was recommended and adopted by the International Commission on Radiological Units, at the Seventh International Congress of Radiology, Copenhagen, in July 1963. (RMB)

Radiation hazard A situation in which persons might receive ionizing radiation in excess of the applicable maximum permissible dose, or in which radiation damage might be done to materials. (RMB)

Radiation hygiene *See* Radiological health. (RMB)

Radiation protection guide (RPG) The total amount of ionizing radiation dose over certain periods of time that may be permitted to persons whose occupation involves exposure to such radiation. It is equivalent to what was formerly called the maximum permissible exposure (MPE). (RMB)

Radiation sickness A self-limited syndrome characterized by nausea, vomiting, diarrhea, and psychic depression, following exposure to appreciable doses of ionizing radiation, particularly to the abdominal region. Its mechanism is unknown, and there is no satisfactory remedy. The period of onset is typically a few hours after an occupational exposure or after treatment (in the case of patients receiving therapeutic doses), and the symptoms may subside within a day. The sickness may be sufficiently severe to necessitate interrupting the treatment series and may incapacitate a patient. In the case of occupational exposure, it may be so serious that the patient requires hospitalization. *See also* Radiation syndrome. (RMB)

Radiation standards Regulations, promulgated by the Nuclear Regulatory Commission or EPA, that establish maximum exposure limits for protection of the public from radioactive materials. (RMB)

Radiation syndrome The complex of symptoms that characterize the disease known as radiation injury, and that result from excessive exposure of the whole (or a large part) of the body to ionizing radiation. The earliest of these symptoms are nausea, vomiting, and diarrhea. They may be followed by loss of hair (epilation), hemorrhage, inflammation of the mouth and throat, and general loss of energy. In severe cases, in which the radiation exposure has been relatively large, death may occur within 2 to 4 weeks. Those who survive 6 weeks after the receipt of a single dose of radiation may generally be expected to recover. *See also* Chernobyl. (RMB)

Radicidation An analog to pasteurization. It refers to the reduction of specific non-spore-forming pathogens, other than viruses, so that none is detectable by standard methods. The level of treatment is typically between those of radurization and radappertization. *See also* Radappertization; radurization. (HMB)

Radioactive cloud An all-inclusive term for the mixture of hot gases, smoke, dust, and other particulate matter that is carried aloft in conjunction with the rising fireball produced by a nuclear blast. *See also* Chernobyl. (RMB)

Radioactive equilibrium The state that prevails when the ratios between the amounts of successive members of a radioactive series remain constant. A condition called secular equilibrium can be considered to exist if a parent element has a very much longer half-life than do succeeding elements, so that there is no appreciable change in its amount of the parent element in the time interval required for the later products to attain equilibrium. In such a case, after equilibrium is reached, equal numbers of atoms of all members of the series disintegrate per unit time. This condition is never actually attained but is very nearly established in such a case as the disintegration of radium and its series, through radium C. The half-life of radium is 1600 years;

that of radon is only 3.83 days; those of each of the subsequent members are only a few minutes. After about a month of observation of any sample of radium, essentially the equilibrium amount of radon is present, together with equilibrium amounts of all other members of the series, through radium C. (RMB)

Radioactive substances Substances that emit radiation. (FSL)

Radioactivity The spontaneous emission of radiation, generally alpha or beta particles, often accompanied by gamma rays, from the nucleus of an unstable isotope. As a result of this emission, the radioactive isotope is converted (or decays) into the isotope of a different element, which may or may not be radioactive. Ultimately, as a result of one or more stages of radioactive decay, a stable (nonradioactive) end product is formed. (RMB)

Radioactivity concentration guide (RCG) The maximum permissible amount of any specified radioisotope that may be allowed to accumulate in the body. It is equivalent to what was formerly called the maximum permissible concentration (MPC). (RMB)

Radiobiology The branch of biology that deals with the effects of radiation on biological systems. (FSL)

Radiography, industrial The use of penetrating radiation, such as X-rays, gamma rays or neutrons, to make visual images of the insides of objects—e.g., to inspect metal castings or welds for internal flaws. Industrial radiography does not include the medical uses of radiation such as in chest or dental X-rays. *See also* Radiology. (RMB)

Radiological health The art and science of protecting humans and the environment from injury or damage by radiation. (RMB)

Radiological survey The evaluation of the radiation hazards incident to the production, use, or existence of radioactive materials or other sources of radiation under a specific set of conditions. Such evaluation customarily includes a physical survey of the disposition of materials and equipment, measurements or estimates of the levels of radiation that may be involved, and a knowledge of processes involving materials that is sufficient to allow prediction of the hazards resulting from expected or possible changes in the materials. (RMB)

Radiology The branch of medicine that deals with the diagnostic and therapeutic applications of radiant energy, including X-rays and the energy produced by radioisotopes. (RMB)

Radiometer An instrument used to demonstrate the transformation of radiant energy into mechanical energy. It consists of an evacuated glass vessel containing vanes that revolve around an axis when exposed to radiant energy. (RMB)

Radionecrosis The destruction of tissue by radiant energy. (RMB)

Radionuclide Radioactive element that is characterized in terms of its atomic mass and atomic number. Radionuclides can be made artificially or can be naturally occurring. They can have a long life in form of soil or water pollutants, and have potentially mutagenic effects on the human body. (RMB)

Radiosensitivity The relative susceptibility of cells, tissues, organs, organisms, or any living substances to the injurious action of radiation. Radioresistance and radiosensitivity are currently employed in a comparative sense, rather than in an absolute one. (RMB)

Radius of vulnerability zone Measured from the point of release of a hazardous substance, the maximum distance at which the airborne concentration could reach the level of concern under specified weather conditions. (FSL)

Radon (Rn) A naturally occurring radioactive gas that is a breakdown product of radium and uranium. Radon has a half-life of 3.8 days and an atomic number of 86. It is highly toxic and emits ionizing radiation. When this material is handled, lead shielding, as well as protective clothing, may be required. In recent years there has been increased scientific and public interest in radon because of possible associations with various types of cancers in humans. EPA has developed regulations pertaining to acceptable levels of radon in various environmental situations. Radon is used in treatments for malignant growths and in industrial radiography. *See also* Radiography, industrial. (RMB)

Radon daughters *See* Radon decay products. (FSL)

Radon decay products A term used to refer collectively to the immediate, short-lived products of the radon decay chain. These include Po-218, Pb-214, Bi-214, and Po-214, which have an average half-life of about 30 minutes. They normally exist as solids and appear in air as free ions or as ions attached to dust particles. The NIOSH recommended exposure limit (REL) for worker exposure in underground mines is based on evidence that a substantial risk of lung cancer is associated with an occupational exposure to radon decay products. EPA has developed other standards for radon in various environmental and residential settings. (FSL)

Radon progeny *See* Radon decay products. (FSL)

Radurization An analog to pasteurization. It involves substantial reduction in the number of spoilage organisms through an appropriate radiation treatment applied to enhance the keeping quality of foods. Usually treatment is in the 0.75–2.5 megarad dose range for various types of food. *See also* Radappertization; radicidation. (HMB)

Rain shadow A dryness effect created on the leeward side of a mountain region when warm moist air cools adiabatically and releases its precipitation as it ascends the wet, windward side of the mountain range. The cold dry air then becomes warm and produces little or no precipitation as it descends the leeward side of the mountain. The rain-shadow effect is especially pronounced in areas in which a north-south mountain range blocks prevailing winds. *See also* Adiabatic. (SMC)

Ramazzini, Bernardino (1633–1714) Considered to be the founder of industrial medicine, Ramazzini wrote a book in 1700 entitled *De Morbis Artificium Diatriba* (titled *The Diseases of Workmen*, in translation). In that book, he described the signs and symptoms of lead and mercury poisoning and promoted the concept that physicians should question patients about their specific occupations. (FSL)

Rancid Refers to a musty, rank taste or smell that is usually associated with fats that have undergone decomposition. (FSL)

Random In a series of events, refers to any or all outcomes that have an equal chance of occurrence. (VWP)

Random number Value with digits selected from 0, 1, 2, . . . , 9 with no specific pattern. (VWP)

Random number table A collection of digits, used to select sampling frequencies, in which each of the digits 0, 1, 2, . . . , 9 has an equal chance of occurrence and the value of any digit is independent of the value of any other digit. (VWP)

Random variable A variable that assumes a unique numerical value for each of the outcomes in the sample space of a probability experiment. (VWP)

Range **(1)** (Biostatistics) The difference between the smallest and largest data values. **(2)** (Radiation) The distance radiation will penetrate a given material before all of the ionizing power of the radiation is spent. (RMB, VWP)

Rank correlation A nonparametric analog to linear correlation, based on ranks rather than data values. (VWP)

Ranked data A series of values arranged in order from smallest to largest. (VWP)

Raoult's law Law that states that equal molar weights of different nonvolatile nonelectrolytes dissolved in a definite weight of a given solvent under the same conditions lower the solvent's freezing point, elevate its boiling point, and reduce its vapor pressure equally, irrespective of the nature of the nonelectrolytes. (DEJ)

Rasp A machine that grinds waste into a manageable material, reduces bulk, and helps prevent odor. (FSL)

Rate A measure of the frequency with which a specified event occurs in a particular population, either at a certain instant or during a particular period. The three major types of rates used in public health practice are the morbidity rate, the mortality rate, and the natality rate. **(1)** *Morbidity rate:* A measure of the frequency of disease in a population. There are two major groups of morbidity rates: incidence and prevalence rates. An *incidence rate* is a measure, in a particular population, of the frequency of cases of disease, the onset of which occurred during a specified period of time. Incidence rates that are calculated for narrowly defined populations (in terms of age, sex, etc.) during intervals of time, as in epidemics, are often called attack rates. Attack rates are usually expressed as a percent. A secondary attack rate is a measure of the frequency of new cases of a disease among close contacts of known cases. Secondary attack rates are usually calculated for household contacts. A prevalence rate is a measure of the frequency of all current cases of a disease (regardless of the time of onset) within a particular population, either at a specified instant (a point-prevalence rate) or during a specified period (a period-prevalence rate). **(2)** *Mortality rate:* A measure of the frequency of deaths within a particular population during a specified interval of time. If deaths from all causes are included, the rate is called a crude death rate; if only deaths from a specified cause are included, the rate is a cause-specific death rate. A case fatality rate is a measure of the frequency of deaths due to a particular disease among members of a population who have the disease—e.g., the percentage of persons with a specific disease who die as a result of that disease. **(3)** *Natality rate:* A measure of the frequency of births in a particular population during a specified period of time. *See also* Disability; OSHA frequency rates; OSHA Log of Occupational Injuries and Illnesses; OSHA recordable injuries and illnesses. (FSL, VWP)

Rated load The manufacturer's recommended maximum load for certain types of equipment. For hoists, this is the manufacturer's maximum allowable operating load for the particular event. For powered industrial platforms, this is the combined weight of workers, tools, equipment, and other material that are permitted to be carried by the working platform at the installation, as indicated on the load rating plate on the equipment. (SAN)

Rat eradication The elimination or extermination of all rats within a building or some other area by the use of poisoning, fumigation, trapping, or ratproofing techniques. (FSL)

Rat harborage Conditions that provide shelter or protection for rats and thus favor their multiplication and existence in, under, or outside of a structure. (FSL)

Ratio A measure of the frequency of one group of events (e.g., the contraction of a specified disease by a certain number of males) relative to the frequency of a different group of events (e.g., the contraction of the specified disease by a certain number of females). (VWP)

Ratio scale data Cardinal data with a fixed zero value on the scale. (VWP)

Ratproofing Construction or technique that is designed to prevent the entrance of rats into structures, or their movement from one structure to another. Openings in exterior walls, ground floors, basements, roofs, and foundations should be constructed of materials that are resistant to gnawing and entry by rats. (FSL)

Raynaud's disease Named after the French physician Maurice Raynaud (1834–1881) and also known as Raynaud's phenomenon and vibration syndrome, a disease that consists of vasospastic, neuromuscular, and arthritic disorders, usually of the hands and upper limbs. It is associated with the operation of hand-held or manually supported or guided machines that produce intense vibration in the frequency range of 10 to 1000 Hz. The condition is manifested by pain in the fingers, loss of manual dexterity, stiffness in the joints, radiographically observable changes in the bones and joints, paroxysmal blanching (whitening) and numbness in one or more fingers of either hand, skin atrophy, and occasionally gangrene. (DEJ)

Reaction As applied to chemistry, the action of two or more substances upon each other whereby new substances are formed from them. (FSL)

Reactivity A measure of the tendency to undergo a chemical change. As defined by EPA, a solid waste is considered reactive if a sample of it has any of the following properties: **(1)** It is normally unstable and readily undergoes violent change without detonation. **(2)** It forms potentially explosive mixtures with water. **(3)** It reacts violently with water. **(4)** When mixed with water, it generates toxic gases, vapors, or fumes in a quantity sufficient to present a danger to human health or the environment. **(5)** It is readily capable of detonation or explosive decomposition or reaction at standard temperature or pressure. **(6)** It is capable of detonation or explosive reaction if it is subjected to a strong initiating source or if it is heated under confinement. **(7)** It is a forbidden explosive as defined in 49 CFR 173.51, or is a Class A explosive as defined in 49 CFR 173.53 or a Class B explosive as defined in 49 CFR 173.88. (FSL)

Reactor, nuclear A device in which nuclear energy may be sustained and controlled in a self-supporting nuclear reaction. The varieties are many, but all incorporate certain features, including fissionable material, or fuel, a moderating material (unless the reactor is operated on fast neutrons), a reflector to conserve escaping neutrons, and provisions for removal of heat, for measuring and controlling elements, and for personnel protection. (RMB)

Reagent A substance that produces a chemical reaction that can be used to detect, measure, or produce another substance. (FSL)

Real time *See* Data automation. (DEJ)

Real-time sampling instrument An environmental sampling device, also known as a direct-reading instrument, that displays the concentration or level of a contaminant instantaneously. *See also* Grab sample. (DEJ)

Reasonably available control technology (RACT) limit The lowest emissions limit that a particular source is capable of meeting by the application of control technology that is reasonably available and technologically and economically feasible. RACT limits are usually applied to existing pollution sources in nonattainment areas and in most cases are less stringent than new source performance standards. (FSL)

Rebuttable presumption against registration (RPAR) *See* Special review. (FSL)

Receiving waters A river, lake, ocean, stream, or other watercourse into which wastewater or treated effluent is discharged. *See also* Sewer; Sewerage. (FSL)

Recharge The process by which aquifers are renewed by the infiltration of water during wet seasons. *See also* Aquifer. (FSL)

Recharge area A land area in which water reaches the zone of saturation through surface infiltration—e.g., an area in which rainwater soaks through the earth and reaches an aquifer. (FSL)

Reclamation A controlled method of sorting and storing material from solid wastes, for future use. (FSL)

Recombinant bacterium A type of microorganism whose genetic makeup has been altered by deliberate introduction of new genetic elements. The offspring of such altered bacteria also contain the new genetic elements. (FSL)

Recombinant DNA (rDNA) The new deoxyribonucleic acid (DNA) that is formed by combining pieces of DNA from different organisms or cells. *See also* Deoxyribonucleic acid. (FSL)

Recommended exposure limits (RELs) Occupational exposure limits developed by the National Institute for Occupational Safety and Health (NIOSH). RELs are recommended to OSHA for promulgation as legally enforceable permissible exposure limits, under the Occupational Safety and Health Act. In general, OSHA has used the threshold limit values, not the recommended exposure limits established by NIOSH, in establishing its permissible exposure limits because RELs are based solely on health criteria, not feasibility considerations. (DEJ)

Recommended maximum contaminant level (RMCL) In drinking water, the maximum level of a contaminant at which no known or anticipated adverse effect on human health would occur. The level designated includes an adequate margin of safety. Recommended levels are nonenforceable health goals. *See also* Maximum contaminant level. (FSL)

Reconstructed source As defined by the EPA, an existing facility in which components are replaced to such an extent that the fixed capital cost of the new components exceed 50% of the capital cost that would be required to construct a comparable, entirely new facility. New source performance standards may be applied to sources that are constructed after the proposal of a standard if it is technologically and economically feasible to meet the standard. (FSL)

Recording thermometer A thermometer with a recording device, used to obtain a permanent record of temperature of treatment—e.g., pasteurization temperature—or storage temperature. Recording thermometers are usually the pressure-spring type, and are actuated by means of a volatile liquid (generally an ether derivative). A single length of tubing extends between the case of the instrument and the temperature-sensing bulb; therefore, any break or leak will make the instrument inoperative. *See also* Indicating thermometer. (HMB)

Record-keeping The OSHA act requires employers to maintain records of work-related injuries and illnesses and inspection reports of high-injury-potential equipment. In addition, many state departments of labor require employers to maintain certain types of reports. Records of accidents and injuries and of the training of the persons involved are essential to the management of efficient and successful safety programs. The records must not only be responsive to regulatory requirements, but must also provide a detailed chronology of the events involved, so that appropriate prevention and control programs can be initiated. (FSL)

Record of decision (ROD) A public document that explains which cleanup alternative(s) will be used at National Priorities List sites for whose cleanup, under CERCLA, Trust Funds must pay. 40 CFR §300.430(f)(5). *See also* Comprehensive Environmental Response, Compensation, and Liability Act; National Priorities List; remedial action. (DEW)

Recovery (radiobiology) The return toward normal of a particular cell, tissue, or organism after radiation injury. (RMB)

Recovery rate (radiation) The rate at which recovery takes place following radiation injury. It may be different for different tissues. Tissues having slower rates of recovery will ultimately suffer greater damage from a series of successive irradiations. This differential effect is taken advantage of in fractionated radiation therapy if the neoplastic tissues have a slower recovery rate than do the surrounding normal structures. (RMB)

Recycling Any process by which reclaimed materials or other materials that would otherwise become waste are collected, separated, or processed, and then reused or returned to use in the form of raw materials or products. (FSL)

Red algae *See* Algae; red tide. (FSL)

Red earth Red clayey soils of the tropics. They usually have very deep profiles and are low in silica and high in sesquioxides. (SWT)

Redistribution, soil The process by which water moves through soil until an equilibrium energy is achieved throughout the soil profile. (SWT)

Redox potential An abbreviation of *oxidation-reduction potential*, an intrinsic factor affecting the survival and growth of microorganisms. *See also* Oxidation-reduction potential. (HMB)

Red squill A rodenticide obtained from the bulbs of a perennial onionlike plant belonging to the lily family and native to the areas around the Mediterranean. The toxic action of red squill depends upon the presence of glycoside and scilliroscide in the plant. A desirable quality of red squill is its natural emetic action, which causes vomiting in humans and most animals other than rats. The selective rodenticidal usefulness of red squill, then, takes advantage of the physiologic peculiarity of the rat's in this respect. Although this rodenticide is safe for use around residential areas, it is irritating to the skin; thus, rubber gloves must be worn during preparation and placement of bait. *See also* Strychnine; warfarin.

Red tide A proliferation of marine plankton that is toxic and often fatal to fish. This natural phenomena may be stimulated by the addition of nutrients. A tide can be called red, green, or brown, depending on the coloration of the plankton. Shellfish, such as oysters, clams, and other bivalves, are unharmed by these plankton, but they may store and concentrate the toxin, which causes paralytic poisoning when the shellfish are consumed by humans. *See also* Paralytic shellfish poisoning. (FSL)

Redundancy As related to systems and operations, a method of reducing the probability of error or failure by providing more than one mode to accomplish an objective. The primary method of redundancy consists of using two or more parallel subsystems or a backup system. Some systems utilize partial redundancy, in that only a portion of the system is protected. Personnel redundancy can be achieved by use of more than one person in a hazardous situation (such as when a safety watch is included during a confined space entry or the "buddy system" is used). (SAN)

Reed, Walter (1851–1902) An American bacteriologist who, as a military physician during the Spanish-American War, was appointed to head a committee to investigate typhoid fever epidemics in army camps. Later, when yellow fever was epidemic in Cuba, he studied the modes of transmission of that disease and was able to demonstrate that the vector involved was the mosquito *Aedes aegypti*. (FSL)

Reentry interval The time between application of a pesticide and the allowable entry of unprotected workers into the field. (FSL)

Reference population The specific group under consideration in an epidemiologic investigation. (VWP)

Reflectance The ratio of the radiant energy reflected by a body to that incident upon it. (DEJ)

Refuse Collective term for all putrescible and nonputrescible solid wastes, except sewage. The term is intended to apply in the broadest sense to include garbage, rubbish, and abandoned equipment, furniture, appliances, and other solids that have been or are intended to be discarded. *See also* Garbage; offal; refuse reclamation; rubbish. (FSL)

Refuse reclamation Conversion of solid waste into useful products—e.g., the composting of organic wastes to make soil conditioners, and the separation of aluminum and other metals for melting and recycling. (FSL)

Regeneration/regenerative process The process by which damaged human, plant, or animals cells are replaced by new ones of the same type. (FSL)

Regional response teams (RRTs) Teams responsible for regional planning and preparedness activities before response actions under a national contingency plan, and for providing advice and support to the on-scene coordinator or remedial project manager when activated during a response. They consist of designated representatives from federal agencies participating with state and (as agreed upon by the state) local government representatives. 40 CFR §300.115. *See also* National contingency plan; national response team. (DEW)

Registered sanitarian (RS) Designation granted to an individual after a credentialing process that was originally developed by the National Association of Sanitarians, now the National Environmental Health Association. The process requires applicants for the RS designation to meet certain educational and experience criteria and in some cases to pass a written examination. Following the successful completion of these requirements, eligible individuals are able to use the RS designation. In addition, many states have sanitarian registration laws and require a similar process for environmental health professionals who wish to practice their specialty areas within the state boundaries. *See also* Environmental engineer; environmental health; environmental health specialist; sanitarian. (FSL)

Registrant Any manufacturer or formulator who obtains registration for a pesticide product or active ingredient in a pesticide. (FSL)

Registration Formal listing with the EPA of a new pesticide before the substance can be sold or distributed in intrastate or interstate commerce. The product must be registered under the Federal Insecticide, Fungicide, and Rodenticide Act. EPA is responsible for registration (pre-market licensing) of pesticides on the basis of data demonstrating that they will not cause unreasonable adverse effects on human health or the environment when used according to approved label directions. *See also* Federal Insecticide, Fungicide, and Rodenticide Act. (FSL)

Registration standards Published reviews by EPA of the data available on the active ingredients currently used in pesticides. (FSL)

Regolith Soil; the unconsolidated mantle of weathered rock and soil material on the earth's surface; earth materials above solid rock. (SWT)

Regression analysis A calculation procedure using an equation that provides values of one variable, given values of another variable. One of its primary objectives is to make predictions. The estimation or prediction of the value of a variable from the given values of two or more variables is known as multiple regression analysis. (VWP)

Regression line The line of best fit for a given set of data points. (VWP)

Rehydration The replacement of water or fluid content in a body or substance that has become dehydrated. (FSL)

Rejection region The range of values of the arithmetic mean for which the null hypothesis is rejected. (VWP)

Relapsing fever Any of a group of similar infectious diseases transmitted to humans by the bites of ticks or lice. The causative organisms belong to the genus *Borrelia*. The disease is characterized by a high fever that usually lasts 2 or 3 days, followed by a relapse of high fever several days later. This pattern may continue through three or four attacks, until the disease has run its course. (FSL)

Relative biological effectiveness (RBE) The ratio of the number of rads of gamma radiation or X-radiation of a certain energy that will produce a specified biological effect to the number of rads of another radiation required to produce the same effect. The appropriate value of biological effectiveness of radiation relative to that of X-rays with an average specific ionization of 100 ionizations per centimeter in water for the particular biological system and effect under consideration and for the condition under which the radiation is received. (RMB)

Relative biological effectiveness dose (RBE dose) The product of the absorbed dose in radiation and an agreed conventional value of RBE with respect to a particular form of radiation effect. The standard of comparison recommended by the International Commission on Radiological Units and Measurements (ICRU) is X-radiation or gamma radiation in water of 3 keV/μ delivered at a rate of about 10 rad per minute. The unit of RBE dose is the rem. The RBE dose equals the number of rads times the RBE. *See also* Radiation absorbed dose. (RMB)

Relative frequency A proportional measure of the frequency of an occurrence. It is calculated by dividing the class frequency by the total number of observations. (VWP)

Relative-frequency histogram A frequency histogram with the vertical scale giving relative frequencies rather than frequency of classes. (VWP)

Relative humidity The amount of moisture in the air as compared with the maximum amount that the air could contain at the same temperature, expressed as a percentage. *See also* Humidity. (DEJ)

Relative plateau slope In radiation detection, the relative increase in the number of counts as a function of voltage, expressed in percentage per 100-volt increase above the Geiger threshold. (RMB)

Release For purposes of CERCLA, any spilling, leaking, pumping, pouring, emitting, emptying, discharging, injecting, escaping, leaching, dumping, or disposing into the environment (including the abandonment or discarding of barrels, containers, and other closed receptacles containing any hazardous substance or pollutant or contaminant). It excludes certain workplace releases, engine exhaust emissions, and specified radioactive releases related to the generation of nuclear power. 42 USC §9601(22). *See also* Comprehensive Environmental Response, Compensation, and Liability Act; contaminant; hazardous substance; pollutant. (DEW)

Relief *See* Safety relief valve. (FSL)

Remedial action (RA) Under CERCLA, an action consistent with a permanent remedy taken instead of or in addition to removal actions, in the event of a release or threatened release of a hazardous substance into the environment. Such an action is undertaken to prevent or minimize the release of hazardous substances so that they do not migrate to cause substantial danger to present or future public health, welfare, or the environment. 42 USC §9604(34). *See also* Comprehensive Environmental Response, Compensation, and Liability Act; national contingency plan; removal design; remedial investigation. (DEW)

Remedial design A phase of remedial action that follows a remedial investigation/feasibility study and includes development of engineering drawings and specifications for site cleanup. *See also* Remedial action; remedial investigation. (FSL)

Remedial investigation (RI) For purposes of a cleanup action under CERCLA, a process undertaken by the lead agency to determine the nature and extent of the problem presented by a release. It emphasizes data collection and site characterization, and is generally performed concurrently and in an interactive fashion with the feasibility study. It includes sampling and monitoring, as necessary, and the gathering of sufficient information to determine the necessity for remedial action and to support the evaluation of remedial alternatives. 40 CFR §300.5. *See also* Comprehensive Environmental Response, Compensation, and Liability Act; feasibility study; release; remedial action; remedial design. (DEW)

Remedial project manager (RPM) The EPA or state official responsible for overseeing remedial action at a hazardous-waste site. (FSL)

Remedial response A long-term action that stops or substantially reduces a release or threat of a release of hazardous substances that is serious but not an immediate threat to public health. *See also* Release; remedial action; removal action. (FSL)

Remedy The means by which a right is enforced or the violation of a right is prevented, redressed, or compensated. *See also* Remedial action. (DEW)

Removal action Under CERCLA, the cleanup or removal of released hazardous substances from the environment, or an action taken in the event of the threat of release. Such actions include monitoring, assessment, and evaluation of the release or threat of release, the disposal of removed material, or other actions to prevent, minimize, or mitigate damage to the public health, or welfare or environment that may otherwise result. Under the Clean Water Act, it refers to containment and removal of the oil or hazardous substances from the water and shorelines or the taking of such other actions as may be necessary to minimize or mitigate damage to the public health or welfare, including, but not limited to, actions involving fish, shellfish, wildlife, public and private property, shorelines, and beaches. 33 USC §1321(a)(8), 40 CFR §300.5. *See also* Clean Water Act; Comprehensive Environmental Response, Compensation, and Liability Act; national contingency plan; remedial action. (DEW)

Repeated sampling The selection of a subset from a population more than one time. (VWP)

Repeated violations *See* Willful violations. (DEJ)

Repellents Chemical substances used to produce avoidance reactions in humans and animals. These materials are commercially available in solid, liquid, and spray formulations. The most common repellents on the market usually contain DEET (diethyl-m-toluamide), which provides satisfactory prevention against bites by mosquitoes, ticks, and chiggers. (FSL)

Reportable quantity (RQ) For purposes of CERCLA and the Clean Water Act, the quantity of specified hazardous substances the release of which must be reported to the National Response Center. 42 USC §9602 and 9603, 40 CFR §300.125. *See also* Comprehensive Environmental Response, Compensation, and Liability Act; Clean Water Act; hazardous substance; release; national response center. (DEW)

Reregistration The reevaluation and relicensing of existing pesticides originally registered prior to enactment of current scientific and regulatory standards. EPA reregisters pesticides through its registration standards program. *See also* Federal Insecticide, Fungicide, and Rodenticide Act; registration. (FSL)

Resazurin reduction test A dye reduction test that is used as a quality control measure to estimate indirectly the number of bacteria in milk. This test is often used as a substitute for the methylene blue test. Resazurin gives a characteristic blue color to fresh milk. Reduction of the dye is indicated by a color change through shades of purple to pink, at which point the dye has been reduced to resorufin. This change is irreversible in milk. A second stage carries the reduction through a color change to white, characteristic of dihydroresorufin, and is reversible. Two such tests are in use. In the triple reading test, the time required (up to 3 hours) to obtain dye reduction to a prescribed color end point is measured. In the one-hour test, the degree of color change is determined after one hour of incubation. *See also* Methylene blue reduction test. (HMB)

Research and Special Programs Administration (RSPA) A program of the federal Department of Transportation that embraces several offices emphasizing safety regulations, emergency preparedness, and research and development associated with the transportation of hazardous materials. (FSL)

Reservoir Any natural or artificial holding area used to store, regulate, or control water. (FSL)

Reservoir of infectious agents Any human being, animal, arthropod, plant, soil, or inanimate matter in which an infectious agent normally lives and multiplies, upon which it depends primarily for survival, and from which it can be transmitted to a susceptible new host. (FSL)

Residual That which remains or is left behind. As applied to the use of disinfectants, the concentration of chemical available for microbiocidal purposes in a given medium—e.g., water. In the case of pesticides, the amount of active material that remains to eliminate a target insect or weed pest. *See also* Residual chlorine. (FSL)

Residual chlorine Chlorination is widely practiced and generally accepted as the most common means of treating public water supplies and wastewater effluent to prevent the spread of waterborne disease. Chlorine is a potent oxidizing agent and possesses strong bactericidal properties. It combines with water to form hypochlorous and hydrochloric acids. It is the only recognized disinfectant that is capable of providing protective "residuals" within a treated system to guard against subsequent bacterial growth or inadvertent contamination. This is of paramount importance in any application of chlorination, whether it be for water treatment or in a food sanitation program as a sanitation measure. Chlorine will react with a wide variety of substances and impurities in water, especially ammonium ions and ammonia, which react with chlorine or hypochlorous acid to form mono-, di-, and trichloramines, depending upon the relative amounts of each and the pH. The amines have some disinfection capability and are therefore important in determining the residual effect of chlorination. It has become common practice to refer to chlorine, hypochlorous acid, and hypochlorite ion as free chlorine residuals. The chloramines are called combined chlorine residuals. Conditions favoring the formation of HOCl over OCl are more effective in disinfection. It has also been shown that a greater concentration of combined chlorine residual than of free chlorine residual is required to accomplish a given kill in a specified time. The reactions are as follows:

$$Cl_2 + H_2O \rightarrow HOCl + H^+ + Cl^-$$

$$HOCl \leftrightarrow H^+ + OCl^-$$

where the double arrow indicates a reversible reaction. *See also* Disinfectant; hypochlorous acid. (HMB)

Residual materials Unconsolidated and partly weathered mineral materials accumulated by disintegration of consolidated rock in place. (SWT)

Residual nuclear radiation Nuclear radiation, chiefly beta particles and gamma rays, that persists for some time following a nuclear explosion. The radiation is emitted mainly by the fission products in the fallout and to some extent by earth and other materials in which radioactivity has been induced by the capture of neutrons. *See also* Fallout; initial nuclear radiation; induced radioactivity. (RMB)

Residual number (RN) A measure of the effectiveness of a counter measure or of a reclamation method. (FSL)

Residual shrinkage The associated decrease in the bulk volume of soil, caused by the loss of water. (SWT)

Residual volume The amount of air remaining in the lungs after a maximum expiratory effort. (DEJ)

Resin, natural A solid, semisolid, or pseudosolid organic material that has indefinite and often high molecular weight; mixtures of carboxylic acids, oils, and terpenes, which occur as exudates on the bark of many trees and shrubs. They are combustible, most are soluble in alcohols, ethers, and carbon disulfide, and are insoluble in water. Examples include rosin and balsam, which are obtained from coniferous trees. The broad category of resins also includes shellac, which is obtained from the secretion of an insect found in India. (FSL)

Resistance The total of host mechanisms that interpose barriers to invasion or multiplication of infectious agents or that prevent damage by the agents' toxic products. Resistance can be of various kinds. **(1)** *Immunity:* Resistance usually associated with possession of antibodies that have an inhibitory effect on a specific microorganism (or its toxin) that causes a particular infectious disease. Passive immunity is acquired either naturally (by maternal transfer) or artificially (by inoculation of specific protective antibodies—e.g., convalescent or immune serum, or immune serum globulin). Passive immunity is of brief duration—generally, days or months. Active immunity lasts months to years and is acquired either naturally (by infection, with or without clinical manifestations) or artificially (by inoculation of fractions or products of the infectious agent or of the agent itself, in killed, modified, or variant form—e.g., vaccines). **(2)** *Inherent resistance:* An ability to resist disease without the action of antibodies or of specifically developed tissue response. Inherent resistance commonly stems from anatomic or physiologic characteristics of the host, and can be either genetic or acquired, permanent or temporary. *See also* Vaccine. (FSL)

Resolving time, counter The minimum time interval between two distinct events that will permit both to be counted. It may apply to an electronic circuit, to a mechanical indicating device, or to a counter tube used in radiation-monitoring programs. (RMB)

Resonance capture An inelastic nuclear collision occurring when the nucleus exhibits a strong tendency to capture incident particles, photons, or particular energies. (RMB)

Resource Conservation and Recovery Act (RCRA) Amendments made in 1976 to the Solid Waste Disposal Act, which is generally now referred to as the Resource Conservation and Recovery Act. Federal law that regulates the disposal of solid waste, including hazardous waste. Subtitle C addresses hazardous waste, with the intent to establish a "cradle-to-grave" regulatory scheme covering the generation, transport, storage, treatment, and disposal of hazardous wastes. Subtitle D addresses the disposal of nonhazardous solid waste. Other provisions include a land disposal ban for some types of wastes and the regulation of underground storage tanks. Substantial amendments to the act in 1984 are referred to as the Hazardous and Solid Waste Amendments (HSWA). 42 USC §6901 *et seq. See also* Corrective action; disposal facility; generator; hazardous waste; land disposal; solid waste; transporter; treatment. (DEW)

Resource recovery The process of obtaining matter or energy from materials formerly discarded. (FSL)

Respirable dust Airborne particulate matter capable of passing through the upper respiratory system and being deposited in the lungs. Such particles typically are less than 10 microns in diameter. (DEJ)

Respirable-mass monitor A device used to measure the amount of respirable dust in air. Direct-reading instruments are based on attenuation of beta radiation or on the change in resonant frequency of a piezoelectric quartz crystal. Both employ prestages to separate respirable from nonrespirable dust. (DEJ)

Respiration Process that involves the oxidation of inorganic or organic molecules, the generation of energy (by means of adenosine triphosphate, ATP) by the transport of electrons and hydrogen ions through an electron transport system, and the donation of electrons to an inorganic electron acceptor. Aerobic respiration occurs when the electron acceptor is molecular oxygen, and anaerobic respiration occurs when the electron acceptor is an inorganic substance other than oxygen, such as a nitrate. *See also* Aerobic; anaerobic. (HMB)

Respirator fit testing Includes a number of test protocols that measure the integrity of the seal between the respirator face mask and the wearer's face. Also measures the integrity of the valves and air filtration capability of a respirator. Fit testing usually involves a challenge exposure using an irritant smoke, saccharin, isoamyl acetate, or a particulate. If the individual responds to the test agent, it is an indication that the respirator is not properly fitted. A qualitative fit test utilizes the wearer's ability to smell or react to an irritant, and provides a nonnumerical gross assessment of respirator fit. A quantitative fit test measures the concentration of the contaminant outside and inside the mask to arrive at a numerical estimate of how well the respirator fits and the extent of leakage. *See also* Protection factors. (DEJ)

Respiratory system The group of organs concerned with the exchange of oxygen and carbon dioxide in organisms. In higher animals, it consists successively of the air passages through the mouth, nose, and throat; the trachea; the bronchi; the bronchioles; and the alveoli of the lungs. (FSL)

Respondeat superior Legal doctrine that states that a master, usually an employer, may be liable for the wrongful acts of his servant or employee when the latter is acting within the scope of his employment. *See also* Negligence. (DEW)

Response action A CERCLA-authorized action involving either a short-term or long-term removal response that may include but is not limited to: removing hazardous materials from a site to an EPA-approved hazardous-waste facility for treatment, containment, or destruction; containing the waste safely onsite; destroying or treating the waste on-site; and identifying and removing the source of ground-water contamination and halting further migration of contaminants. (FSL)

Response variable A characteristic of particular interest involving each individual element of a population or sample. (VWP)

Restoration Measures taken to return a waste disposal site to pre-violation conditions. (FSL)

Restricted propellant A propellant that is used in explosives and a portion of whose individual propellant grains have been treated to control burning. (FSL)

Restricted use When a pesticide is registered, some or all of its uses may be classified (under FIFRA standards) for restricted use if the pesticide requires special handling because of its toxicity. Restricted-use pesticides may be applied only by trained, state-certified applicators or those under their direct supervision. (FSL)

Restriction enzymes Enzymes that recognize certain specific regions of a DNA molecule and cut the DNA into smaller pieces. (FSL)

Retentivity profile, soil A graphic representation of the retaining capacity of soil as a function of depth. The retaining capacity may represent water at any given tension, cations, or any other substances held by soils. (SWT)

Reticulate mottling A network of streaks of different colors that are commonly found in the deeper profiles of lateritic soils. (SWT)

Retrieval line A line or rope secured to a worker by a chest–waist or full-body harness, or wristlets. The other end is secured to either a lifting device or an anchor point located outside the entry portal. *See also* Lifeline. (SAN)

Retrospective study (case control, case history) A method of identifying causal associations between disease occurrence and either host or environmental characteristics. These causal associations are identified by comparison of the exposure histories of two defined population groups: one whose members have the disease and one whose members do not. An epidemiologic investigation in which two subgroups of the population are identified: the subgroup that is diagnosed (the cases) and the subgroup that does not have the disease under investigation (the control). The purpose of the study is to establish a relationship between prior health habits and current disease status. *See also* Epidemiology. (FSL, VWP)

Reverberation The persistence of sound after direct reception of the sound has stopped. (DEJ)

Reverse osmosis A water treatment process used in small water systems by addition of pressure to force water through a semipermeable membrane. Reverse osmosis removes most drinking-water contaminants. Also used in wastewater treatment. Large-scale reverse-osmosis plants are now being developed. *See also* Primary wastewater treatment; secondary wastewater treatment. (FSL)

Revetment A facing of masonry, brick, or stones used to protect an embankment against erosion. *See also* Riprap. (FSL)

Rhizobia A type of bacteria capable of living symbiotically in the roots of legumes. The association often results in an increase in the availability of nitrogen for use by plants. (SWT)

Rhizocylinder A plant root and the adjacent soil that is influenced by it. (SWT)

Rhizome A horizontal underground or subsurface plant stem. Rhizomes usually root at the nodes and are commonly used in vegetative propagation. (SMC)

Rhizosphere The area in soil in which plant roots and microorganisms interact. (SWT)

R-Horizon *See* Soil horizons. (SWT)

Ribonucleic acid (RNA) A component of all cells, RNA is found in both the nucleus and cytoplasm and is a natural polymer that consists of long chains of alternating phosphate and d-ribose units with the bases adenine, guanine, cytosine, and uracil bonded to the 1-position of the ribose. RNA fractions are identified by location, form, or function. *See also* Deoxyribonucleic acid. (FSL)

Ricin A toxin, contained in the seeds of the castor bean (*Ricinus communis*), the inhalation or ingestion of which can cause adverse effects in humans. (FSL)

Rickets A nutritional disease caused by a deficiency of Vitamin D in the diet. *See also* Scurvy. (FSL)

Rickettsia A class of microbial agents, named for the pathologist Howard T. Ricketts (1871–1910) and resembling small bacteria that multiply by simple fission, but only within a living cell. (FSL)

Rickettsialpox A disease caused by the microorganism *Rickettsia akari*, which is transmitted by mites. Also known as Kew Garden spotted fever, it is a mild febrile disease of 7 to 10 days' duration. Outbreaks of the disease have occurred in tenement buildings and other dwellings infested with mice. (FSL)

Rigging A general term that refers to the use of ropes, chains, and slings to lift loads. Riggings form the interface between the hoisting equipment and the load. Usually, there is a hook at the end of a hoist or crane wire rope. Slings with loops on both ends are wrapped around the item to be raised. Ground rigging is a method of suspending a working platform from a safe surface to a point of suspension above the surface. (SAN)

Rill A small, intermittent water course with steep sides, usually only several centimeters deep. Rills are not considered obstacles to tillage operations. (SWT)

Ringlemann chart A series of shaded illustrations, scored from 0 to 5, that are used to measure the opacity of air pollution emissions. The chart ranges from clear (indicating no pollution) to light gray, through black. The readings are used to establish and enforce emissions standards. The chart has been replaced by more-accurate smoke-measuring devices. (FSL)

Riparian Refers to land that borders a river, stream, lake, or coast. Such areas have a high density, diversity, and productivity of plant and animal species relative to nearby uplands. (FSL)

Riparian rights Rights that accrue naturally to the owners of land on the banks of waterways and that include the use of the water. (DEW)

Riprap Materials, such as broken rocks, cobbles, or boulders, that have been placed on earth surfaces such as the face of a dam or on a stream bank for protection against the erosive action of waves. *See also* Revetment. (FSL)

Risk (disease) The likelihood that a person having specified characteristics (e.g., age, sex, immune status) will acquire a particular disease. (FSL)

Risk assessment The qualitative and quantitative evaluation performed in an effort to define the risk posed to human health and/or the environment by the presence or potential presence and/or use of specific chemicals. (FSL)

Risk communication An interactive process of exchange of information and opinion among individuals, groups, and institutions. It involves multiple messages regarding the nature of risk and other messages that are not strictly about risk and that express concerns, opinions, or reactions to risk messages or to legal and institutional arrangements for risk management. (FSL)

Risk management The process of evaluating alternative regulatory and nonregulatory responses to risk and selecting among them. The selection process necessarily requires the consideration of legal, economic, and social factors. (FSL)

River basin The land area drained by a river and its tributaries. (FSL)

River wash Course, textured, barren alluvial land exposed along water courses at low water flows. These areas are subject to shifting during high water. *See also* Accretion. (SWT)

Rock A naturally formed aggregate or mass of mineral matter. (FSL)

Rock land Land areas that contain many rock outcrops and shallow soils. The rock outcrops usually occupy 25 to 90% of the total land area. (SWT)

RODAC plate A device used in an agar contact method of estimating the sanitary quality of surfaces. Such plates should be used on surfaces that have been previously cleaned and sanitized. A total plate count of microorganisms may be obtained using a nonselective medium, or, if more-qualitative data are desirable, a differential medium may be selected. *See also* Agar contact plate. (HMB)

Rodent Any mammal belonging to the order Rodentia, which includes gnawing or nibbling mammals such as beavers, mice, rats, and squirrels. *See also* Rodenticide; Rodent control. (FSL)

Rodent control Pest control is an essential part of any sanitation program in food-service or food-processing operations. The control of rodents, especially mice and rats, is an important part of this process. Proper sanitation is the most fundamental and effective way of controlling the presence of these animals. Proper storage and handling of foods within a facility can aid in eliminating their food sources. Use of proper containers with tight-fitting lids, and quick removal of scraps will contribute to this effort. Harborages should be eliminated also. Rodents are attracted to areas in which garbage and other solid waste is located. Overstocked storage areas and poor housekeeping practices also contribute to the problem. All opportunities for entry into the structure should be eliminated. Screens should be used over drains and vent areas. Building foundations and all masonry should be repaired. Holes or gaps around pipes should be secured. Screens should be in good repair and doors should be tight-fitting. A multitude of other preventive measures can be taken to control these pests, which are aesthetically unacceptable, destroy and contaminate supplies and facilities, and are potential carriers of disease to humans. *See also* Rat eradication; rat harborage; rat proofing. (HMB)

Rodenticide A pesticide, such as red squill, strychnine, or warfarin, used to kill rodents, generally through ingestion. *See also* Fumigation; pesticide; red squill; rodent control; strychnine; warfarin. (HMB)

Roentgen (R) Named after Wilhelm K. Roentgen (1845–1923), this is a unit of exposure dose of gamma radiation or X-radiation. It is defined as the quantity of gamma radiation or X-radiation that will produce (in 0.001293 gram of air [1 cc] at 0°C and 760 mm of Hg pressure) ions carrying one electrostatic unit quantity of electricity of either sign. From the accepted value for the energy lost by an electron in producing a positive–negative ion pair in air, it is estimated that 1 roentgen of gamma radiation or X-radiation results in the absorption of about 87 ergs of energy per gram of air. (RMB)

Roentgen equivalent, man (or mammal) (REM) A unit of biological dose of radiation in which the number expressing the relative biological effectiveness of radiation is equal to the number of rads absorbed, multiplied by the RBE of the given radiation (for a specified effect). The quantity of ionizing radiation of any type that, when absorbed by a human or other mammal, produces a physiological effect equivalent to that produced by the absorption of 1 roentgen of X-rays or gamma rays. *See also* Radiation absorbed dose; relative biological effectiveness; RBE dose. (RMB)

Roentgen equivalent physical (REP) An obsolete term. A unit of absorbed dose of radiation. Basically, the REP is intended to express roentgens of gamma radiation or X-radiation. It is estimated to be about 97 ergs per gram of tissue, although the actual value depends on certain experimental data that are not precisely known. The REP is thus defined, in general, as the dose of any ionizing radiation that results in the absorption of about 97 ergs of energy per gram of soft tissue. For soft tissue, the REP and the RAD are essentially the same. *See also* RAD; roentgen. (RMB)

Rollover protection A structure that surrounds the operator of a powered industrial truck, tractor, bulldozer, or any similar type of equipment, and that offers protection from injury in the event the unit turns over. Many powered industrial trucks have a high center of gravity, which makes them susceptible to rolling over. Since many of these vehicles, such as forklifts, carry loads, they require counterweights for stability. Operating these vehicles on rough, irregular, and sloping surfaces can add to their instability. *See also* Overhead protection; powered industrial trucks. (SAN)

Root-mean-square sound pressure The square root of the mean of the squares of instantaneous sound pressure levels. Often used to measure sound pressures in sound-pressure-level meters. (DEJ)

Rope grab A deceleration device that travels on a lifeline and automatically engages the lifeline frictionally and locks it so as to arrest the fall of an employee. A rope grab usually employs the principle of inertial locking, cam/lever locking, or both. *See also* Deceleration device; fall protection; self-retracting lanyard. (SAN)

Rosaniline [$HOC(C_6H_4NH_2)_2C_6H_3(CH_3)NH_2$] A substance derived from coal tar and used as a basis for dyes and stains, and as a fungicide. (FSL)

Rotameter A device that is used for measuring flow rates and that consists of a precision-made tapered transparent tube with one or more solid floats inside. The tube is marked with graduations to delineate different flow rates, with the reading usually being taken at the top meniscus of the ball float. (DEJ)

Rotating-vane anemometer A device used to measure air flow through large supply and exhaust openings and consisting of a propeller or revolving vane connected through a gear train to a readout unit that displays linear feet per minute. It is used to determine airflow over a given period of time (usually one minute). (DEJ)

Rotenone ($C_{23}H_{22}O_6$) Also known as tubatoxin, this is the most active of several rotenoids produced by legumes such as species of *Derris* and *Lonchocaycus*. Although in its use as an insecticide, it has been replaced by other chemicals, rotenone is now used primarily as a fish poison. (FSL)

Rough fish Fish not prized for eating, such as carp, gar and suckers. Most are more tolerant of changing environmental conditions than are game species. (FSL)

Route of entry/exposure *See* Exposure. (FSL)

Rubbish Term that includes waste paper, cartons, boxes, wood, tree branches, yard trimmings, furniture, appliances, metals, cans, glass, crockery, dunnage, and/or similar materials. *See also* Construction/demolition wastes; solid waste. (FSL)

Rule Under the Administrative Procedure Act, a federal rule is the whole or part of an agency statement of general or particular applicability and future effect, designed to implement, interpret, or prescribe law or policy or describing the organization, procedure, or practice requirements of an agency. This includes the approval or prescription of the future of rates, wages, corporate or financial structures, or reorganizations thereof, as well as related prices, facilities, appliances, services, allowances, valuations, costs, accounting, or practices. General notice of proposed rules must be published in the *Federal Register*. Final rules are codified in the Code of Federal Regulations, 5 USC §§551, 553. (DEW)

Rulemaking Under the Administrative Procedure Act, the agency process for formulating, amending, or repealing a rule. 5 USC §551(5). (DEW)

Runaway reactor **(1)** A chemical reactor that has become overpressured as a result of a reaction that is out-of-control; a violent release of materials results. To prevent this from occurring, chemical reactors are equipped with relief valves that allow them to release material gradually, thus relieving pressure and possibly preventing reactor blow-up. **(2)** A nuclear reactor that is out of control as a result of too-rapid rearrangement of nuclei by either fission or fusion. The reaction can cause overheating of the vessel and potentially a meltdown of the reactor. *See also* Safety relief valves; Chernobyl. (SAN)

Runoff That portion of the precipitation (storm water), on an area, that is discharged from a watershed through stream channels. Precipitation not entering the soil is called surface runoff. That which enters the soil before reaching a stream is called ground water runoff or seepage flow. In soil science *runoff* usually refers to the water lost by surface flow; in geology and hydraulics the term usually refers to both surface and subsurface flow. (SWT)

Rupture disks Portions of a pressure relief valve that are designed to release at an established pressure in order to discharge the contents of a gas cylinder. *See also* Safety relief valve. (FSL)

S

Sabin, Albert Bruce (1906–1993) American virologist noted for the discovery of an effective oral vaccine against poliomyelitis. This vaccine, when taken orally, stimulates the production of antibodies in the digestive system as well as in other body systems. *See also* Vaccine. (FSL)

Safe Drinking Water Act Federal law that sets standards for drinking water quality for public water systems by establishing primary and secondary drinking water regulations. Also controls the underground injection of fluids that could result in the contamination of drinking water supplies. 42 USC §300f *et seq. See also* Maximum contamination level goal; maximum contamination level; primary drinking water regulation; secondary drinking water regulation. (DEW)

Safety The development of systems and techniques to ensure that individuals in residential, recreational, and occupational settings and their environment are relatively free from conditions that could cause unnecessary morbidity or mortality from injuries or disease. (FSL)

Safety and health audits A comprehensive safety and health survey intended to determine compliance of a facility with OSHA regulations and in-house safety and health policies. A regular schedule for repeating routine audits should be in place in order to ensure that changes in conditions and activities do not create new hazards and that controls remain effective and in place. (SAN)

Safety belt A device, also called a body belt, that is usually worn around the waist and that, by reason of its attachment to a lanyard, lifeline, or deceleration device and an anchorage, will prevent a worker from falling.

OSHA requires that all safety belts be capable of withstanding a tensile loading of 4000 pounds without cracking, breaking, or becoming permanently deformed. Safety belts should not be used for any purpose other than guarding workers. *See also* Anchorage; fall protection; deceleration device; lanyard, lifeline. (SAN)

Safety color codes In the ANSI standard (Z53.11971), certain color schemes have been developed for the identification and location of fire systems, first-aid kits, traffic aisles, and tripping hazards. Red is usually used to identify fire protection equipment and apparatus, danger warnings (e.g., lights, safety cans), and emergency stops (signs, bars, buttons). Orange is the basic color for designating dangerous parts of machines or energized equipment that may cut, crush, shock, or otherwise injure. Yellow is the basic color for designating caution and is used to mark stumbling, falling, and tripping hazards. Green is the basic color for safety instructions and equipment such as safety bulletin boards, first-aid kits, and dispensers. Purple (magenta) is used to designate radiation hazards. Black, white, or a combination of the two is used for designating traffic and housekeeping markings. (SAN)

Safety engineers Engineers trained in the recognition and control of safety hazards, such as those involving machine guarding, equipment, controls, fire systems, and electrical systems. They apply scientific and engineering principles and methods to the elimination and control of occupational hazards. *See also* Certified safety professional. (SAN)

Safety Factor Also referred to as the "uncertainty factor," a number by which the "no observed effect level" (NOEL) is divided to obtain the acceptable daily intake of a chemical, for regulatory purposes. This factor presumably reflects the uncertainties involved in the process of extrapolating data about toxic exposures and allows for differences in sensitivity between test animals and humans. (FSL)

Safety fuse A flexible cord containing an internal burning medium by which fire is conveyed at a continuous and uniform rate for the purpose of detonating blasting caps. *See also* Explosion. (SAN)

Safety harness Also called a body harness, a design of straps that may be secured about the employee in a manner to distribute the fall arrest forces over portions of the body, such as the thighs, pelvis, waist, chest, and shoulders, with a means of attachment to other components of a personal fall arrest system. *See also* Deceleration device; fall protection; lanyard; lifeline. (SAN)

Safety Nets OSHA requires safety nets be provided when workplaces are more than 25 feet above the ground or are above water or other surfaces for which the use of ladders, scaffolds, catch platforms, temporary floors, safety lines, or safety belts is impractical. The nets must extend 8 feet beyond the edge of the work surface where workers are exposed and should be installed as close below the work surface as practical, but not more than 25 feet below the work area. Nets used for worker safety must meet certain construction requirements. The mesh size cannot exceed 6 inches by 6 inches and nets should have a label certifying a minimum impact resistance of 17,500 foot-pounds. (SAN)

Safety relief valve A device, often backed up with rupture disks, that is installed in tanks containing pressurized liquids, vapors, and gases. Such valves are designed to release small amounts of material. In the event that excessive pressure causes the valve to fail, the rupture disk beneath it will allow for larger quantities to be released. *See also* Pressure vessel; rupture disk. (SAN)

Safety Shoes Devices worn on the feet for protection against a variety of hazards, including impact, electrical shock, heat, and cold. Certain types of safety shoes are equipped with steel toes to protect against heavy falling objects and crushing, and/or may be equipped with flexible steel shanks to protect against penetration by nails or other sharp objects through the soles. Other types are constructed to eliminate the possibility of static electricity discharge in explosive atmospheres. Still others are well insulated to protect the wearer from heat and cold. Safety shoes must meet certain minimum standards. (DEJ)

Safety system A process for the elimination of or control of hazardous events through engineering design, education, management policies, and supervisory control of work site conditions and practices. (SAN)

Safety watch *See* Attendant; fire watch; hole watch; standby person. (SAN)

Salic horizon A mineral soil horizon that is enriched with secondary salts that are more soluble in cold water than is gypsum. A salic horizon is 15 cm or more in thickness and contains at least 20 g/kg salt. The product of the thickness in centimeters and the amount of salt by weight is more than 600 g/kg. *See also* Gypsum requirement. (SWT)

Saline A solution of salt (sodium chloride) in water. (FSL)

Salinization A process in which soluble salts accumulate in soils. (SWT)

Saline seep A saline discharge that is intermittent or continuous, exists at or near the soil surface under dryland conditions, and reduces or eliminates crop growth. It is differentiated from other saline soil conditions by its recent and local origin, a shallow water table, a saturated root zone, and sensitivity to cropping systems and precipitation. (SWT)

Saline-sodic soil A soil that contains both sufficient soluble salt and exchangeable sodium to affect crop production adversely under most soil and crop conditions. (SWT)

Salinity The percentage of salt in a solution, determined by measuring the density of the solution, using a hydrometer. (FSL)

Salk, Jonas Edward (1914–) American physician who developed a vaccine for the prevention of poliomyelitis. The vaccine is a preparation of three types of killed polioviruses and is given in a series of injections. *See also* Vaccine. (FSL)

Salt-affected soil Soil that has been adversely affected as a medium for plant growth by the presence of soluble salts, exchangeable sodium, or both. (SWT)

Salt balance The quantity of soluble salt that has been removed from an irrigated area in drainage water, minus the quantity of salt delivered in the irrigation water. (SWT)

Salts Minerals that water may pick up as it passes through the air, and over and under the ground, and as it is used in industry and households. (FSL)

Salt tolerance The ability of plants to resist or tolerate the adverse effects of excessive soluble salts in the soil. (SWT)

Salt water intrusion The invasion of fresh surface or ground water by salt water. If the salt water comes from the ocean, the process may be called sea water intrusion. (FSL)

Salutary Pertaining to health or to a state of well-being. *See also* Health. (FSL)

Salvage The utilization or reclamation of waste materials. *See also* Scavenge. (FSL)

Sample **(1)** (Statistics) A subset of a test human, animal, or plant population. **(2)** (Environmental) A process consisting of the withdrawal, isolation, or concentration of a fractional part of a whole as related to environmental media such as air, water, soil, or other liquid, solid, or gas, for analysis by acceptable procedures. Samples can be either cumulative (i.e., collected over a given period of time), in which case the result is expressed as a time-weighted average, or instantaneous. (DEJ, VWP)

Sample space The set of all possible outcomes of an experiment. (VWP)

Sampling distribution The spread of values for a sample statistic obtained from all possible samples of a population. (VWP)

Sampling frame A list of the items belonging to the population from which a sample will be drawn. (VWP)

Sampling plan *See* Sample design. (VWP)

Sand A soil particle between 0.05 and 2.0 mm in diameter; any one of five soil separates—namely: very coarse sand, coarse sand, medium sand, fine sand, and very fine sand; a separate soil class. (SWT)

Sand filters Devices that remove some suspended solids from sewage. Air and bacteria decompose additional wastes filtering through sand, so cleaner water drains from the bed. *See also* Primary wastewater treatment; secondary wastewater treatment. (FSL)

Sandflies Flies of the genus *Phlebotomus* that are vectors in the transmission of phlebotomus (sandfly) fever and leishmaniasis. (FSL)

Sandhog A worker performing tunneling operations that require atmospheric pressure control. *See also* Caisson. (FSL)

Sand test One of the major tests used in the United States to determine the performance of an explosive. It consists of detonation of a known explosive agent in sand that consists of a single grain size fraction. The strength of the explosive is determined by the amount of the sand that passes a fine-meshed sieve following fragmentation. (FSL)

Sandwich wall In a structure, a non-weight-bearing wall whose inner faces enclose insulating core materials such as styrofoam. (FSL)

Sanguine Found in the presence of, or associated with blood. *See also* Biomedical waste; medical waste. (FSL)

Sanitarian A public health or sanitation expert. A person who is involved in the maintenance of healthful and hygienic conditions in the environment. The term is synonymous with *environmentalist. See also* Sanitation. (HMB)

Sanitary landfill A site where putrescible solid wastes are disposed of by means of placement of an earth cover onto them. The site is managed by engineering methods that include spreading the wastes in thin layers, compacting them to the smallest practical volume, placing an earth cover onto them, and using such other measures as are necessary to protect human health and the environment. *See also* Landfill; solid waste. (FSL)

Sanitary science The body of knowledge concerned with the prevention and control of diseases through environmental control measures and the application of techniques for achieving standards of living that are necessary and economically attainable for the physiological needs of people in their particular environmental setting. *See also* Sanitarian. (HMB)

Sanitary sewers A system of underground pipes that convey only domestic sewage or nonhazardous industrial waste, and not storm water (except in the case of a combined sewer system). *See also* Sewage; sewer. (FSL)

Sanitary survey An on-site review of the water sources, facilities, equipment, operation, and maintenance of a public water system undertaken to evaluate the adequacy of those elements for producing and distributing safe drinking water. Can also refer to a plant or facility survey. Such surveys are to be comprehensive and critical and must address all phases of food production and food service, including cleaning and sanitizing to ensure product safety, quality, and freedom from contamination. *See also* Safety and health audit. (FSL, HMB)

Sanitation The formulation and application of measures designed to protect public health. Sanitation in the food industry refers to either processing or food service and is the creation and maintenance of healthful and hygienic conditions. Effective sanitation refers to the mechanisms that help accomplish these goals. *See also* Sanitarian. (HMB)

Sanitizer A chemical agent that reduces the microbial contaminants on a precleaned surface to safe levels, as determined by public health requirements. *See also* Disinfection. (HMB)

Sap The common name for the vascular tissues in plants. *See also* Phloem; xylem. (SMC)

Saponification The chemical process whereby an oil or fat is converted into soap by placement in combination with an alkali compound such as sodium hydroxide (NaOH). (FSL)

Saprists Soils that have a high content of plant materials that have decomposed, such that the original plant structure cannot be determined. Unless they are artificially drained, saprists are saturated with water for periods long enough to limit their use for most crops. (SWT)

Saprophytes Organisms, such as bacteria or fungi, that can survive on decaying or dead organic material. *See also* Garbage; offal; refuse. (FSL)

Saprophytic competence The ability of symbiotic or pathogenic microorganisms to survive as saprophytes in soil. (SWT)

Saranex® A multilayer laminate of polyethylene and polyvinylidene chloride used as a nonpermeable coating on various articles of protective clothing. *See also* Penetration; permeation. (FSL)

SARA Title III *See* Emergency planning; Community Right-to-Know Act. (DEW)

Sarcoma Malignant neoplasm composed of cells imitating the appearance of supportive and lymphatic tissues. *See also* Carcinogenic; oncogenic. (FSL)

Sarcopter A genus of mites that includes the itch mite (*Sarcopter scabies*), which causes the disease scabies in people. (FSL)

Sarin [$(CH_3)_2CHO(CH_3)FPO$] Methylphosphonofluoride acid isopropyl ester, an organic phosphate compound that has been developed for use as a nerve gas and designated by the military as GB. Sarin is a cholinesterase inhibitor and is toxic by inhalation and skin absorption. It was originally incorporated into various types of munitions (bombs) that would explode on impact and disseminate the gas. *See also* Phosgene; soman; tabun. (FSL)

Sassolite *See* Boric acid.

Saturation **(1)** The impregnation of one substance by another to the greatest possible extent; the action of causing something to be filled, charged, or supplied with the maximum that it can absorb—for example, the filling of all voids between individual soil particles with a liquid. **(2)** Preparation of the most concentrated solution. In chemistry, refers to the ability to dissolve the maximum amount of a gas, liquid, or solid in a solution at a given temperature and pressure. (FSL)

Saturation percentage The amount of water contained in a saturated soil paste, expressed as a dry weight percentage basis. (SWT)

Saturation zone A subsurface area in which all pores and cracks are filled with water under pressure equal to or greater than that of the atmosphere. (FSL)

Saturnism The medieval term for lead poisoning, or plumbism. *See also* Lead; lead poisoning. (FSL)

Scaffold A temporary metal or wooden framework for supporting workers and materials during the erection or repair of a building or other structure. According to OSHA, scaffolds must be designed and constructed to carry a work load not to exceed the following limits: **(1)** light duty, 25 pounds per square foot; **(2)** medium duty, 50 pounds per square foot; **(3)** heavy duty, 75 pounds per square foot. *See also* Scaffold, tube and coupler; scaffold, two-point suspension. (SAN)

Scaffold, tube and coupler The most common portable scaffold in use on work sites. An assembly consisting of tubing that serves as posts, bearers, braces, ties, and runners, a base supporting the posts, and special couplers that serve to connect the uprights and to join the various members. Couplers must be made of materials such as drop-forged steel, malleable iron, or structural grade aluminum. Platforms are made of either wood planking or metal grating planks that hook to the tubes. OSHA requires that these scaffolds be erected in a certain configuration, although their height, width, and length can vary. The runners consist of tubing that is installed along the length of the scaffold and located on both the inside and the outside posts at even height. Bearers are tubing installed transversely between posts and securely coupled to posts. Cross bracings are installed across the width of the scaffold at certain specific intervals. Longitudinal diagonal bracings on the inner and outer rows of poles must be installed at a 45° angle, from near the base of the first outer post upward to the extreme top of the scaffold. All tube and coupler scaffolds must be constructed and erected to support four times the maximum intended load. *See also* Scaffold. (SAN)

Scaffold, two-point suspension A scaffold the platform of which is supported by wire or rope hangers at two points and suspended from overhead supports so as to permit the raising or lowering of the platform to the desired working position by tackle or hoisting machines. This type of scaffold is used as a surface to work upon or as a means of providing access to vertical sides of structures on a temporary basis. It is also known as a swinging scaffold. (SAN)

Scale A buildup of mineral deposits on industrial equipment, in boilers, or on other equipment surfaces, largely as a result of the use of hard water in processing. Appropriate selection of a cleaning compound is essential in minimizing and removing this buildup. The effectiveness of sanitizers is also affected by scaling that could result in the survival of microorganisms on surfaces. It is therefore critical that an effective maintenance, cleaning, and sanitizing program be carried out to minimize this potential. *See also* Hard waters. (HMB)

Scaler As used in radiation safety work, an electronic device that produces an output pulse whenever a prescribed number of input pulses have been received. (RMB)

Scaling laws The methods by which nuclear weapons effects may be estimated for varying yields. For blasts, it has been found that such factors as overpressure and time of arrival scale as the cube root of the yield ($W^{1/3}$). For thermal effects, it is found that the total thermal energy is directly proportional to the weapon's yield (W). Similar approximations may be made for initial nuclear radiation estimates. However, all scaling laws must be modified depending upon the conditions of the burst. These conditions include height of burst, the yields themselves (when widely divergent yields are compared), and the weapon design. (RMB)

Scalping Forest soil preparation whereby ground vegetation and roots are removed to expose mineral soil. (SWT)

Scat A generic term for feces. (SMC)

Scatermia The type of toxemia wherein chemical toxins are absorbed through the intestines. (FSL)

Scatology The scientific study and analysis of feces for physiologic and diagnostic purposes. (FSL)

Scatter diagram A plot of all ordered pairs of bivariate data on a coordinate axis system. (VWP)

Scattered radiation Radiation that, during its passage through a substance, has been deviated in direction. It may also have been modified by an increase in wavelength; one form of secondary radiation. *See also* Bremsstrahlung. (RMB)

Scattering The diversion of radiation, either thermal or nuclear, from its original path as a result of interactions (or collisions) with atoms, molecules, and larger particles in the atmosphere or other medium between the source of the radiations. As a result of scattering, radiation (especially gamma rays and neutrons) will be received at a point at some remote distance from a nuclear explosion from many directions instead of only from the direction of the source. (RMB)

Scavenge The uncontrolled picking and removal of waste materials from discarded solid waste items at disposal sites, such as dumps and landfills. *See also* Salvage. (FSL)

SCBA *See* Self-contained breathing apparatus. (DEJ)

Scintillation counter The combination of a phosphor (which converts ionizing particle energy to a light pulse), a photomultiplier (which converts a light pulse to many electric pulses), and associated circuitry for counting electric pulses. The instruments are used for the routine detection of radioactive material in various media. (RMB)

Scintillation detector The combination of phosphor, photomultiplier tube, and associated electronic circuits for counting light emissions produced in the phosphor by ionizing radiation. *See also* Scintillation counter. (RMB)

Score variable A quantity introduced to represent some particular numerical attribute of a group. (VWP)

Scoria land Area that is characteristic of burned-out coal beds and that contains slaglike clinkers, burned shale, and fine-grained sandstone. Such areas may support a sparse cover of vegetation, but are not usually valuable for agricultural use. (SWT)

Scouring compound A chemical abrasive, usually silica, feldspar, or a material such as calcium carbonate, that is used to provide mild cleaning and polishing action. *See also* Cleanser. (HMB)

Scram Refers to the sudden closure of a nuclear reactor, usually by the dropping of safety rods. (RMB)

Scrap Materials, discarded from manufacturing operations, that may be suitable for reprocessing. *See also* Recycling; salvage. (FSL)

Screefing Forest soil preparation that consists of moving the humus layer to expose mineral soil. (SWT)

SCUBA *See* Self-contained breathing apparatus. (DEJ)

Scrubber An air pollution device that uses a spray of water or reactant or a dry process to trap pollutants in emissions. (FSL)

Scurvy A disease caused by the lack of ascorbic acid (vitamin C) in the diet. It is characterized by the presence of anemia, edema, ulceration of gums, and hemorrhages into the skin and from the mucous membranes. *See also* Rickets. (FSL)

Sealed source Radioactive material that is encased in, and is to be used in, a container in a manner intended to confine the material. For example, a calibration source such as ^{57}Co or ^{60}Co, such as that used in industrial radiography to check the integrity of welds. (RMB)

Secondary attack rate *See* Rate. (FSL)

Secondary blast An explosion undertaken primarily to reduce the size of rocks resulting from the primary detonation. *See also* Primary blast. (FSL)

Secondary clarifiers In a sewage treatment plant, the settling basins that receive wastewater after biological treatment. They are identical in principle to primary clarifiers and, depending upon the degree of treatment, may also be called final clarifiers. *See also* Clarifiers; clarification; primary clarifiers. (HMB)

Secondary drinking water regulation Under the Safe Drinking Water Act, a regulation that sets a maximum contaminant level for contaminants that adversely affect the odor or appearance of water or otherwise adversely affect the public welfare. *See also* Maximum contaminant level; Safe Drinking Water Act. (DEW)

Secondary food contamination A source of food contamination that may come from animals, humans, fomites, or feed additives. Animals—e.g., rodents and invertebrates—are familiar sources of secondary food contamination. Infected humans may be the source of contamination at any step in the food chain, but most commonly are the cause of problems associated with preparation of foods for the table—e.g., in restaurants and homes. Fomites (animal feed antibiotics), growth stimulators, or toxic chemicals accidentally added to food can be sources of secondary food contamination.

See also Carrier; insect infestation; personal hygiene; rodent. (HMB)

Secondary radiation Radiation originating as the result of absorption of other radioactivity in matter. It may be either electromagnetic or particulate in nature. (RMB)

Secondary wastewater treatment The second step in most publicly owned waste treatment systems in which bacteria consume the organic parts of the waste. It is accomplished by bringing together waste, bacteria, and oxygen in trickling filters or in the activated sludge process. This type of treatment removes floating and settleable solids and about 90% of the oxygen-demanding substances and suspended solids. Disinfection is the final stage of secondary treatment. *See also* Primary wastewater treatment. (FSL)

Second bottom The first terrace above the normal floodplain of a stream. (SWT)

Second quartile A number such that at most one-half of the data are smaller in value than the number and one-half of the data are larger. The same value as the median. (VWP)

Secure maximum contaminant level Maximum permissible level of a water contaminant delivered to the free-flowing outlet of the consumer, the ultimate user of a water supply. (FSL)

Sedimentary rock **(1)** A rock formed from materials that were deposited from a suspension or precipitated from solution. Principal sedimentary rocks are sandstones, shales, limestones, and conglomerates. **(2)** Those rocks that consist of chemical precipitates and of rock fragments deposited by water, ice, or wind. They include deposits of gravel, sand, silt, clay, the hardened equivalents of these (sandstone, siltstone, slate, gravel, sand) and deposits of gypsum and salt. (SWT, FSL)

Sedimentation **(1)** The process whereby sediments are deposited by wind or water. **(2)** As applied to water treatment, the process of settling by gravity and deposition of heavy suspended material in water. This settling action can be accomplished in a pond or properly constructed basin. (SWT, FSL)

Sedimentation tanks Holding areas for wastewater, in which floating wastes are skimmed off and settled solids are removed for disposal. (FSL)

Sediments Soil, sand, and minerals washed from land into water, usually after rain. They pile up in reservoirs, rivers, and harbors, destroying fish-nesting areas and holes of water animals, and cloud the water so that needed sunlight may not reach aquatic plants. Careless farming, mining, and building activities can expose sediment materials, allowing them to be washed off the land after rainfalls. (SWT)

Seepage pit A covered pit with a lining designed to permit sewage effluent from a septic tank from seeping into the surrounding soil. This type of subsurface disposal system is now outlawed in most areas because of the high potential for soil and/or groundwater contamination. *See also* Cesspool. (FSL)

Seepage spring The type of spring in which the water oozes out of sand, gravel, or other material that contains many small interstices. Such springs are generally free from pathogenic bacteria, but they are susceptible to contamination by surface runoff that collects in valleys or depressions. (FSL)

Seiche A variation in the level of the water in a lake or pond. (FSL)

Selective localization (radiation) As applied to radioisotopes, the accumulation of a particular isotope to a significantly greater degree in certain cells or tissues. *See also* Differential absorption ratio. (RMB)

Selective pesticide A chemical designed to affect only certain types of pests, leaving other plants and animals unharmed. *See also* Insecticides; rodenticides. (FSL)

Selectivity coefficient A conditional equilibrium coefficient for an ion exchange reaction that is expressed in terms of concentration variables for the exchangeable ions and either concentration variables or activities of the ions in solution. (SWT)

Self-absorption Absorption of radiation emitted by radioactive atoms by the matter in which the atoms are located; in particular, the absorption of radiation within a sample being assayed. (RMB)

Self-contained breathing apparatus A device, worn by an individual, that provides a containerized source of high-pressure breathing air to a pressurized mask through a regulator. Such devices usually worn in atmospheres containing extremely high levels of air contaminants or unknown concentrations of hazardous substances. They are often used by firefighters, emergency response workers, and some miners. When used underwater, such a device is known as a SCUBA (acronym for *self-contained underwater breathing apparatus*). *See also* Pressure-demand respirator. (DEJ)

Self-mulching soil A soil that does not crust or seal under the influence of raindrops because of the amount of surface soil aggregation. (SWT)

Self-retracting lanyard A deceleration device, used in the work environment, that contains a drum-wound line that may be slowly extracted from, retracted from, or retracted onto the drum under slight tension during normal employee movement, and that after onset of a fall automatically locks the drum and stops the fall. Can also be attached to a lifeline. *See also* Deceleration device; fall protection; rope grab. (SAN)

Semelparous Describes populations in which individuals reproduce only once during their lifetime. Examples are annual plants and the Pacific salmon. (SMC)

Semi-confined aquifer An aquifer that is partially confined by a soil layer (or layers) of low permeability through which recharge and discharge can occur. (FSL)

Senescence The aging process; the term is sometimes used in describing bodies of water in advanced stages of eutrophication. *See also* Eutrophication. (FSL)

Sensitive ingredient Any food ingredient historically associated with a known microbiological hazard. *See also* Hazard analysis critical control point. (HMB)

Sensitive volume The portion of a counter tube or ionization chamber that responds to a specific radiation source. (RMB)

Sensitivity The probability that a symptom is present given that the affected person has a disease. (VWP)

Sensitization Immunization, especially involving antigens (immunogens) not associated with infection; the induction of acquired sensitivity or of allergy. *See also* Vaccine. (FSL)

Sensorineural hearing impairment Usually an irreversible hearing loss involving the auditory nerve and the loss of hair cells in the organ of Corti in the ear, caused by exposure to excessive noise levels, viruses, congenital defects, and drug toxicity. (DEJ)

Sensors Portions of an instrument that rely on some physical property of an environmental contaminant to produce a response that is subsequently detected in another part of the instrument, amplified or otherwise treated, and then displayed and interpreted by the instrument user. All sensors must be calibrated in media with known concentrations of the substance to be measured. Examples of areas in which sensors are used include aerosol photometry, chemiluminescence, combustion studies, electrical conductivity and thermal conductivity studies, coulometry, flame ionization studies, potentiometry, polarography, and electron capture studies. Some sensors are physically separated from the instrument, and rely on transmission to measure remote phenomena. (DEJ)

Sepsis The presence of pathogenic microorganisms or their toxins in the blood or other tissue. *See also* Infection. (FSL)

Septic tank An underground storage tank for wastes from homes having no sewer line to a treatment plant. The waste goes directly from the home to the tank, where the organic waste is decomposed by bacteria and the sludge settles to the bottom. The effluent flows out of the tank into a subsurface disposal network of drains. The accumulated sludge in the tanks is pumped out periodically. Local laws or ordinances usually regulate the placement of septic tanks and subsurface disposal systems and generally prohibit their location in places such as parking lots or other paved areas or in proximity to water sources, structures, etc. *See also* Cesspool; seepage pit; soil absorption system; sub-surface disposal system. (FSL)

Sequestrant A compound, sometimes called a chelating agent, that, in aqueous solution, combines with metal ions to form a water-soluble product. Sequestrants are used in combination with cleaning compounds to inactivate the water-hardness metals (calcium and magnesium) and iron and manganese, which would otherwise result in precipitation deposits. *See also* Chelating agent; water softening. (HMB)

Sere A stage in the succession of a plant community in a particular habitat. (FSL)

Serial distribution As related to subsurface sewage disposal systems, a distribution achieved by arranging individual trenches in the absorption system so that each trench is forced to pond or flood to the full depth of the gravel fill before liquid flows into the succeeding trenches. This is the accepted technique in most areas for the installation of soil absorption fields used in conjunction with septic tanks. *See also* Septic tank. (FSL)

Serious violation In OSHA proceedings, a term used to describe a situation in which there is substantial probability that death or serious physical harm could result from a condition in the workplace. *See also* Willful violation. (DEJ)

Serotyping The characterization of a microorganism by the identification of the antigens possessed by that organism; a means of strain differentiation. (FSL)

Serpentine **(1)** A type of rock high in iron, magnesium, nickel, chromium, and cobalt, and low in calcium, potassium, sodium, and aluminum. Soils that are derived from serpentine parent rocks are chemically distinct from soils derived from granite, limestone, or other rocks, and they support different ecological communities than do the latter soils. The biomass of communities on serpentine soils is generally lower than that of nearby communities on other soils, and serpentine communities usually have a high number of endemic species. **(2)** As relates to asbestos, serpentine is the mineral chrysotile, a magnesium silicate. *See also* Asbestos. (FSL)

Service connection A physical link between water lines allowing delivery of water from a public or semipublic system to the piping system of a single structure. The service connection is usually joined by a valve and/or meter and is located in a curb box on the property. Service connections composed of lead piping may pose a health hazard. (FSL)

Sesquan A cutan that is composed of sesquioxides (such as those of iron or aluminum). *See also* Cutan. (SWT)

Set pressure The point at which a pressure relief valve on a dishwasher or hot water heater is set to discharge. *See also* Safety relief valve. (FSL)

Settleable solids Material heavy enough to sink to the bottom of a wastewater treatment tank. (FSL)

Settlement agreement Negotiated agreement, in settlement of a dispute, that is not entered as an order of the court. *See also* Consent decree. (DEW)

Settling chamber A device used to remove particulates from air or water by reducing the conveying velocity of the fluid stream to a level at which the particles are no longer being transported. Space requirements for such devices are usually substantial, and the devices are sometimes incapable of removing significant percentages of certain contaminants (such as respirable dust when there are eddy currents). Settling chambers are often used as coarse precleaners, and their use may be followed by that of more-efficient cleaners. (DEJ)

Settling tank A wastewater holding area in which heavier particles sink to the bottom for removal and disposal. (FSL)

Severity rate A formula applied exclusively to OSHA recordable injuries that result in lost time from the workplace.

$$\text{Severity rate} = \frac{\text{(Days aways from work) (200,000)}}{\text{Number of hours of exposure}}$$

See also Rate. (SAN)

Seveso A village near Milan, Italy, that was the site of an incident involving the herbicide 2,4,5-T in July 1976. As the result of an explosion at a factory that was manufacturing this herbicide, numerous workers were exposed to the chemical and vegetation in the area was destroyed. The workers developed chloracne, a skin condition known to be associated with exposure to specific chlorinated compounds. *See also* Chloracne. (FSL)

Sevin *See* Carbaryl. (FSL)

Sewage Excremental or other discharges from the bodies of humans or animals; the wastes and wastewater produced by residential and commercial establishments that are ultimately discharged into sewers. (FSL)

Sewage sludge Sludge produced at publicly owned treatment works as a result of the sewage treatment process. The disposal of such sludge is regulated under the Clean Water Act. The sludge from a disposal plant is useful as a soil conditioner. However, the material may contain minute concentrations of heavy metals, such as lead and cadmium, and other contaminants potentially detrimental to the environment. (FSL)

Sewer A channel or conduit that carries wastewater and storm water runoff from the source to a treatment plant or receiving stream. Sanitary sewers carry household, industrial, and commercial waste; storm sewers carry runoff from rain or snow; combined sewers are used for both purposes and are prone to flooding. (FSL)

Sewerage The entire system of sewage collection, treatment, and disposal. (FSL)

Shade numbers Filter lenses or plates used in welding helmets and goggles to protect the eyes from the ultraviolet and infrared rays given off by both flames and arcs. The lenses are tinted either green or brown and are available in shade intensities (expressed as numbers). OSHA requires specific shade numbers for specific operations, but allows shades more dense to be used if desired by the worker. The higher the number, the more dense (dark) is the shade. Filter lenses must also conform to the ANSI requirements for eye protection for welding (ANSI Z87.1-1968). Approved lenses carry a shade number and manufacturer's mark. *See also* Flash burn. (SAN)

Shakeout That part of a foundry process in which the mold (usually consisting of treated sand) is separated from the metal form produced. The shakeout area of a foundry is often characterized by high exposures to silica dust, carbon monoxide, heat, and other stressors. (DEJ)

Shaly Describes a soil that contains a large amount of shale fragments. (SWT)

Shaver's disease A disease of the lungs, found in workers exposed to fumes containing aluminum oxide and silica in the process of smelting bauxite in the manufacture of corundum, a form of aluminum oxide. This disease, a type of pneumoconiosis, is marked by opacities on X-ray film, and is sometimes accompanied by decreased lung function. *See also* Pneumoconiosis. (DEJ)

Sheet erosion *See* Erosion. (SWT)

Shielding Any material or obstruction that absorbs radiation and thus tends to protect personnel or materials from the effects of nuclear radiation. A moderately thick layer of any opaque material will provide satisfactory shielding from thermal radiation, but a considerable thickness of material of high density may be needed for nuclear radiation shielding. (RMB)

Shigellosis An illness caused by bacteria of the genus *Shigella. Shigella spp.* are aerobic, non-spore-forming, nonmotile, Gram-negative bacilli of the family Enterobacteriaceae. The genus consists of four species: *S. dysenteriae*, *S. flexneri*, *S. boydii*, and *S. sonnei*, referred to as subgroups A, B, C, and D, respectively. Shigellosis is often thought of as a waterborne illness, but is frequently foodborne as well. Large outbreaks of foodborne *Shigella sonnei* gastroenteritis have been reported by the Centers for Disease Control and Prevention. Nine outbreaks in the U.S. in 1987 involved 6494 individuals. The most frequently reported species is probably *S. sonnei.* The incubation period is from 1 to 7 days, and usually less than 4 days. Symptoms of the disease include fever, diarrhea, vomiting, abdominal cramps, and tenesmus (urgent need to urinate or defecate). Blood and mucus in the stool occurs in severe cases. Foods implicated in outbreaks are varied and often involve contamination from food handlers (humans are the major reservoir for the organism). Identified vehicles include salads, vegetables, poultry, shellfish, and stews. (HMB)

Shock-sensitive chemicals Refers to those chemicals for which mechanical shock can serve as an initiator of an explosive reaction. Organic peroxides, nitro compounds, organic nitrates, and azides are shock-sensitive materials. Organic peroxides have unusual stability problems that make them among the most hazardous chemicals normally used in laboratories. They are sensitive to heat, friction, impact, and light. All organic peroxides are highly flammable. Peroxides present as contaminants in reagents or solvents can alter the nature of a planned reaction. *See also* Picric acid; unstable material. (FSL)

Shock wave The intense compression wave produced by the detonation of an explosive. *See also* Explosive. (FSL)

Shoring Structures, such as metal hydraulic, mechanical, or timber reenforcement systems, that support the sides of an excavation and are designed to prevent cave-ins of excavations. Cross braces are the horizontal members of a shoring system and are installed perpendicular to the sides of the excavation, the ends of which bear against either uprights or wales. Wales are horizontal members of a shoring system and are placed parallel to the excavation face, whose sides bear against the vertical members of the shoring system or earth. Uprights are the vertical members of a trench shoring system that are placed in contact with the earth and positioned so that individual members do not come into contact with each other. Uprights placed so that individual members are closely spaced, in contact with or interconnected to each other, are often called sheeting. *See also* Sloping. (SAN)

Short-term exposure limit (STEL) The concentration to which workers may be exposed continuously for a short period of time without suffering adverse health effects. The STEL is defined by the American Conference of Governmental Industrial Hygienists and by OSHA as a 15-minute time-weighted average that should not be exceeded at any time during the workday even if the 8-hour time-weighted average is within acceptable limits. Exposures to the STEL should not be longer than 15 minutes, should not occur more than four times each day, and should be separated by at least 1 hour between successive exposures. Also known as TLV-STEL. *See also* Permissible exposure limit. (DEJ)

Shot anchor A device used to hold explosive charges in a bore hole so that the charges will not be blown out by the detonation of other charges. (FSL)

Shotgun A nonscientific term that refers to the process of breaking up the DNA derived from an organism and then moving each separate and unidentified DNA fragment into a bacterium. (FSL)

Shoulder The convex slope component at the top of an erosional sideslope. *See also* Angle of repose. (SWT)

Shredding The process by which solid waste is cut, torn, or ground into small pieces for final disposal or further processing. (FSL)

Shrub A bush; a woody perennial plant, usually branching from the base and having several main stems, that grows to a height of less than 5 meters. (SMC)

Sierra Club A national organization founded in 1892 and devoted to the preservation of natural resources and the maintenance of the integrity of the ecosystem. (FSL)

Sigmoid curve S-shaped curve, often characteristic of a dose-effect curve in radiobiological studies. (RMB)

Sign The objective evidence of a disease as demonstrated on examination of a patient. *See also* Syndrome; symptom. (FSL)

Signal words As applied to the labeling of pesticides or other containers of chemicals, signal words represent the three main categories of toxicity, based on the contents of the material. ''Danger'' indicates highly toxic material (oral LD_{50} up to 50 mg/kg), ''warning'' (oral LD_{50} of 50–500 mg/kg) for moderately toxic material, and ''caution'' (oral LD_{50} of 500–5000 mg/kg) for slightly toxic chemicals. (FSL)

Significant deterioration Pollution resulting from a new source in previously ''clean'' areas. (FSL)

Significant level The probability of committing the type-one error. *See also* Error, type-one. (VWP)

Significant municipal facilities Publicly owned sewage treatment plants that discharge a million or more gallons per day and are therefore considered by states to have the potential for substantially affecting the quality of receiving waters. (FSL)

Significant violations Violations, by point-source dischargers, of sufficient magnitude and/or duration to be a regulatory priority. (FSL)

Silage Chopped corn stalks and other green fodder, also known as ensilage, stored in a silo, allowed to ferment, and used ultimately as a food source for cattle. *See also* Silo-filler's disease. (FSL)

Silica (SiO_2) A family of minerals composed of silicon dioxide and nearly ubiquitous in the earth's crust. Silica can refer to amorphous (noncrystalline) silica or crystalline silica, such as quartz (sand) and silicates (including clay). The crystalline forms, such as quartz, tridymite, cristobalite, are known to cause silicosis, a lung disease found in workers employed in foundry work, glass manufacturing, granite cutting, mining, and tunneling. Silicosis is also known as miner's asthma, grinder's consumption, miner's phthisis, potter's rot, and stonemason's disease. Silicosis is marked by a loss of elasticity of the lung tissue, and development of silica-containing nodules in the lungs, resulting in decreased lung function, shortness of breath, enlargement of the heart, and often lung cancer. *See also* Anthracosis; black lung disease; Collier's disease; pneumoconiosis. (DEJ)

Silica–alumina ratio The ratio of units of silicon dioxide (SiO_2) per unit of aluminum oxide (Al_2O_3) in clay minerals or in soils. (SWT)

Silica–sesquioxide ratio The ratio of units of silicon dioxide (SiO_2) per unit of aluminum oxide (Al_2O_3) plus ferric oxide (Fe_2O_3) in clay minerals or soils. (SWT)

Silicon (Si) The second-most-abundant element in the earth's crust. Silicon is a major component of silica and silicates (rocks, quartz, sand, clays, etc.). It is used as a semiconductor in solid-state devices (transistors, computer circuitry, photovoltaic cells, etc.), as an alloying agent in steels, aluminum, and copper, and as a deoxidizer in the manufacture of steel. It is used in the manufacture of glass and in various types of water-repellent finishes. (FSL)

Silicosis *See* Silica. (DEJ)

Silo-filler's disease Pulmonary inflammation due to the inhalation of oxides of nitrogen and other chemicals liberated in farm silos during the fermentation process associated with fresh silage. *See also* Silage. (FSL)

Silt **(1)** A textural class of soil that consists of particles between 0.05 mm and 0.002 mm in diameter. (SWT)

Silting Deposition of waterborne sediments in stream channels, lakes, reservoirs, or on flood plains. Deposition is usually the result of a decrease in the velocity of the water. (SWT)

Silt loam A textural class of soils that contain a relatively large amount of silt and small quantities of sand and clay. (SWT)

Silty clay A textural class of soils that contain a relatively large amount of silt and clay and a small amount of sand. (SWT)

Silty clay-loam A textural class of soils that contain a large amount of silt, a lesser quantity of clay, and a still smaller quantity of sand. (SWT)

Silviculture Management of forest land for timber; may contribute to water pollution as a result of erosion following clear-cutting. (FSL)

Single-service articles Utensils used in food-service operations, such as cups, containers, lids, closures, plastic plates, knives, forks, and spoons, stirrers, paddles, straws, napkins, wrapping materials, toothpicks, and similar articles intended for one-time, one-person use and then discarded. (HMB)

Sinking Controlling oil spills by using an agent to trap the oil and sink it to the bottom of a body of water, where the agent and the oil are biodegraded. (FSL)

Site In ecology, an area that is described or defined by its biotic, climatic, and soil conditions as related to its capacity to produce vegetation. Also refers to the location of a facility such as a water treatment plant or waste-disposal operation. *See also* Dump; sanitary landfill. (SWT)

Site index A quantitative assessment of soil productivity for forest growth in an existing or a specified environment. The height in meters of the dominant forest vegetation is taken at or calculated to an index age, usually 50 to 100 years. (SWT)

Site inspection Under the Comprehensive Environmental Response, Compensation, and Liability Act, an on-site investigation to determine whether there is a release or potential release and the nature of the associated threats. The purpose is to augment the data collected in the preliminary assessment and to generate, if necessary, sampling and other field data to determine whether further action or investigation is appropriate. 40 CFR §300.5. *See also* Comprehensive Environmental Response, Compensation, and Liability Act; preliminary assessment; release. (DEW)

Siting The process of choosing a location for a facility such as a sewage treatment plant or waste-disposal operation. (FSL)

Skeleton grains Individual grains of soil that are stable and not readily translocated, concentrated, or reorganized by soil-forming processes. Skeleton grains include mineral grains and resistant siliceous and organic bodies that are larger than colloidal size. (SWT)

Skewness State of skewing of data distribution. In the case of positively skewed data (data skewed to the right), if sample points are graphed, the points above the median are generally farther from the median in absolute value than are the points below the median. The arithmetic mean is larger than the median. In the case of negatively skewed data (data skewed to the left), if sample points are graphed, the points below the median are generally farther from the median in absolute value than are the points above the median. The arithmetic mean is smaller than the median. (VWP)

Skimming The use of a machine to remove oil, scum, or other floating debris from a water surface. (FSL)

Skin dose (radiology) Dose at the center of an irradiation field on skin. It is the sum of the air dose and backscatter. (RMB)

Skyshine Radiation, particularly gamma rays from a nuclear explosion, reaching a target from many directions as a result of scattering by the oxygen and nitrogen in the intervening atmosphere. (RMB)

Slag Nonmetallic residue resulting from the smelting of metallic ores. (FSL)

Slash Downed timber (boles, branches, and whole small trees) left where it has fallen. Slash usually refers to the woody material left on site after a logging operation, but it can also refer to timber felled by natural events, such as tornados and ice storms. Slash does not usually refer to branches or single trees that have fallen because of senescence. (SMC)

Slaty Describes soils that contain a considerable quantity of slate fragments. (SWT)

Slickens Fine-textured materials separated during mining and ore mill operations. Slickens may decrease plant growth unless confined in specially constructed basins. (SWT)

Slickensides Polished and grooved soil surfaces caused by the sliding of one mass past another. Slickensides are common in vertisols. *See also* Vertisols. (SWT)

Sloping A method of protecting employees from cave-ins by excavating to form sides of an excavation that are inclined away from the area of disturbed earth. The angle of incline required to prevent a cave-in varies with differences in many factors, such as the soil type, environmental conditions of exposure, and the application of surcharge loads. The maximum allowable slope is the steepest incline of an excavation face that is acceptable for the most favorable site conditions as protection against cave-ins, and is expressed as the ratio of horizontal distance to vertical rise. *See also* Angle of repose; distress; excavations; terrace. (SAN)

Slough A swamp, bog, or marsh, especially one that is part of an inlet or backwater. *See also* Wetland. (FSL)

Slow sand filtration Treatment process involving the passage of raw water through a bed of sand at low velocity, which results in the substantial removal of chemical and biological contaminants. (FSL)

Sludge The accumulated settled solids that are deposited from sewage and that contain a high proportion of water and form a semiliquid mass. The material that is removed as a result of chemical treatment, coagulation, flocculation, sedimentation, flotation, and/or biological oxidation of water or wastewater. *See also* Activated sludge; sewage treatment. (HMB)

Sludge bed An area constructed of natural or artificial layers of porous material (e.g., sand) upon which digested or stabilized sludge is dried by evaporation and drainage. A sludge bed is usually open to the atmosphere, but may be covered with a greenhouse-like structure to facilitate drying. Often called a sludge drying bed. *See also* Secondary wastewater treatment. (HMB)

Sludge blanket An accumulation of sludge that is hydrodynamically suspended within an enclosed body of water or wastewater—e.g., solids that accumulate in a primary or secondary clarifier during wastewater treatment. (HMB)

Slug A bar-shaped piece of material prepared especially for insertion in a nuclear reactor. The term often refers to the fuel unit of a natural uranium–graphite reactor. (RMB)

Slurry A water mixture of insoluble matter that results from some pollution control techniques; a thin, watery mixture of a fine insoluble material such as clay, cement, or soil. (FSL)

Small-hose system System employing hose ranging in diameter from $\frac{5}{8}$ inch to $\frac{1}{2}$ inch. Such a system is generally for use by employees and provides a means of controlling and extinguishing incipient-stage fires. (SAN)

Small-quantity generator As pertaining to hazardous waste, and as defined by the EPA, a facility that produces, in a calendar month: more than 100 kg (2.2 lbs.), but less than 1000 kg, of nonacutely hazardous waste; or less than 100 kg of waste resulting from cleanup of any residue or contaminated soil, water, or other debris involving the cleanup of an acutely hazardous waste; or less than 1 kg of an acutely hazardous waste. *See also* Generator. (FSL)

Smelter A facility that melts or fuses ore, often with an accompanying chemical change, to separate the metal. Emissions from these facilities are known to cause pollution. *See also* Slag. (FSL)

Smog A combination of atmospheric smoke and fog (thus the name ''smog'') that is composed of a complex mixture of gases, dust, and droplets that appear as atmospheric haze. It constitutes atmospheric contamination by aerosols from a combination of natural and anthropometric (people-related) sources. In urban areas, *smog* often refers to the complex mixture of air contaminants produced during peak traffic hours from the photochemical reaction of vehicular emissions, including hydrocarbons, oxides of nitrogen and carbon, and ozone. (DEJ)

Smoke Small gasborne particles resulting from incomplete combustion that consist mainly of carbon and other combustible materials. (DEJ)

Smudge pot A smoke generator used to create dense smoke to prevent ground frost or frost that might damage fruit trees or forest in a specific area. (FSL)

Snag Standing dead trees or boles; dead trees, shrubs, or branches that are partially or completely underwater in a lake or river. (SMC)

Snakehole A hole drilled in a slightly downward direction from a horizontal plane into the floor elevation of a quarry face. (FSL)

Snaphook A connector, made up of a hook-shaped member with a normally closed keeper or similar arrangement, that may be opened to permit the hook to receive an object and that may be released to close automatically and retain the object. Snap hooks are generally of two types: a locking type with a self-closing, self-locking keeper that remains closed and locked until pressed open for connection or disconnection; and a nonlocking type with a self-closing keeper that remains closed until pressed open for connection or disconnection. *See also* Connector, D-ring; fall protection. (SAN)

Sneeze shields Shields designed of transparent material to protect displays of unpackaged foods offered for sale to the consumer. The shields intercept the direct line between a customer's mouth and the displayed food and minimize contamination by the customer. Various designs are acceptable if they satisfy these basic requirements. (HMB)

Snood A small netlike device for keeping hair in place. It may be considered a hair restraint, and is required by regulations for workers in food-handling operations. It may be worn over beards for the same purposes. *See also* Hair restraint. (HMB)

Snow, John (1813–1858) English epidemiologist who collected facts about drinking water sources and related them to the incidence of cholera in London. He published several books, including *On the Mode of Transmission of Cholera, On Chloroform and Other Anaesthetics*, and *On the Inhalation of the Vapour of Ether in Surgical Operations.* (FSL)

Soap The product formed by the saponification of fats, oils, waxes, rosin, or their acids with organic or inorganic bases. (HMB)

Sodic soil A soil that is nonsaline but that contains sufficient exchangeable sodium to affect crop production and soil structure adversely under most conditions. (SWT)

Sodication A process that increases the exchangeable sodium content of a soil. (SWT)

Sodium (Na) A highly reactive, soft, silver-white metal that reacts violently with water to form sodium hydroxide and hydrogen gas. This element represents a severe fire risk when in contact with water in any form. Sodium is used in the production of sodium peroxide and sodium hydride, in nonglare lighting for highways, and as a heat transfer agent in solar-powered electric generators. It has an atomic weight of 22.9898 and an atomic number of 11. (FSL)

Sodium adsorption ratio (SAR) test A soil test used to predict the exchangeable sodium percentage equilibrated with a given solution, based on the relation between soluble sodium and soluble divalent cations. (SWT)

Sodium chlorate ($NaClO_3$) A colorless, odorless chemical generally derived from a concentrated acidified solution of sodium chloride that has been heated and electrolyzed. Sodium chlorate is a dangerous fire risk, and in contact with organic materials may cause fire. It is used as an oxidizing agent and bleach and in the manufacture of matches, explosives, and flares. It is also used as a herbicide that can persist in the soil for months after application. (FSL)

Sodium hydroxide (NaOH) A strongly alkaline substance used as the cleaning agent in some detergents. Also known as caustic soda, lye, and or white caustic, it is the eighth-highest-volume chemical produced in the United States. Sodium hydroxide is corrosive to tissues and is a strong irritant to the eyes, skin, and mucous membranes, and is toxic by ingestion (TLV, 2 mg/m^3). (FSL)

Sodium sulfate (Na_2SO_4) White, crystalline or powdered, odorless substance that is among the highest-volume chemicals produced in the United States. Sodium sulfate is used in the manufacture of paper, glass, detergents, and ceramic glazes; as a laboratory reagent; and as a food additive. (FSL)

Sodium sulfite (Na_2SO_3) White crystalline or powdered material derived from the reaction of sulfur dioxide with soda ash and water, with treatment of the resulting sodium bisulfite with soda ash. Sodium sulfite is occasionally used illegally to retard the discoloration of fresh meat. Its use for that purpose has resulted in deaths and illness among humans who have consumed the contaminated product. Sodium sulfite is also used in water treatment as an antioxidant and as a bleaching agent. (FSL)

Softness A relative specification of the quality or penetrating power of X-rays. In general, the longer the wavelength, the softer is the radiation. (RMB)

Soft water *See* Water softening; zeolite. (HMB)

Soil **(1)** Dirt, dust, and organic materials with discrete particles that could be encountered in a food-service or food-processing facility. Examples may include fat deposits, lubricant deposits, and other organic deposits on equipment. Soils can be classified according to the method of removal from an object being cleaned: soils soluble in water (or other solvents) containing no cleaner or detergent; soils soluble in a cleaning solution that contains a ''solubilizer'' or detergent; and soils insoluble in a cleaning solution. **(2)** Unconsolidated mineral material on the immediate surface of the earth that serves as a growth medium for land plants. This material has been subjected to and influenced by genetic and environmental factors such as parent material, climate (including moisture and temperature effects), macro- and microorganisms, and topography. All of these factors act over a period of time and produce soil that differs in physical, chemical, biological, and morphological properties from the material from which it was derived. *See also* Dirt. (HMB, SWT)

Soil absorption field A subsurface area containing a trench or bed with clean stones and a system of distributed piping through which treated sewage may seep into the surrounding soil for further treatment and disposal. *See also* Septic tank. (FSL)

Soil absorption system Any subsurface sewage disposal system that uses soil for absorption of the treated sewage—e.g., an absorption trench, seepage bed, or seepage pit. *See also* Cesspool; seepage pit; septic tank. (FSL)

Soil air The soil volume that is not occupied by solids or liquids; the gaseous phase within the soil. (SWT)

Soil amendment Any material, such as fertilizer or mulch, that is added to soil for the purpose of improving physical or chemical properties. (SWT)

Soil association A map unit used in soil surveys that is composed of delineations that show the size, shape, and location of a landscape unit composed of two or more kinds of component soils or component soils and miscellaneous land areas. (SWT)

Soil auger A tool used to remove a small core of soil for field or laboratory observation. (SWT)

Soil biochemistry The branch of soil science that is concerned with the study of vital processes, enzymes, reactions, and products of soil microorganisms. (SWT)

Soil characteristics Soil properties, such as color, temperature, water content, pH, and structure, that are described or measured by field or laboratory observations. (SWT)

Soil chemistry The division of soil science that involves the study of the chemical composition, properties, and reactions of soils. (SWT)

Soil classification system (OSHA) A method of categorizing excavation soil and rock deposits in a hierarchy. The method consists of grouping soils into three major categories (types A, B, and C) according to specific criteria—e.g., the unconfined comprehensive strength (UCS) in tons per square foot (TSF). Type A soils, for example, have a UCS of 1.5 TSF. Similar technical considerations apply to the other categories of soils. Knowledge of this information is invaluable in the development of a safety program to prevent injuries and deaths in excavation-related accidents. *See also* Excavations; shoring; sloping; soil structure grade; trenching. (SAN)

Soil complex A map unit, used in soil surveys, that is composed of delineations, each of which shows the size, shape, and location of a landscape unit made up of two or more kinds of component soils or component soils and a miscellaneous land area. (SWT)

Soil conditioner Any material that will improve the physical characteristics of soil as a plant growth medium. (SWT)

Soil conservation Protection of soil from losses by erosion or overuse of fertilizers; a system of management practices and land use methods that safeguard soil against depletion or deterioration such as that caused by overfarming. (SWT)

Soil extract A soil solution that has been separated from a soil suspension or from a soil by methods such as filtration, centrifugation, suction, or pressure. (SWT)

Soil formation factors The natural variables that are responsible for soil formation. The five major factors of soil formation are parent rock, climate, organisms, topography, and time. (SWT)

Soil gas Gaseous elements and compounds that occur in small spaces between particles of soil or rock. Such gases can move through or leave the soil or rock, depending on changes in pressure. *See also* Radon. (FSL)

Soil genesis The formation or origin of soil, with special reference to the soil-forming factors responsible for the development of the solum, or true soil, from the unconsolidated parent materials. (SWT)

Soil geography A division of physical geography concerned with the distributions of soil types. (SWT)

Soil horizon A layer of soil or soil material that is closely parallel to the land surface and differs from adjacent genetically related layers in physical, chemical, and biological properties or characteristics such as color, structure, texture, consistency, kinds and numbers of organisms present, and degree of acidity or alkalinity. (SWT)

Soil interpretations Quantitative and/or qualitative predictions of soil behavior in response to specific uses or management, based on inferences from soil characteristics and qualities such as erodibility or productivity. (SWT)

Soil loss tolerance Maximum average annual soil loss that will allow continuous cropping and maintenance of soil productivity without requiring additional management techniques. (SWT)

Soil management The total of all soil management practices, including tillage operations, cropping practices, and fertilizer, lime, and other treatments conducted on or applied to a soil for the production of plants. (SWT)

Soil management groups Taxonomic soil units with similar management requirements for one or more specific purposes, including adapted crops or crop rotations, drainage practices, fertilization, forestry, and highway engineering. (SWT)

Soil map A map that indicates the soil types or soil mapping units in relation to physical and cultural features of the earth's surface. (SWT)

Soil mechanics and engineering A branch of soil science that is concerned with the effect of forces on the soil and the application of engineering principles to problems involving soil. (SWT)

Soil microbiology The branch of soil science concerned with soil-inhabiting microorganisms and their functions and activities. (SWT)

Soil mineral A mineral that occurs as a part of soil or that is incorporated into soil. A natural inorganic compound with definite physical, chemical, and crystalline properties that occurs in the soil. (SWT)

Soil mineralogy The specialty area of soil science that involves the study of the inorganic materials found in soil to the depth of weathering or sedimentation. (SWT)

Soil monolith A vertical section of a soil profile that has been removed from the soil and mounted for display or study. (SWT)

Soil morphology The physical constitution or structural properties of a soil profile, as characterized by the types, thickness, and arrangement of the horizons and by the texture, structure, consistency, and porosity. (SWT)

Soil organic matter The organic fraction of soil, including the remains of animals and plants that have decomposed to a point at which they are unrecognizable; microbial residues; the relatively stable end products of decomposition (humus). (SWT)

Soil organic residue Animal and vegetative materials that have been added to soil and that are recognizable as to origin. (SWT)

Soil piping or tunneling A type of accelerated erosion that results in subterranean voids and tunnels. (SWT)

Soil population The total of all organisms that live in soil. (SWT)

Soil pores The portions of the soil bulk volume that are not occupied by particles. (SWT)

Soil productivity The capacity or ability of a soil to produce crops or other plants, with optimum management. (SWT)

Soil reaction *See* pH. (SWT)

Soil salinity The amount of soluble salts contained in a soil. Soil salinity is measured by means of electrical conductivity tests of a saturation extract. (SWT)

Soil science The science that deals with soils as a natural resource on the surface of the earth and includes such topics as soil formation, classification, and mapping, and physical, chemical, biological, and fertility properties of soils, and their properties in relation to crop production. (SWT)

Soil separates Mineral particles less than 2.0 mm in diameter. The names and size limits of separates recognized in the United States are: very coarse sand, 2.0 mm to 1.0 mm; coarse sand, 1.0 mm to 0.5 mm; medium sand, 0.5 mm to 0.25 mm; fine sand, 0.25 to 0.10 mm; very fine sand, 0.10 to 0.05 mm; silt, 0.05 to 0.002 mm; and clay, less than 0.002 mm. (SWT)

Soil series The lowest category of the United States' system of soil taxonomy. Soil series are commonly used to name soil groupings represented on detailed soil maps. The soil series serve as a method of transferring soil information and research knowledge from one area to another. (SWT)

Soil structure grade A grouping or classification of soil structure based on inter-and intra-aggregate adhesion, cohesion, or stability within the profile. There are four grades of structure, which are designated from 0 to 3 as follows. **Grade 0:** structureless; no observable aggregation or no definitive and orderly arrangement of natural lines of weakness; massive if coherent; single-grain if noncoherent. **Grade 1:** weak; poorly formed indistinct peds, barely observable in place. **Grade 2:** moderate; well-formed, distinct peds, moderately durable and evident, but not distinct in undisturbed soil. **Grade 3:** strong; durable peds that are quite evident in undisturbed soil, adhere weakly to one another, withstand displacement, and become separated when the soil is disturbed. *See also* Soil classification system (OSHA). (SWT)

Soil structure types A soil structure classification based on the shape of the aggregates or peds and their arrangement in the soil profile. (SWT)

Soil survey The systematic examination, description, classification, and mapping of soils in an area. (SWT)

Soil suspending agent An ingredient used in detergents to aid in suspending and dispersing soil as part of washing operations. The ability of a cleaning compound to remove soil from a surface is related to its ability to disperse the soil in water. A material such as carboxymethylcellulose is an aid in keeping soil suspended and dispersed during a washing cycle. *See also* Detergent; dispersing agent; soil; surfactant. (HMB)

Soil test A chemical, biological, or physical analysis used to determine the suitability of soil to support plant growth. (SWT)

Soil texture Relative proportions of various separates in a soil, as described by the classes of soil texture. (SWT)

Solclime Soil climate; temperature and moisture conditions of the soil. (SWT)

Solder A metallic compound used for joining metal surfaces together by filling a joint or covering a junction. The most commonly used solders contain tin and as much as 50% lead. The solder acts as an adhesive and does not form an intermetallic solution with the metals being connected. *See also* Plumbism. (FSL)

Sole-source aquifer An aquifer that supplies 50% or more of the drinking water to an area. (FSL)

Solidification and stabilization Removal of wastewater from a waste or the changing of it chemically to make the waste less permeable and susceptible to transport by water. (FSL)

Solid waste **(1)** Putrescible and nonputrescible waste, except water-carried body waste (excrement) and materials destined for recycling, and including garbage, rubbish (paper, cartons, boxes, wood, tree branches, yard trimmings, furniture, appliances, metal, tin cans, glass, crockery, or dunnage), ashes, street refuse, dead animals, sewage sludges, animal manures, industrial waste (waste materials generated in industrial operations), residue from incineration, food-processing waste, construction waste, and any other waste material in a solid or semisolid state. **(2)** As defined by the Resource Conservation and Recovery Act, any garbage, refuse, or sludge from a waste treatment plant, water supply treatment plant, or air pollution control facility, or any other discarded material, including solid, liquid, semisolid, or contained gaseous material resulting from industrial, commercial, mining, and agricultural operations and from community activities. Solid waste does not include solid or dissolved material in domestic sewage, solid or dissolved materials in irrigation return flows or industrial discharges (which are point sources subject to permits under the Clean Water Act), or source, special nuclear, or by-product material as defined by the Atomic Energy Act. 42 USC §6903(27). *See also* By-product material; compost; garbage; hazardous waste; source material; special nuclear material; Resource Conservation and Recovery Act. (FSL, DEW)

Solid waste disposal The final placement of refuse that is not salvaged or recycled. *See also* Garbage; refuse. (FSL)

Solid Waste Disposal Act *See* Resource Conservation and Recovery Act; Medical Waste Tracking Act. (DEW)

Solid waste handling The storage, collection, transportation, treatment, utilization, processing, or disposal of solid wastes, or any combination of these activities. (FSL)

Solid waste handling facility Any location where storage, collection, transportation, treatment, utilization, processing, or disposal of solid waste, or any combination of these activities occurs. *See also* Landfill. (FSL)

Solid waste management Supervised handling of waste materials from their source through recovery processes to disposal. (FSL)

Solum The upper and most weathered part of the soil profile; the A and B soil horizons. (SWT)

Solvent Substance (usually liquid) capable of dissolving or dispersing one or more other substances. Solvents can be divided into two major categories: polar (having a high dielectric constant) or nonpolar. Water, a universal solvent, is strongly polar. Examples of organic solvent groups are esters, ethers, and chlorinated hydrocarbons. Many solvents are flammable and toxic and are significant fire hazards. *See also* Methanol; trichloroethylene. (FSL)

Soman [$(CH_3)_3CCH(CH_3)OPF(O)CH_3$] Methylphosphonofluoric acid, trimethylpropyl ester. One of a family of volatile liquid anticholinesterase nerve agents with the military designation GD. This is a colorless, odorless gas highly toxic by ingestion, inhalation, and skin absorption. A fatal dose for humans is 0.01 mg/kg. *See also* Sarin; tabun. (FSL)

Somatic cells Body cells, usually having two sets of chromosomes, as opposed to germ cells. (FSL)

Sombric horizon A type of soil horizon that is common in cool, moist soils of high altitudes in tropical regions. The sombric horizon is described as a subsurface mineral horizon that is darker in color than the overlying horizon but that lacks the properties of a spodic horizon. *See also* Spodic horizon. (SWT)

Soot Carbon dust formed by incomplete combustion. (FSL)

Sorption The action of absorbing or attracting substances; a process used in many pollution control systems. *See also* Absorption. (FSL)

Sound Any pressure variation in air, water, or other media that the human ear can detect. Sound is characterized by both frequency and pressure. *See also* Noise. (DEJ)

Sound intensity The average rate at which sound energy is transmitted through a unit area normal to the direction of sound propagation, measured in joules per square meter per second ($J/m^2/sec$) or in decibels referenced to 10^{-12} watts/m^2. (DEJ)

Sound-level meter An instrument used for measuring sound pressure levels in decibels, referenced to 0.0002 microbar. Readings can also be made in specific octave bands ranging from 75 Hz to 10,000 Hz. (FSL)

Sound power The total sound energy radiated by the source per unit time, measured in decibels, referenced to 10^{-12} watt. (DEJ)

Sound pressure The root-mean-square value of the pressure changes above and below atmospheric pressure, normally described in decibels referenced to 2×10^{-5} newtons/m^2. Sound pressure level can also be measured in dynes per square centimeter, microbars, or pascals. (DEJ)

Sound velocity The speed of propagation of a sound wave through a medium. The velocity of sound is 1130 ft/s in air, 4700 ft/s in water, 13,000 ft/s in wood, and 16,500 ft/s in steel. (DEJ)

Source A discrete amount of radioactive material or radiation-producing equipment. The term usually refers to radioactive material specifically packaged for scientific or industrial use. (RMB)

Source individual Under the OSHA Bloodborne Pathogen Standard (29 CFR 1910.1030), any individual, living or dead, whose blood or other potentially infectious materials may be a source of occupational exposure to the employee. Examples include, but are not limited to, hospital and clinic patients; clients in institutions for the developmentally disabled; trauma victims; clients of drug and alcohol treatment facilities; residents of hospices and nursing homes; human remains; and individuals who donate or sell blood or blood components. *See also* Exposure control plan. (FSL)

Source of contamination A person, animal, or inanimate substance responsible for the presence of a biologic, chemical, or radiologic vehicle that may or may not have the capability to transmit such an agent. (FSL)

Source of infection The person, animal, or substance from which an infectious agent is transmitted to a susceptible host. (FSL)

Source material As defined by the Atomic Energy Act, uranium, thorium, or any other material so determined by the NRC; or ores containing one or more of the foregoing materials in concentrations determined by the NRC. 42 USC §2014(z). (DEW)

Source reduction **(1)** The reduction or elimination of hazardous waste at the source, usually within a process. Source reduction measures include process modifications, feedstock substitutions, improvements in feedstock purity, housekeeping and management practices, increases in the efficiency of machinery, and recycling within a process. Source reduction includes any action that reduces the toxicity or the amount of waste exiting from a process. **(2)** The elimination of either insect- or rodent-breeding areas by the use of environmental control measures, such as draining swamps or reworking rodent harborage. (FSL)

Sovereign immunity Legal doctrine that a government cannot be sued unless it has consented to be sued on a particular matter. For example, many environmental statutes contain provisions that explicitly allow the United States government to be sued to enforce provisions of the law. *See also* Administrative Procedure Act; citizen suits; Federal Tort Claims Act. (DEW)

Spearman rank correlation coefficient (r_s) A nonparametric alternative to the linear correlation coefficient. Only rankings are used to calculate this value:

$$r_s = 1 - \frac{6 \sum d_i^2}{n(n^2 - 1)}$$

where d_i is the difference in rankings and n is the number of pairs of data. The value of r_s will range from -1 to $+1$. (VWP)

Special nuclear material As defined by the Atomic Energy Act, plutonium, uranium enriched in the isotope 233 or in the isotope 235, on any other material so designated by the NRC; any material artificially enriched by any of the foregoing substances. 42 USC §2014(aa). (DEW)

Special review Formerly known as rebuttable presumption against registration (RPAR), this is the regulatory process through which existing pesticides suspected of posing unreasonable risks to human health, nontarget organisms, or the environment are referred for review to the EPA. The review requires an intensive risk–benefit analysis with opportunity for public comment. If the risk of any use of a pesticide is found to outweigh social and economic benefits, regulatory actions—ranging from label revisions and use-restriction to cancellation or suspended registration—can be initiated. (FSL)

Species The smallest, most exclusive taxon in the classification hierarchy (excluding subgroups like subspecies, variants, races). A species is a group of organisms that share a common gene pool and (for the plant and animal kingdoms) can potentially interbreed. (SMC)

Specific activity The total radioactivity of a given isotope per gram, given in Ci/g or d/s-g. (RMB)

Specific ionization Number of ion pairs per unit length of path of the ionizing particle in a medium—e.g., per centimeter of air or per micron of tissue. (RMB)

Specificity The probability that a given clinical symptom is not present, given that the person does not have an active case of the disease under investigation. (VWP)

Spectral distribution The distribution of energy throughout the electromagnetic spectrum. For thermal effects on people, the spectrum from nuclear weapons is thought to be quite similar to that from the sun. (RMB)

Spill prevention control and countermeasures plan (SPCC) Plan covering the release of hazardous substances as defined in the Clean Water Act. (FSL)

Splash zone In food sanitation, this refers to surfaces, other than in the food zone or from food contact surfaces, subject to routine splash or other food soiling during normal use. (HMB)

Spodic horizon A mineral soil horizon that is characterized by an illuvial accumulation of amorphous materials that are composed of aluminum and organic carbon with or without iron. The spodic horizon has a certain minimum capacity of extractable carbon plus iron and aluminum in relation to its clay content. (SWT)

Spodosols Mineral soils that have a spodic horizon or a placic horizon overlying a fragipan. (SWT)

Spoil Dirt or rock that has been removed from its original location, destroying the composition of the soil in the process, as with strip mining or dredging. Spoils are generally excavated materials and include rock, soil, and debris. Employees and others must be protected against the possible falling back of these materials into trenches and other excavations. All materials such as disturbed soil and excavation equipment must be kept at least 2 feet from the edge of the excavation in order to comply with OSHA requirements. *See also* Excavation. (SAN)

Spoil banks Dumps of rocks and soil resulting from the excavation of ditches and strip mines. (SWT)

Spontaneous combustion Starting of a fire without an external source such as a flame, spark, ember, or heat. A frequent cause is heat buildup from decomposition of organic material. *See also* Autoignition. (SAN)

Sporadic case In an investigation, a case that has, as far as is known, no epidemiologic relationship to any other case. (FSL)

Sporocides Chemical agents that kill spores. *See also* Disinfectants. (HMB)

Sprawl Unplanned development of open land. (FSL)

Spring *See* Gravity spring. (FSL)

Sprinkler alarm A device installed so that any water flow from a sprinkler system equal to or greater than that from a single automatic sprinkler will result in an audible alarm signal. *See also* Sprinkler system. (SAN)

Spurious count In a radiation-counting device, a count caused by any agent other than radiation. (RMB)

Squib A generic term for a small explosive device that is similar in appearance to a detonator but is loaded with low-order explosives to produce an output that is mainly heat or flash. *See also* Explosive. (FSL)

Stabilization Conversion of the active organic matter in sludge into inert, harmless materials. (FSL)

Stabilization pond *See* Lagoon. (FSL)

Stabilizers Additives used in substances and compounds to keep them in a stable state, retard deterioration, act as an antioxidant, etc. (FSL)

Stable air A mass of air that is not moving normally, so that it holds rather than disperses pollutants. (FSL)

Stable isotope A nonradioactive isotope of an element. (RMB)

Stable materials Chemicals that have the ability to resist changes in their composition despite exposure to water, air, pressure, heat, or shock. (FSL)

Stack A chimney or smokestack, a vertical pipe that discharges used air and other materials. (FSL)

Stack effect Movement of used air upward, as in a chimney, because the air is warmer than the surrounding atmosphere. (FSL)

Stack gas *See* Flue gas. (FSL)

Stagnation In a mass of air or water, lack of motion, which tends to hold and concentrate pollutants. (FSL)

Standard deviation **(1)** *Population (σ):* The square root of the population variance.

$$\sigma = \sqrt{\sum_{i=1}^{k} [(x_i - \mu)^2 \cdot p(x_i)]}$$

where x_i is a population data value, μ is the population mean, and $p(x_i)$ is the probability of x_i occurring, and $i = 1, 2, \ldots, k$. **(2)** *Sample (s):* The square root of the sample variance.

$$s = \sqrt{\sum_{i=1}^{n} \frac{(x_i - \bar{x})^2}{n - 1}}$$

where x_i is a sample data value, $\bar{x}$ is the sample mean, and n is the number of sample values. (VWP)

Standardized morbidity ratio The standardized ratio that is applied for nonfatal conditions. *See also* Rate; standardized mortality ratio. (VWP)

Standardized mortality ratio (SMR) A quantity equal to 100%, times the number of observed deaths in a select population, divided by the expected number of deaths in the population under the assumption that the mortality rates for the select population are the same as those for the general population. *See also* Mortality rate; rate. (VWP)

Standardized variable *See* Z-score. (VWP)

Standard normal distribution A normal distribution where $\mu = 0$ and $\sigma = 1$. (VWP)

Standard of performance Under the Clean Water Act, a standard for the control of pollutant discharges that reflects the greatest degree of effluent reduction. This is determined by the Administrator of EPA to be achievable through application of the best available demonstrated control technology, processes, operating methods, or other alternatives, including, where practicable, a standard permitting no discharge of pollutants. 33 USC §1316(a)(1). *See also* Clean Air Act; discharge of pollutants. (DEW)

Standard operating procedures (SOPs) The OSHA process safety standard 1910.119 requires that employers develop and implement written SOPs that provide clear instructions for safely conducting activities involved in each process. SOPs should address the following: **(1)** steps for each operating phase, such as initial startup, normal operation, temporary operations, emergency operations (including emergency shutdowns, and who may initiate these procedures), normal shutdown, and startup following a turnaround or after an emergency shutdown; **(2)** operating limits, the consequences of deviation, steps required to correct and/or avoid deviation, and safety systems and their functions; **(3)** safety and health considerations such as those relating to the properties of and hazards presented by the chemicals used in the process, precautions necessary to prevent exposure (including administrative controls, engineering controls, and personal protective equipment), measures to be taken if physical contact or airborne exposure occurs, safety procedures for opening process equipment (such as pipeline breaking), quality control of raw materials and hazardous-chemical inventory levels, and concerns relating to any special or unique hazards. *See also* Process safety. (SAN)

Standards Prescriptive norms that govern action and actual limits on the amount of pollutants or emissions produced. The EPA and OSHA, under most of their responsibilities, establish minimum standards for specific chemicals. (FSL)

Standard score *See* Z-score. (VWP)

Standing Generally the legal right to bring a lawsuit. In federal court, standing turns on whether there is an actual case or controversy between the parties, in which the plaintiff has suffered personal harm by the actions of the defendant. *See also* Negligence. (DEW)

Standpipe systems A class I standpipe system in a building consists of a $2\frac{1}{2}$-inch hose connection and is for use by fire departments and those trained in handling heavy firefighting equipment. A class II standpipe system is a $1\frac{1}{2}$-inch hose system and provides a means for the control or extinguishment of incipient-stage fires. A class III standpipe system is a combined system of hose; it is designed for employees trained in the use of hose operations and is capable of furnishing effective water discharge during the more advanced stages of fire in the interior of workplaces. Hose outlets are available for both $1\frac{1}{2}$-inch and $2\frac{1}{2}$-inch hose. *See also* Incipient-stage fire; interior structural firefighting. (SAN)

Staphylococcal intoxication An illness caused by the bacterium *Staphylococcus aureus*. This organism is a member of the Family Micrococcaceae and is a Gram-positive, nonmotile, non-spore-forming, facultative anaerobe that grows singly, in pairs, and in tetrads, and may form irregular clusters. The various staphylococcal species can be identified by conventional schemes for the identification of human-adapted species and animal-adapted species. *Staphylococcus aureus* has been confirmed to be a major agent of foodborne illness. When it or its enterotoxins are identified in foods, poor sanitation is usually suspected. The bacterium is widely distributed and humans are a natural reservoir. It may be present in healthy individuals. It is a bacterium that is commonly found on the skin, and can be isolated from throat and nasal swabs and from skin lesions. It is apparent that foods handled by infected individuals are among the most common sources of contamination. This fact emphasizes the critical importance of personal hygiene as well as sanitation programs in food-processing and food-service operations. At least five antigenically distinct staphylococcal protein toxins, designated A through E, have been characterized, and are demonstrated to be quite heat resistant. Foodborne outbreaks of staphylococcal intoxication have occurred in which enterotoxin is present but no viable staphylococci are apparent. Once a food has become contaminated with the preformed toxin, little can be done to the product to make it ''safe'' because of the highly stable nature of the toxin.

Onset of symptoms may be abrupt, and there are sometimes violent reactions. The incubation period is short, usually 2 to 3 hours after ingestion of the contaminated food. Vomiting, nausea, abdominal cramps, diarrhea, and prostration are typical. There is generally no fever or chills, and diarrhea is less frequent than is vomiting. The symptoms are usually most acute during the first few hours and the illness runs its course typically in 1 to 2 days. Mortality is rare, especially in healthy individuals, but the severity is greater than is perhaps generally believed, and many victims require hospitalization. Foods implicated as vehicles of the disease are varied, and many times are products that require handling. High-protein foods, such as ham, beef, pork, poultry, dairy products, custards, and potato salad, have been implicated in outbreaks. Prevention of staphylococcal intoxication can be most successful through good sanitation, good personal hygiene, and prevention of the growth of the bacterium, so that the enterotoxin cannot be produced in the food. *See* Foodborne disease; foodborne intoxication. (HMB)

Stare decisis Legal policy in which courts abide by or adhere to previously decided cases. (DEW)

State emergency response commission (SERC) Commission appointed by each state governor according to the requirements of the Superfund Amendments and Reauthorization Act, Title III. The commissions designate emergency planning districts, and appoint local emergency planning committees and supervise and coordinate their activities. (FSL)

State Health Planning and Development Agency (SHPDA) Agency created at the state level, as required by the National Health Planning and Resources Development Act of 1974 (Public Law 93-641), to conduct health planning functions and review plans for medical care facilities. (FSL)

State implementation plan (SIP) EPA-approved state plan for the establishment, regulation, and enforcement of air pollution standards. (FSL)

Static pressure The potential pressure exerted in all directions by a fluid at rest. For a fluid in motion, static pressure is measured in a direction normal to the direction of flow, and can be thought of as a measure of the fluid's ability to burst a pipe or duct. Usually measured in inches of water gauge mercury. On the suction side of a fan, static pressure is always negative with respect to atmospheric pressure; on the exhaust side, static pressure is always positive with respect to atmospheric pressure. (DEJ)

Stationary growth phase The phase at which population growth rate is static—that is, when the number of growing, multiplying organisms is equal to the number of dying organisms. During the exponential growth phase (log phase) of a microbial population, large amounts of organic acids and other toxic substances may accumulate in the culture medium. The accumulation of toxic materials and the depletion of essential nutrients are largely responsible for the development of the stationary phase of the growth cycle. The stationary phase is followed by the death phase of the population. *See also* Death phase; lag phase; log phase. (HMB)

Stationary source A nonmoving producer of pollution. Refers mainly to power plants and other facilities using industrial combustion processes. (FSL)

Statistic A numerical characteristic of a sample. (VWP)

Statistical software package A selection of statistical computer programs that describe data and perform statistical tests on the data. Some widely used packages include SAS, SPSS, BMBP, and Minitab. (VWP)

Statute A state or federal law. In the United States, federal laws are proposed as bills by either the House of Representatives or the Senate and are referred to by number (for example, *S. 1234*, which means *Senate Bill number 1234*). Bills passed by the House and Senate and signed by the President are referred to as public laws and are numbered according to the number of the Congress plus an additional number (for example, *Public L. 101-12*, which means *Public Law 12 passed under the 101st Congress*). Laws are codified into the United States Code, which is organized by subject matter. Codified laws are cited by title number and section (for example, *5 USC §552* means *Title 5 of the United States Code, section 552*). (DEW)

Statute of limitations Statute limiting the period of time by which a legal action must be brought. (DEW)

Stay A court order to stop a judicial proceeding. Usually, a stay is temporary and allows time for some action to take place. (DEW)

Stay time The period during which personnel are allowed to remain in a radiation and/or contaminated area before accumulating their permissible dose. (RMB)

Steam sanitizing *See* Thermal sanitizing. (HMB)

Stem-and-leaf diagram A display of data that uses the leading digit (S), called the stem, and trailing digit (s), called the leaf, to organize and sort data. (VWP)

Stem flow Precipitation intercepted by plants that flows down the stem to the ground. (SWT)

Stemming In mining terminology, use of inert materials, such as mud or clay, to plug a borehole into which an explosive charge has been placed. Often called "tamping." (FSL)

Sterilant A chemical or physical agent used to destroy microorganisms. *See also* Disinfectant. (FSL)

Sterility (biological) The temporary or permanent incapability to reproduce. (FSL)

Sterilization The destruction of all microorganisms in or on an object, using heat, steam, chemical agents, ultraviolet radiation, or a combination of these. In pest control, sterilization is the use of radiation and chemicals to damage body cells needed for reproduction. *See also* Disinfectant; preservatives; sanitizer. (FSL)

Stibialism Antimony (Sb) poisoning. Antimony is a silvery, white, brittle semimetallic chemical element of crystalline structure that is used to harden alloys and increase their resistance to chemical action. Antimony must be used with adequate ventilation and has a TLV of 0.5 mg/m^3 of air. (FSL)

Stokes' law An equation that relates the terminal settling velocity of a smooth rigid sphere in a viscous fluid of known density and viscosity to the diameter of the sphere when subjected to a known force field. This technique is used in particle-size analysis of soils by the pipette hydrometer or centrifuge method. (SWT)

Stones Rock fragments greater than 25 cm in diameter if rounded, and greater than 38 cm along the greater axis if flat. (SWT)

Stony Describes an area that contains stones in large enough number that they interfere with or prevent tillage. (SWT)

Stopping power The reciprocal of the thickness of a solid that absorbs the same amount of alpha radiation as does one centimeter of air. (RMB)

Storage The temporary holding of waste, pending treatment or disposal. Storage methods include the use of containers, tanks, waste piles, and surface impoundments. *See also* Part B permit. (FSL)

Storm sewer A system of pipes (separate from sanitary sewers) that carry only water runoff from buildings and land surfaces. (FSL)

Stratified Refers to soils arranged in or composed of strata, or layers. (SWT)

Stratosphere The portion of the atmosphere that is 10 to 25 miles above the earth's surface. (FSL)

Stray radiation Radiation not serving any useful purpose. It includes direct radiation and secondary radiation from irradiated objects. (RMB)

Streptococci Nonmotile, non-spore-forming, aerobic to facultatively anaerobic, Gram-positive bacteria. These organisms occur regularly in the mouth and intestines of humans and other animals, and in food and dairy products; some species are pathogenic for humans. (FSL)

Strict liability Legal principle on whose basis a person can be held liable for an injury without a showing of fault, negligence, bad faith, or knowledge of the wrong. Applicable under common law in limited situations or can be prescribed by statute. (DEW)

Stridor A high-pitched sound, resembling the blowing of wind, that is caused by obstruction of air passages in the lungs. (FSL)

Strip cropping A practice that entails growing crops that require different types of tillage, such as row and sod, in alternate strips along contours or across the prevailing direction of wind. (SWT)

Strip mining A process that uses machines to scrape soil or rock away from mineral deposits under the earth's surface. (FSL)

Strip planting A method of simultaneous tillage and planting in isolated bands of varying width that are separated by bands of erect residues undisturbed by tillage. (SWT)

Strontium (Sr) A pale-yellow, soft, reactive metal that resembles calcium. This element has an atomic weight of 87.62 and an atomic number of 38. There are several isotopes of strontium, including Sr-90 and strontium-89. Strontium-90 is a deadly radioactive isotope present in the fallout of nuclear explosions. Strontium reacts with water and releases hydrogen, is spontaneously flammable in powder form, and ignites when heated above its melting point. Strontium is used primarily in alloys and in fireworks. (FSL)

Structural charge In a clay mineral, a negative charge that is the result of isomorphic substitution within the layer. (SWT)

Strychnine ($C_{21}H_{22}N_2O_2$) A highly toxic alkaloid derived from seeds of the *Strychnos* nux vomica plant. Strychnine stimulates all parts of the central nervous system. The symptoms first noted are tremors and twitching, which progress to severe convulsions and respiratory arrest. Strychnine has been used as an effective rodenticide; only a single dose is required to eliminate rats. Its use for that purpose has been replaced by that of multiple-dose poisons, such as the anticoagulates, which have a much greater safety factor for nontarget animals. *See also* Red squill; rodenticide; warfarin. (FSL)

Stubble mulch Crop residues left in place as a surface cover before and during seedbed preparation and for some time during the growing of a succeeding crop. (SWT)

Study of the Efficacy of Nosocomial Infections Control (SENIC) A study conducted by the Centers for Disease Control from 1976 through 1978 of approximately 350 hospitals in the United States to determine whether the risk of nosocomial infections could be lowered by infection control programs. (FSL)

Subclinical case Case in which there is an absence of any clinical signs of a disease or the presence of a very mild form of a disease. *See also* Apparent infection; typhoid mary. (FSL)

Sublethal injury (microbiology) When microorganisms are subjected to environmental stress, such as heat, freezing, adverse pH, chemicals, and sanitizers, many cells experience metabolic injury resulting in their inability to grow on selective media that an uninjured organism can tolerate. Plating organisms on both selective and nonselective media results in recovery of a larger number from the nonselective medium, because growth on that medium is that of both injured and uninjured cells. The difference in the two represents the number of injured cells in the population. Sublethally injured cells in foods are of great significance in recovery not only of potential pathogens but also of spoilage organisms. Sublethally injured cells can repair, given adequate time and nutrients in a ''recovery medium.'' The same recovery phenomenon may occur in food, resulting in the growth of organisms—i.e., pathogens might grow that were previously undetected because of their failure to form colonies on the selective medium employed. Failure to detect either spoilage organisms or pathogens during a routine interpretation of results in a sanitation program may not give a true picture of the program's effectiveness and could also have an impact on the safety and quality of the product. (HMB)

Sublimation The process of converting a solid into a gas without passage through a liquid state, as, for example, with dry ice, which sublimates as carbon dioxide. (FSL)

Submerged inlet The lack of an air space or air gap between the free-flowing discharge end of a water supply pipe and an unpressurized receiving vessel. For example, a submerged inlet may occur when a section of hose is attached to a faucet and extends downward into a dishwashing sink. Backflow of contaminated water from the filled sink may now occur, depending upon the pressure differential between the two systems. *See also* Air gap; backflow; back-siphonage; cross-connection. (HMB)

Subpoena A court order to require a witness to appear and give testimony. (DEW)

Subpoenal duces tecum A subpoena requiring a person in possession or control of documents relevant to a proceeding to provide those documents to the court or the adverse party. (DEW)

Subsidence Sinkage, fall to the bottom, or settling, as in the case of sediment, soils, rocks, or gravel; any sinkage to a lower level. *See also* Excavation; shoring; sloping. (FSL)

Subsoiling A tillage operation that uses narrow tools designed to loosen soils below the depth of normal tillage without inversion and with a minimum mixing of the soil. (SWT)

Substrate The particular material on which a microbial agent grows—e.g., culture media or body fluid. (FSL)

Subsurface sewage disposal system A system for the treatment and disposal of domestic sewage by means of a septic tank in combination with a soil absorption system. *See also* Septic tank; soil absorption system. (FSL)

Subsurface tillage A tillage operation that uses a special sweeplike plow or blade that is drawn beneath the soil surface and that cuts plant roots and loosens the soil without inversion or without incorporating residues on the surface of the soil. (SWT)

Succession Changes in plant community composition through time. Primary succession is begun with only a bare substrate, such as exposed granite or a newly emerged volcanic island, and no organic material. Secondary succession is initiated when a disturbance such as fire or a hurricane wipes out a later successional stage. Secondary succession begins with organic material such as soil and often a seed bank or root stocks. (SMC)

Sulfate The ion that is formed when sulfur trioxide reacts with water to form sulfuric acid in the atmosphere or that is present in other chemical compounds, such as ammonium sulfate. *See also* Sodium sulfate. (FSL)

Sulfite The ion formed in the reaction of sulfur dioxide with water to form sulfurous acid. *See also* Sodium sulfite. (FSL)

Sulfur (S) An element also known as brimstone and found in sedimentary rocks along with gypsum and limestone. It has an atomic weight of 32.064 and an atomic number of 16. Native sulfur is the main source of material from which sulfuric acid is derived. Sulfur is also used in the manufacture of pulp and paper, carbon disulfide, detergents, insecticides, rodent repellents, soil conditioners, fungicides, etc. Sulfur in the finely divided form is a fire and explosion risk. (FSL)

Sulfur cycle In the soil, a sequence of transformation in which sulfur is oxidized or reduced through organic and inorganic products. (SWT)

Sulfur dioxide (SO_2) A component of products of combustion, particularly diesel fuels and coal. It is used as an index of air quality. A colorless gas or liquid with a sharp pungent odor, sulfur dioxide has many properties that are similar to those of salts of sulfites, bisulfites, and metabisulfites. SO_2 has been used as a food preservative in the past, but is not recognized as a common one today. At low pH, it is bacteriostatic against acetic and lactic acid bacteria at lower concentrations and bactericidal at higher concentrations. It has been shown to be inhibitory to *C. botulinum* and to inhibit growth of salmonellae and other Enterobacteriaceae as well. SO_2 has some application as an adjunct in other preservation methods.

SO_2 is used in the manufacture of paper pulp and soybean protein, as a bleaching agent for oils and starch, and for disinfection and fumigation. It is toxic by inhalation, and is a strong irritant to eyes and mucous membranes. It is a dangerous air contaminant and a constituent of smog. The TLV is 2 ppm in air. (FSL, HMB)

Sulfuric acid (H_2SO_4) Also known as hydrogen sulfate, battery acid, and electrolytic acid, this is one of the most widely used industrial chemicals. Sulfuric acid is very corrosive and reactive and will dissolve most metals. It is a strong irritant to tissues, and is used in the manufacture of fertilizers, dyes, pigments, and steel. (FSL)

Sulfur trioxide (SO_3) An oxidizing substance also known as sulfuric anhydride, sulfur trioxide represents a fire risk when it comes in contact with organic materials. This highly toxic chemical is used in the sulfonation of organic compounds, especially detergents, and in solar energy collectors. (FSL)

Summary judgment Judgment that resolves a lawsuit or particular issue in a lawsuit in response to a motion by either party, and that assumes that the facts of the case are not in controversy. The court can therefore render a judgment applying the law to those facts. (DEW)

Summation notation (Σ) The summation of a set of addends, $x_1 + x_2 + x_3 \ldots + x_n$, also denoted Σx, Σx_i, or $\sum_i x_i$.

Sum of squares **(1)** (Corrected) The sum is given by

$$\sum_{i=1}^{n} x_i^2 - \frac{\left(\sum_{i=1}^{n} x_i\right)^2}{n}$$

(2) (Raw) The sum is given by

$$\sum_{i=1}^{n} x_i^2$$

where x_i is the sample value and n is the number of sample values. (VWP)

Summit The highest point of any landform remnant, hill, or mountain. (SWT)

Sump A pit or tank that catches liquid runoff for drainage or disposal. *See also* Cesspool; septic tank; soil absorption field. (FSL)

Sump pump A mechanical means for removing water or wastewater from a sump or wet well. (FSL)

Superaural protector A form of hearing protector that seals the external ear canal, using a soft rubberlike material held in place by a light headband. *See also* Noise. (DEJ)

Superchlorination In a water supply system, a technique for overcoming and simplifying the problem of insufficient chlorine contact in a water supply system. Chlorine is added to the water in increased amounts until a minimum residual of 3.0 mg/L is reached for a contact period of at least 5 minutes. *See also* Chlorination. (FSL)

Supercriticality A term used to describe the state of a given fission system when the specified conditions are such that a greater than critical mass of active material is present. This is the condition for increasing the level of operation of a reactor or for bringing about the explosion of a fission bomb. The fission neutron production rate will exceed all neutron losses, and the neutron population will increase at a controlled rate in a reactor and very rapidly and uncontrollably in a bomb. *See also* Critical mass; criticality. (RMB)

Superfund The trust fund established by CERCLA to finance hazardous substance cleanup and related activities under the statute. Programs operated under Superfund include establishing the National Priorities List, investigating sites for inclusion on the list, and conducting and/or supervising cleanup, removal, and other remedial actions. *See also* Comprehensive Environmental Response, Compensation, and Liability Act. (DEW)

Superfund Amendments and Reauthorization Act (SARA) Law passed in 1986 that reauthorized and amended the Comprehensive Environmental Response, Compensation, and Liability Act of 1980. Other provisions amended the Solid Waste Disposal Act and the Internal Revenue Code. Also enacted as free-standing provisions of law: the Emergency Planning and Community Right-to-Know Act (also known as SARA Title III) and the Radon Gas and Indoor Quality Research Act (Public Law 99-499). *See also* Comprehensive Environmental Response, Compensation, and Liability Act;

Emergency Planning and Community Right-to-Know Act; Solid Waste Disposal Act. (DEW)

Supernatant The liquid found above a layer of a precipitated insoluble substance. (FSL)

Supplied-air respirators A device that supplies breathing air to an individual through hoses carrying a continuous flow of the air, usually from a breathing-air compressor. A series of regulators and valves is typically included to provide air at the proper flow rate. Often used in atmospheres that contain relatively high levels of air contaminants. *See* Pressure-demand respirator; respirator fit testing. (DEJ)

Supreme Court of the United States *See* Federal courts. (DEW)

Surface charge density The excess of charge (negative or positive) per unit of surface area of soil or soil mineral. (SWT)

Surface dose The dose of radiation measured at the surface of a body. It is the sum of the air dose plus the dose due to backscatter from the object. (RMB)

Surface impoundment Treatment, storage, or disposal of liquid hazardous wastes in ponds. *See also* Oxidation pond; waste stabilization pond. (FSL)

Surface sealing The process in which soil becomes impermeable to water, due to the orientation and packing of dispersed particles in the immediate surface layer of the soil. (SWT)

Surface soil The uppermost part of the soil, usually moved in tillage, or its equivalent soils. It ranges in depth from 7 to 20 cm. Frequently designated as the plow layer, the A_p layer, or the A_p horizon. (SWT)

Surface water All water naturally open to the atmosphere (rivers, lakes, reservoirs, streams, impoundments, seas, estuaries, etc.). Also refers to springs, wells, or other collectors that are directly influenced by surface water. *See also* Ground water. (FSL)

Surfactant A complex molecule that, when added to water, reduces the surface tension of the water to permit closer contact between a soil deposit and a cleaning medium. This additive is responsible for emulsification action but is one part of a complete, blended compound. May also be used in conjunction with chemical sanitizers as well. (HMB)

Surveillance of disease The continuing scrutiny of all those aspects of the occurrence and spread of a disease that are pertinent to effective control. *See also* Epidemiology. (FSL)

Surveillance of persons The practice of close medical or other observation of the health of contacts (without restricting their movements) to promote prompt recognition of infection or illness in these contacts if it occurs. (FSL)

Surveillance system A series of monitoring devices designed to determine environmental quality or a series of data-gathering mechanisms to monitor all aspects of the occurrence and spread of disease or adverse health conditions. (FSL)

Survey instrument (radiation) A portable instrument, such as a Geiger-Mueller counter, used for detecting and measuring radiation under various physical conditions. (RMB)

Susceptible Describes a person or animal lacking sufficient resistance to prevent disease if or when exposed to a particular pathogenic agent. (FSL)

Suspect A person whose medical history, symptoms, and possible exposure to a source of infection suggest that an infectious disease may be developing in the person. *See also* Sign; symptom. (FSL)

Suspended solids Small particles of solid pollutants that float on the surface of, or are suspended in, sewage or other liquids. They resist removal by conventional means. *See also* Sewage, sludge; total suspended solids. (FSL)

Suspension The act of stopping the use of a pesticide when the EPA deems it necessary to do so in order to prevent an imminent hazard resulting from the continued use of the pesticide. An emergency suspension takes effect immediately; under an ordinary suspension, a registrant can request a hearing before the suspension goes into effect. Such a hearing process might take 6 months. *See also* Special review. (FSL)

Suspension culture (microbiology) Growth of individual cells or small clumps of cells in a liquid nutrient medium. (FSL)

Swab test A surface contact method of microbiological examination of surfaces, usually used to assess the effectiveness of cleaning and sanitizing procedures. It is described as one of several environmental monitoring procedures used in food-processing or food-service operations. One of two swab methods is used. In the first, sterile nonabsorbent cotton swabs are rubbed thoroughly over a known surface area to remove microorganisms, which are subsequently enumerated. In the second method, a sterile cellulose sponge of standard size is used instead of a cotton swab; this method is suitable for evaluation of larger surface areas. A standard method is used to conduct the swab tests and calculate the numbers of organisms present. For adequately cleaned and sanitized food-service equipment on which an area of approximately 250 square centimeters have been swabbed, residual bacterial counts not exceeding 500 (i.e., averaging 2 colonies per square centimeter) are satisfactory. Acceptable residual bacterial counts as determined by swab tests must be established by each operation. (HMB)

Swamp Any of a wide variety of wetland ecosystems, usually one with saturated soil or some areas of standing water and characteristic woody vegetation. *See also* Bog; fen; marsh. (SMC)

Sweep device A single or double arm (rod), attached to the upper die or slide of a powered industrial press, that moves the operator's hands to a safe position as the dies close, if the hands are within the point of operation. *See also* Machine guarding. (SAN)

Swimming pool granuloma A chronic infection caused by the organism *Mycobacterium marinum* as a result of contamination occurring in a swimming pool, whirlpool, spa, or hot tub. The organism is very resistant to chlorine and can be eliminated by initiating the appropriate maintenance procedures and by the use of quaternary-ammonium-based disinfectants. (FSL)

Swinging-vane anemometer A device that measures air velocity and static pressure by utilizing the compression of a calibrated spring by the moving airstream. Also known as a velometer. (DEJ)

Symbiosis A mutually beneficial cohabitation of two dissimilar organisms. (SWT)

Symmetric distribution The situation in which the spread of sample points on each side of the sample mean is the same. In this situation, the arithmetic mean is approximately the same as the median. (VWP)

Sympatric Occurring in the same geographic area; refers especially to animal species or plant populations whose ranges overlap. *See also* Allopatric. (SMC)

Symptom Subjective evidence of a disease, as indicated by a departure from the normal in structure, function, or sensation, as experienced by a patient. *See also* Sign; syndrome. (FSL)

Synchrocyclotron A cyclotron in which the radio frequency of the electric field is frequency-modulated to permit acceleration of particles to relativistic energies. (RMB)

Synchrotron A device for accelerating particles, ordinarily electrons, in a circular orbit in an increasing magnetic field applied to synchronize with the orbital motion. (RMB)

Syndrome The combined set of signs and symptoms that typify a particular disease. *See also* Sign; symptom. (FSL)

Synecology The ecology of organisms or populations in relation to each other and their biotic environment.

This term is now nearly synonymous with and usually replaced by "community ecology." (SMC)

Synergism The cooperative interaction of two or more chemicals or other substances or phenomena that produces a total effect greater than that of the sum of the individual effects. *See also* Potentiation. (FSL)

Systemic Relating to the body as a whole. (FSL)

Systemic pesticide A chemical that is absorbed from the soil or through the surface and carried through the vascular system of the organism being protected, making the organism toxic to pests. *See also* Insecticide; pesticide. (FSL)

T

Tabun $[(CH_3)_2NP(O)(C_2H_5O)(CN)]$ Dimethylphosphoramidocyanidic acid, ethyl ester. A volatile, liquid anticholinesterase nerve agent with the military designation GA. Tabun reacts irreversibly with the enzyme cholinesterase. This chemical is very toxic by inhalation; a fatal dose for humans is 0.01 mg/kg. *See also* Sarin; soman. (FSL)

TAG Technical assistance grant, issued under the Comprehensive Environmental Response, Compensation, and Liability Act. (DEW)

Tag A notification device used as a means of preventing accidental injury or illness to employees who are exposed to hazardous or potentially hazardous conditions, equipment, or operations in which the hazards are extraordinary, unexpected, or not readily apparent. Tags are usually made of card, paper, pasteboard, or plastic. "Danger" tags are used in hazard situations in which an immediate hazard presents a threat of death or serious injury. "Caution" tags are used in minor hazard situations, in which a potential hazard or unsafe practice presents less of a threat of employee injury. "Warning" tags may be used to represent a hazard level between those of "Caution" and "Danger." Tags need not be used where signs, guarding, or other positive means of protection are being used. *See also* Lockout/tagout. (SAN)

Tagout device A prominent warning device, such as a tag, that can be securely fastened to an energy-isolating device in accordance with an established procedure, to indicate that the energy-isolating device and the equipment being controlled may not be operated until the tagout device is removed. *See also* Energy-isolating device; lockout device; lockout/tagout. (SAN)

Taiga *See* Boreal forest. (SMC)

Tailings Residue of raw materials or waste separated out during the processing of crops or mineral ores. (FSL)

Taking The appropriation of private property for public use. The Fifth Amendment of the U.S. Constitution requires that just compensation be provided for a taking. Courts have also determined that government action in which property is not acquired outright through the power of eminent domain can be considered a taking for which just compensation is required if it has an impact on a protected property interest of the owner. *See also* Eminent domain. (DEW)

Talud A short, steep slope that is formed gradually at the downslope margin of a field, by decomposition, against a hedge, stone wall, or similar barrier. (SWT)

Talus An accumulation of soil material and rock fragments at the base of a cliff. (SWT)

Target population The cohort under study, from which a random sample is selected. (VWP)

Taxon (plural, *taxa*) **(1)** A level in the classification hierarchy or a group of organisms in the same taxonomic category—e.g., phylum, genus. **(2)** In the context of a soil survey, a class at any categorical level in the United States system of soil taxonomy. *See also* Classification. (SMC, SWT)

Taxonomy The science of classification; for example, in biology, a system of arranging animals and plants into natural, related groups based on some common factors, such as structure, embryology, and biochemistry. (FSL)

TCDD *See* Dioxin. (FSL)

TD_{50} A dose of a substance that when administered to all animals in an experimental study, produces a toxic response in 50% of the test population. *See also* LD_{50}. (FSL)

Technical assistance grants Under CERCLA, grants awarded by EPA to any group of individuals who may be affected by a release or threatened release at a National Priorities List site. These grants can be used to obtain technical assistance in interpreting information on the nature of the hazard or to obtain a record of decision or remedial investigation and feasibility study. They can also be used to carry out remedial design, selection and construction of remedial action, operation, and maintenance, or removal action at the site. 42 USC §9617(e). *See also* Comprehensive Environmental Response, Compensation, and Liability Act; feasibility study; National Priorities List; release; remedial investigation. (DEW)

Technology-based standards Effluent limitations, applicable to direct and indirect sources, that are developed on a category-by-category basis, using statutory factors and not including water-quality effects. (FSL)

Teleology The study of final causes or end results; a belief that natural phenomena are determined not only by mechanical causes but by an overall design or purpose in nature, and the study of evidence for this belief. (FSL)

Temperature danger zone A temperature range, usually given as between 45°F and 140°F (7.2°C to 60°C), within which most bacteria thrive in foods. Given the right environment, moisture, nutrients, and time, organisms will multiply rapidly within this zone. At temperatures of about 110°F many bacteria begin to die. At a temperature of 140°F, most will have been killed. Metabolic activity slows with a decrease in temperature. Some bacteria (psychrophiles) thrive at lower temperatures (i.e., refrigeration temperatures), whereas others grow very slowly. At freezing temperatures not all bacteria are killed, but growth stops. Food-service and food-processing regulations specify the temperatures that must be maintained for all foods to ensure safety. Foods that will not be heated or cooked—e.g., salads—must be maintained at temperatures of less than 45°F at all times. Likewise, foods that are cooked must be maintained at temperatures above 140°F to ensure safety. The foods will thus have been kept outside the temperature danger zone at which the microorganisms might reproduce rapidly and, if pathogenic, cause an outbreak of foodborne illness. *See also* Maximum growth temperature; minimum growth temperature. (HMB)

Temporary restraining order An order of the court forbidding a defendant to take a threatening action. May be issued without notice to the adverse party if circumstances warrant. (DEW)

Tenosynovitis A disorder of a tendon, produced by a swelling of the synovial sheath, which surrounds it. The surfaces of the tendon become irritated, rough, and bumpy, and movement is impeded. Often related to long-term exposure to repetitive motion, inappropriate posture or positioning of machinery, or exposure to vibration from equipment. Operations such as buffing, grinding, polishing, sanding, and turning controls can produce this condition. *See also* De Quervain's syndrome; trigger finger; and cumulative trauma disorder. (DEJ, SAN)

Tensiometer An instrument that measures the soil-water matrix potential (or tension) of water in soil *in situ.* The tensiometer consists of a porous, permeable ceramic cup connected through a water-filled tube to a manometer, vacuum gauge, pressure transducer, or other pressure-measuring device. (SWT)

Teratogen A substance capable of producing physical defects *in utero*. This may result in fetal or embryonic mortality or in the birth of offspring with defects. *See also* Carcinogenicity; mutagen. (FSL)

Teratology The branch of science concerned with the production, development, anatomy, and classification of malformed fetuses. (FSL)

Terminal disinfection Measure taken to render the personal effects and the physical environment of infected individuals free from the potential of transmitting the infectious agent to susceptible individuals. This process usually involves a complete wipedown of the area, with approved disinfectants, bagging of personal effects for disinfection, and autoclaving and/or incineration of those materials prior to disposal. (FSL)

Terrace **(1)** A level, narrow plain that borders a river, lake, or the sea. Rivers may be bordered by terraces at different levels. **(2)** A raised horizontal strip of earth constructed on or close to a contour and designed to make the land suitable for tillage and to prevent accelerated erosion. **(3)** An embankment with the uphill side sloping toward and into a channel that conducts water from areas above the terrace at a regulated rate of flow, preventing the accumulation of large volumes of water on the downslope side of a cultivated field. *See also* Sloping. (SWT)

Terrestrial Of or pertaining to land or, especially, organisms that live on land (nonaquatic organisms) and the environments in which they live. (SMC)

Tertiary wastewater treatment Advanced wastewater cleaning that goes beyond the secondary or biological stage. It removes nutrients such as phosphorus and nitrogen and most dissolved oxygen and suspended solids. *See also* Primary wastewater treatment; secondary wastewater treatment. (FSL)

Testimony Evidence given by a competent witness under oath or affirmation, as distinguished from evidence derived from writings and other sources. *See also* Expert witness. (DEW)

Tetraethyl Lead An organic form of lead used for many years as a fuel additive. Leaded gasoline was one of the main sources of widespread lead contamination until it was largely eliminated from the United States in the 1970s. Lead attacks nearly every organ in the body, and is known to cause retardation, birth defects, anemia, hypertension, convulsions, neurophysiological problems, decreased growth, reduced nerve conduction velocities, altered vitamin D metabolism, elevated erythrocyte protoporphyrin levels, miscarriages, stillbirths, many other adverse health effects, and death. Tetraethyl lead is also known for its ability to be absorbed through the skin. *See also* Galena; lead poisoning. (DEJ)

Tetradotoxin A potent neurotoxin found in the liver and ovaries of numerous species of the Japanese puffer fish (*Tetradontoidea*). When ingested, the toxin can almost immediately cause respiratory distress, convulsions, and death. (FSL)

Thallium (Tl) A bluish-white, leadlike metal, with an atomic weight of 204.37 and an atomic number of 81. Thallium forms toxic compounds on contact with moisture and is soluble in nitric and sulfuric acids. Previously used as a rodenticide, it has been replaced by other chemicals. Rodent populations often developed "bait shyness" to the material, and this factor, coupled with the potential for human toxicity, was what led to the virtual elimination of thallium for rodent control purposes. Persons receiving large exposures often exhibit alopecia (hair loss). Thallium has a TLV of 0.1 mg/m^3 of air. *See also* Rodenticide. (FSL)

Therapeutic index A clinical pharmacology term for the ratio of the toxic level of a drug to the therapeutic level. This is calculated as the toxic concentration (TC) of a drug, divided by the effective concentration. This is expressed as TC_{50}/EC_{50}, and is the point at which 50% of patients have a toxic reaction to the drug to be monitored. The lower the therapeutic index, the more difficult it is to titrate the dosage of a drug in a patient. (FSL)

Therapeutics The branch of medicine concerned with the treatment of disease. (FSL)

Therapy The treatment of disease by the use of a variety of methods. (FSL)

Thermal analysis The measurement of changes in properties (physical or chemical) of materials, as a function of temperature change, usually brought about by heating or cooling at a uniform rate. (SWT)

Thermal death time Time required to achieve sterility under a given set of circumstances. *See also* Sterilization. (JHR)

Thermal pollution Raising of the temperature of a part of the environment by the discharge of substances whose temperature is higher than the ambient levels, such as the discharge of hot water from cooling towers into a lake or stream. (FSL)

Thermal sanitizing Use of heat treatment to kill microorganisms, which can be destroyed if they are heated long enough at the correct temperature. Thermal sanitizing is not efficient in terms of energy costs, but it is still used in some applications today, particularly in food-service operations. The two main methods of thermal treatment involve use of hot water and steam. The immersion of small utensils in water at a temperature of at least 80°C or higher is effective. Steam may be applied to larger pieces of equipment. (HMB)

Thermoduric Describes microorganisms that are highly resistant to heat. Some vegetative cells and many types of spores are thermoduric. These organisms may be of importance in food processing in which products are heat treated but not sterilized—e.g., pasteurization. *See also* Mesophiles; psychrophiles; thermophiles. (HMB)

Thermogenic soils Soils in which the properties have been influenced primarily by high temperature as the dominant soil-formation factor. (SWT)

Thermophiles A physiological group of bacteria identified by their temperature range of growth. Thermophiles have a minimum growth temperature of about 35°C, an optimum temperature between 55° and 65°C, and a maximum growth temperature as high as 75–90°C. *See also* Mesophiles; psychrophiles. (HMB)

Thesaurosis A condition caused by the storage of abnormal amounts of natural or foreign substances in the body. (FSL)

Third quartile A number such that at most three-fourths of the data are smaller in value than the number and at most one-fourth of the data are larger. (VWP)

Thixotropic Refers to the ability of certain gels to become less viscous when shaken or subjected to shearing forces, and to return to the original viscosity upon standing. (FSL)

Threatened species Under the Endangered Species Act, any species likely to become an endangered species within the foreseeable future throughout all or a significant portion of its range. 42 USC §1543(20). *See also* Endangered Species Act; endangered species. (DEW)

Threshold dose The dose of a chemical below which an adverse effect does not occur. (FSL)

Threshold limit value (TLV) Refers to airborne concentration of a substance and represents conditions under which the American Conference of Governmental Industrial Hygienists (ACGIH) believes that nearly all workers may be repeatedly exposed day after day without adverse health effects. The ACGIH adds that individuals may be hypersusceptible or otherwise unusually responsive to some chemicals, and may not be adequately protected from adverse health effects at concentrations below the TLV, in which case an occupational physician should evaluate the individual. TLVs are not legal standards and do not carry the force of law, although OSHA has twice adopted them as permissible exposure limits. Some localities and communities have also used them as a basis for air pollution control. TLVs do not serve as the divisions between safe and dangerous concentrations, nor are they a relative index of toxicity. They are revised annually and are to be used in the practice of industrial hygiene. *See also* Ceiling exposure limit; occupational exposure limits; short-term exposure limit; skin notation; time-weighted average. (DEJ)

Threshold moisture content The minimum moisture condition, measured in terms of moisture content or moisture stress, at which biological activity becomes measurable. (SWT)

Threshold planning quantity A quantity, designated for each chemical on the list of extremely hazardous substances, that triggers notification by facilities to the State Emergency Response Commission that such facilities are subject to emergency planning under SARA Title III. *See also* Superfund Amendments Reauthorization Act. (FSL)

Throughfall That portion of precipitation that falls or drips through the plant canopy to the soil below. (SWT)

Tidal flats Areas consisting of nearly flat, barren mud that is periodically covered by tidal waters. Normally the material has an excess of soluble salts. (SWT)

Tidal marsh *See* Estuary. (SMC)

Tidal volume (TV) The volume of gas inspired or expired during each respiratory cycle of the lungs. (DEJ)

Tile drain Concrete or ceramic pipes placed in the soil or subsoil to provide drainage from the soil. *See also* Soil absorption system. (SWT)

Till **(1)** Unstratified glacial drift that has been deposited by ice and that consists of clay, silt, sand, gravel, and boulders, intermingled in any proportion. **(2)** To prepare the soil for seeding; to seed or cultivate the soil. (SWT)

Tillage The manipulation, generally mechanical, of soil properties for any purpose; usually restricted to improving soil conditions for crop production. One or more of the following operations may be involved. **(1)** *Cultivation:* A shallow tillage operation that improves soil conditions (such as aeration and infiltration), water conservation, or weed control. **(2)** *Harrowing:* A tillage operation that pulverizes, smooths, and firms the soil in seedbed preparation, and controls weeds. **(3)** *Incorporation:* The mixing of materials, such as pesticides, fertilizers, or plant residues, into the soil. **(4)** *Lifting:* The separation of roots or other crops and their elevation to the soil surface. **(5)** *Listing* (middle breaking): A tillage and land-forming operation that uses a tool that turns two furrows laterally in opposite directions to produce beds or ridges. **(6)** *Scarifying:* The loosening of topsoil aggregates, using an implement that rakes the soil surface with a set of sharp teeth. **(7)** *Clean Tillage:* A process of plowing and cultivation whereby all residues are incorporated and all growth of vegetation is prevented except that of the particular crop desired during the growing season. **(8)** *Conservation Tillage:* Any type of tillage operation whose purpose is to minimize or reduce loss of soil and water. **(9)** *Minimum Tillage:* A tillage system that minimizes the soil manipulation necessary for crop production or for meeting tillage requirements under the existing soil and climatic conditions. **(10)** *No-tillage System:* A method of crop planting in which crops are planted directly into the soil without preparatory tillage; usually a special planter is necessary to prepare a narrow, shallow seedbed immediately surrounding the seeds being planted . **(11)** *Chemical Fallow:* A method of fallowing (placing the soil in a state of dormancy) whereby all vegetative growth is killed or prevented by the use of chemicals. **(12)** *Contour Tillage:* A tillage system in which all operations and planting are on the contour within a given range. **(13)** *Crop Rotation:* The planting of a sequence of crops in recurring succession on the same area or land, as opposed to continuous culture of one crop. (SWT)

Tilth Soil condition as it is related to use of the sod as a growth medium and to ease of tillage. (SWT)

Times Beach A small Missouri town that was evacuated and eventually abandoned in 1983 as the result of legal action by the EPA. High levels of dioxin were found in the soil on the unpaved streets of this town, as a result of placement of contaminated oil there earlier to control dust. The community is now being considered as a possible location for an incinerator, which will be used to burn the contaminated soil from the streets, roadsides, and other locations. *See also* Dioxin. (FSL)

Time-weighted average (TWA) An exposure averaged over a given time period, often an 8-hour workday. Threshold limit values and permissible exposure limits are often based on 8-hour time-weighted averages. Instantaneous or short-term exposures may be permitted to exceed the exposure limit to a certain extent, if there is a corresponding period of time in which the exposure is below the limit, such that the overall time-weighted average is within acceptable limits. If the work shift is longer than normal, a corresponding reduction in the TLV is usually determined. Also known as TLV-TWA. *See also* Occupational exposure limits. (DEJ)

Tip-up mound The mound created by the roots when a tree tips over. Soil in the mound is usually highly fertile top soil that contains much organic material, especially directly on the roots of the fallen tree. The pit left when the roots and surrounding soil are removed is usually less fertile subsoil with little organic material. When both the mound and the pit are at the surface, the plants that become established in the two areas can differ significantly because of the difference in fertility. (SMC)

Tissue A group of cells with similar structure and function. There are four basic tissues in the body: muscle, nerve, epithelium, and connective tissues (including blood, bone, and cartilage). (FSL)

Titration A chemical procedure whereby a measured amount of a solution of one reagent is added to a solution of another until a reaction is complete and the second reagent has been consumed. (FSL)

TLV *See* Occupational exposure limits; threshold limit value. (DEJ)

TLV-C A threshold limit value with a ceiling exposure limit (to be distinguished from a threshold limit value with a time-weighted average). The ceiling limit is a definite boundary that exposures should not be permitted to exceed, even instantaneously, during any part of the workday. Usually applies to substances that are fast-acting. This limit is a concentration, set by the American Conference of Governmental Industrial Hygienists, to which most, but not all, workers may be exposed for a working lifetime without suffering adverse health effects. *See also* Occupational exposure limits; threshold limit value. (DEJ)

TLV-TWA *See* Occupational exposure limits; threshold limit value. (DEJ)

TLV-STEL A Threshold Limit Value with a short-term exposure limit, a concentration to which most, but not all, workers may be continuously exposed over a short period of time without suffering from irritation, chronic or irreversible tissue damage, or narcosis. The TLV-STEL is a 15-minute time-weighted average exposure limit that should not be exceeded at any time during the workday, even if the 8-hour time-weighted average (TWA) is acceptable. Exposures up to the

STEL should not be longer than 15 minutes and should not occur more than four times in one day, and there should be at least 60 minutes between exposures. *See also* Occupational exposure limits; threshold limit value. (DEJ)

Tolerance **(1)** The permissible residue levels for pesticides in raw agricultural produce and processed foods. Whenever a pesticide is registered for use on a food or a feed crop, a tolerance for exemption from the tolerance requirements must be established. EPA establishes the tolerance levels, which are enforced by the Food and Drug Administration and the Department of Agriculture. **(2)** In the Superfund program, the existence of a contamination concentration in the environment high enough to warrant action or trigger a response under SARA and the National Oil and Hazardous Substances Contingency Plan. The term can be used similarly in other regulatory programs. *See also* Food, Drug, and Cosmetic Act; special review. (DEW, FSL)

Topdressing The practice of applying fertilizer on established crops on the soil surface. *See also* Runoff. (SWT)

Topography The physical features of a surface area, including relative elevations and the position of natural and man-made features. (FSL)

Toposequence A sequence of related soils that differ mainly because of difference in topography. (SWT)

Topsoil The layer of cultivated soil, the A1 horizon; fertile soil that is used as a topdressing. (SWT)

Tort A private or civil wrong or injury. It requires three elements: legal duty owed to plaintiff by defendant, breach of that duty, or damage to plaintiff that was proximately caused by breach of duty. One who commits a tort may be referred to as a tort-feasor. (DEW)

Tortuosity The term that describes the curving nature of soil pores. (SWT)

Total coliform count An enumeration of all the microorganisms in a sample that belong to the coliform group. *See also* Coliforms. (HMB)

Total flooding system A fixed fire-suppression system that is arranged to automatically discharge a predetermined concentration of agent into an enclosed space for the purpose of fire extinguishment or control. (SAN)

Total pressure The algebraic sum of the velocity pressure (always positive) and the static pressure (which may be either positive or negative). It is a measure of the energy content of the airstream. Total pressure declines as airflow proceeds downstream through a duct, and rises across a fan. (DEJ)

Total suspended solids (TSS) A measure of the suspended solids in wastewater, effluent, or water bodies. Suspended matter in these types of media can be divided into settleable and nonsettleable solids. Settleable solids make up the portion that settles readily upon standing; fresh sewage, for example, will have a higher proportion of settleable solids than will stale sewage because disintegration and decomposition is greater in the latter. Nonsettleable suspended solids make up about one-half of the suspended solids in sewage. They consist of finely divided suspended material in the form of colloidal particles. (FSL)

Toxic Describes a chemical in any of the following categories. **(1)** A median lethal dose (LD_{50}) of more than 50 mg per kg but not more than 500 mg per kg of body weight when administered orally to albino rats weighing between 200 and 300 g each. **(2)** A median lethal dose (LD_{50}) of more than 200 mg per kg but not more than 1000 mg per kg of body weight when administered by continuous contact for 24 hours (or less if death occurs within 24 hours) on the bare skin of albino rabbits weighing between 2 and 3 kg each. **(3)** A median lethal concentration (LC_{50}) in air of more than 200 parts per million but not more than 2000 parts per million by volume of gas or vapor, or more than 2 mg per liter but not more than 20 mg per liter of mist, fume, or dust, when administered by continuous inhalation for 1 hour (or less if death occurs within 1 hour) to albino rats weighing between 200 and 300 g each. *See also* Toxic, highly. (FSL)

Toxic, Highly Describes a chemical in any of the following categories. **(1)** A medial lethal dose (LD_{50}) of 50 mg or less per kg of body weight when administered orally to albino rats weighing between 200 and 300 g each. **(2)** A median lethal dose (LD_{50}) of 200 mg or less per kg of body weight when administered by continuous contact for 24 hours (or less if death occurs within 24 hours) with the bare skin of albino rabbits weighing between 2 and 3 k each. **(3)** A median lethal concentration (LC_{50}) in air of 200 parts per million by volume or less of gas or vapor, or 2 mg per liter or less of mist, fume, or dust, when administered by continuous inhalation for 1 hour (or less if death occurs within 1 hour) to albino rats weighing between 200 and 300 g each. *See also* Toxic. (FSL)

Toxicant A poisonous agent that kills or injures animals or plants. *See also* Herbicide; insecticide. (FSL)

Toxic chemical release inventory Compilation of data reported by industry to the EPA pursuant to the Emergency Planning and Community Right-to-Know Act (also known as SARA Title III). Contains information on releases of specified toxic chemicals, such as: maximum amount of chemicals stored on site; estimated quantity emitted into the air, discharged into bodies of water, injected underground, or released to land; methods used in waste treatment and their efficiency; and data on the transfer of chemicals off site for treatment/disposal, either to publicly owned treatment works or elsewhere. Information is publicly available in printed form or as an on-line database. Available through the National Library of Medicine Toxicology Data Network (TOXNET). 42 USC §§11001 *et seq.* *See also* Emergency Planning and Community Right-to-Know Act. (DEW)

Toxic cloud Airborne mass of gases, vapors, fumes, or aerosols containing toxic materials. (FSL)

Toxicity The state of being poisonous; the capacity of a substance to induce damage to living tissue, impair the central nervous system, or cause illness or death following exposure (by ingestion, inhalation, or skin absorption). Toxicity can be related to time—e.g., *acute toxicity* refers to those situations in which there is a sudden onset that occurs for a short period of time, and *chronic toxicity* is noted by long or permanent duration. Further, toxicities can be determined by the site of action. *Local toxicity* can refer to the site of application or exposure, and *systemic toxicity* to cases in which the toxicant has been absorbed and distributed to the susceptible organs. *See also* Acute exposure; chronic exposure; occupational exposure. (FSL)

Toxicity assessment A statistical evaluation of the toxicity of a chemical, based on the available data from human and animal experimentation. (FSL)

Toxicity characteristic leaching procedure (TCLP) A laboratory test (Method 1311) developed in 1990 by EPA to replace the extraction procedure toxicity characteristic (EPTC) for use in the identification of hazardous waste. This particular procedure is designed to determine the mobility of both organic and inorganic contaminants present in liquid, solid, and multiphasic wastes. When a complete analysis of a hazardous waste demonstrates that individual contaminants are not present in the waste, or that they are present at such low concentrations that the appropriate regulatory thresholds could not possibly be exceeded, this test would ordinarily not be performed. *See also* Comprehensive Environmental Response, Compensation, and Liability Act; extraction procedure toxicity characteristic; hazardous waste; Resource Conservation and Recovery Act. (FSL)

Toxicology The science and study of chemicals and their adverse systemic effects, including source, chemical composition, action, tests, and antidotes. *See also* Pharmacology. (FSL)

Toxicology Information Online (TOXLINE) The online, interactive collection of toxicological information at the National Library of Medicine. It contains references to published materials and research in progress in the following areas: adverse drug reactions, air pollution, animal venom, antidotes, carcinogens via chemicals, chemically induced diseases, drug evaluation, drug toxicity, environmental pollution, food contamination, mutagenesis, occupational hazards, pesticides, radiation, teratogenesis, toxicological analysis, waste disposal, and waste treatment. *See also* Medlars. (FSL)

Toxic pollutants Poisonous materials released into the environment. Under the Clean Water Act, EPA must establish effluent standards for the discharge of listed toxic pollutants. The pollutants currently regulated are aldrin/dieldrin, DDT, DDD, DDE, endrin, toxaphene, benzidine, and polychlorinated biphenyls (PCBs). 33 USC §1317(a), 40 CFR Part 129. *See also* Clean Water Act. (DEW)

Toxic Substances Control Act of 1976 Federal law, administered by the EPA, that regulates the manufacture, process, use, and distribution of chemical substances for commercial purposes. Section 4 sets out a mechanism for the testing of chemical substances and mixtures. Section 5 details requirements for notification of intent to manufacture and process chemical substances by filing a pre-manufacture notice for a new substance or significant new use of a substance (SNUR). Section 6 provides authority for the regulation and control of hazardous chemical substances. Section 8 sets out reporting requirements. 15 USC §2601 *et seq.*, 40 CFR Parts 700–799. *See also* Chemical substance; interagency testing committee. (DEW)

Toxic tort Generally, a tort action arising from an injury allegedly caused by exposure or threatened exposure to a hazardous or toxic substance. The alleged injury may take a variety of forms, such as property damage or loss of value, personal injury, mental distress, and fear of future illness. Requested remedies may include abatement of the exposure or threat of exposure, compensation for actual damages, costs of medical monitoring, and punitive damages. (DEW)

Toxin Proteins or conjugated protein substances that are lethal to other organisms. They are produced by some higher plants, certain animals, and pathogenic bacteria. The high molecular weight and antigenicity of toxins differentiates them from simple chemical poisons and vegetable alkaloids. Endotoxins are found within bacterial cells but not in cell-free filtrates of intact bacteria. They are released from a bacterium when its cell wall is ruptured. Exotoxins are released by the bacterial cell into the culture medium (or host) and hence are found in cell-free filtrates and cultures of intact bacteria. (FSL)

Toxoid A neutralized toxin that is capable of producing active immunity when injected into the body. Tetanus toxoid is used, for example, to prevent the disease tetanus. *See also* Vaccine. (FSL)

Trace elements Copper, zinc, boron, and other chemicals that are necessary in small amounts for the proper functioning of plants and animals. (FSL)

Transcript Official verbatim written record of a proceeding. (DEW)

Transduction The transfer of genetic material by bacteriophages from one bacterium to another. *See also* Bacteriophage. (FSL)

Transfer facility Any transportation-related facility, including loading docks, parking areas, storage areas, and other similar areas where shipments of waste are held (where they come to rest or are managed) during the course of transportation. For example, a location at which regulated waste is transferred directly between two vehicles is considered a transfer facility. (FSL)

Transfer station A facility used to transfer solid waste from one vehicle to another for transportation to a disposal site or processing operation. (SWT)

Transformation The process of placing new genes into a host cell, thereby inducing the host cell to exhibit functions encoded by the DNA. (FSL)

Transitional soil A soil with properties that are between those of two different soil types. (SWT)

Translocation The movement or transfer of a chemical from one location to another. For example, after the administration of a chemical agent to an animal by mouth, the material will pass from the gastrointestinal tract to the circulatory system and then to individual cells. Following the application of herbicides, the material will move from the point of application (leaves), through the stem, to the roots. (FSL)

Transmission The direct (contact- or droplet-spread) or indirect (vectorborne, vehicleborne, or airborne) transfer of an infectious agent from a reservoir to a susceptible host. (FSL)

Transmission loss A characteristic of a sound barrier, defined as 10 times the logarithm of the ratio of the incident acoustic energy to the acoustic energy transmitted through the barrier. Information on transmission losses is essential in the design of work spaces in which noise may be a factor in employee health. (DEJ)

Transpiration The process by which water vapor is lost to the atmosphere from living plants. The term can also be applied to the quantity of water thus dissipated. (FSL)

Transporter A person engaged in the off-site transportation of waste by air, rail, highway, or water. Transporters of solid waste under the Resource Conservation and Recovery Act must comply with regulations regarding the transport of such waste and requiring an RCRA manifest by the generator. Regulations set out requirements for the manifest system, record-keeping, reporting, and responses to discharges of hazardous wastes. Department of Transportation regulations under the Hazardous Materials Transportation Act may also apply. 42 USC §6923, 40 CFR Part 263. *See also* Generator; Hazardous Materials Transportation Act; hazardous waste; Resource Conservation and Recovery Act; solid waste. (DEW)

Transport Canada (TC) The governmental agency in Canada that has jurisdiction over the transportation and shipment of compressed gases. (FSL)

Treatment, Storage, and Disposal Facilities Under the Resource Conservation and Recovery Act, facilities must have a permit to treat, store, or dispose of hazardous waste. Facilities that were in existence when the law was passed were accorded interim status and allowed to continue to operate if they filed notification and Part A of the permit application and complied with applicable regulations. Facilities that are granted Part B permits must operate in compliance with more stringent standards. 42 USC §6924, 40 CFR 264, 265. *See also* Interim status; loss of interim status; Resource Conservation and Recovery Act. (DEW)

Treble damages provision Statutory provision that allows the amount of damages found to be owed to be tripled in amount. (DEW)

Trench A narrow excavation (in relation to its length). In general, the depth is greater than the width, but the width (measured at the bottom) should not be greater than 15 feet according to the OSHA definition. If forms or other structures are installed or constructed in an excavation that reduce the dimension measured from the forms or structure to the side of the excavation to 15 feet or less (measured at the bottom of the excavation), the excavation is also considered to be a trench. *See also* Trench box; shoring; excavation. (SAN)

Trench box A structure able to withstand the forces imposed on it by a cave-in and protect employees within the structure. Such structures can be permanent or able to be moved along as work progresses. Trench boxes can be premanufactured or job-built. Trench boxes are also referred to as trench shields or shield systems. *See also* Shoring; trench; excavation. (SAN)

Triazine herbicides Selective heterocyclic nitrogen compounds, such as atrazine, simazine, and cyanazine, that contain aromatic ring compounds of both carbon and nitrogen atoms. Atrazine is one of the most widely used herbicides on corn crops. Simazine, the least water-soluble of all the triazines, is registered for use on alfalfa, artichokes, many orchard crops, and wood ornamental plants. Cyanazine, which is less persistent than atrazine, provides control of a broad spectrum of weeds. The triazines are low in acute toxicity to mammals (simazine, for example, has an oral LD_{50} of 5000 mg/kg). The main environmental concerns revolve around their moderate persistence in soil and potential damage to nontarget plants. *See also* Pesticide. (FSL)

Trichloroethylene (TCE, $CHCl:CCl_2$) A stable, low-boiling colorless liquid with a chloroform-like odor. TCE is toxic by inhalation and is used as a solvent and metal degreasing agent, and in other industrial applications. The TLV is 50 ppm in air. *See also* Solvent. (FSL)

Trickling Filter A device used in an aerobic biological process in the secondary treatment of wastewater. The filter beds are usually constructed of crushed stones of a size that provides a maximum surface area and allows a maximum flow through the filter. Sewage that typically has received primary treatment is applied at the surface of the filter, using rotary distributors consisting of two or more arms that turn by the force of sewage being sprayed from fixed nozzles located along the length of the arms. This rotary motion allows for the even distribution of sewage over the surface of the filter bed. The sewage then trickles down through the bed, collecting in underdrains in the bottom of the bed. Different media are commercially available and may be used to replace stones in the bed. Redwood slates and plastic media (polyvinyl chloride) in a vertical-sheet packing arrangement have been used in so-called biological towers. The principle of operation remains the same, however: the creation of a filter bed with a medium of maximum surface area and sufficient void volume to facilitate sewage flow and passage of air that supplies oxygen. Sewage sprinkled over the surface of the fixed media produces a biological film that coats the surface. This film is composed of a mixed microflora that utilizes organic matter and dissolved oxygen from the wastewater flow and releases metabolic wastes. Although this biological film is quite thin, it becomes anaerobic near the media surface and generally more anaerobic throughout, toward the bottom of the filter. Therefore, most aerobic degradation of organic matter occurs in the upper dimensions of the filter bed. Although the process is considered aerobic, in fact aerobic, facultative anaerobic, and anaerobic microorganisms are involved in the process. Trickling filters are of two types: conventional and so-called high-rate filters. They differ in size and loading rates. They may be used as a single unit or in series. They are particularly adapted to smaller plants and offer relative freedom from operating problems. *See also* Secondary treatment; waste stabilization pond. (HMB)

Trigger finger An occupational disorder in which a tendon in the index finger is compressed from exterior swelling. This condition can be produced by repeatedly pulling a trigger or by using hand tools that have sharp edges that press into the tissue or whose handles are so far apart that only the end segments of the user's hand grasp them, while the middle segments remain straight. *See also* Cumulative trauma disorder. (SAN)

Trihalomethane (THM) One of a family of organic compounds, named as derivatives of methane. THMs are generally a by-product of chlorination of drinking water that contains organic material. *See also* Methane. (FSL)

Trimodal distribution A set of data with three modes. (VWP)

Trituration Reduction to powder by friction or grinding. (FSL)

Tropism A response to an external stimulus, in which the organism usually turns toward the direction of the most intense stimulus. *See* Geotropism and Phototropism. (SMC)

Troposphere The lower atmosphere, the portion between 7 and 10 miles from Earth's surface, where clouds are formed. (FSL)

Truncated Describes soil that has lost all or part of the upper horizons. *See also* Erosion. (SWT)

Trust fund (CERCLA) A fund set up under the Comprehensive Environmental Response, Compensation, and Liability Act (CERCLA) to help pay for the cleanup of hazardous waste sites and for legal action to force those responsible for the sites to remediate conditions there. (FSL)

T test A hypothesis test for a normally distributed population that considers the population mean when the variance of the underlying distribution is unknown.

(1) *One sample:* Used to test the hypothesis $H_0: \mu = \mu_0$, versus $H_1: \mu \neq \mu_0$, with significance level α, assuming that the variance is unknown and is the same under both hypotheses. **(2)** *Two samples:* Used to test the hypothesis $H_0: \mu = \mu_0$, versus $H_1: \mu \neq \mu_0$, with significance level α, for two normally distributed populations in which the variance is assumed to be the same for both populations. (VWP)

Tubular springs Springs commonly called ''bold'' springs because the water is released freely from one or more large openings. Water reaching the openings by percolation through sand or other fine-grained material is usually free from contamination. If water is received either directly or indirectly from the effluent of subsurface sewage disposal systems or other contamination, the spring water must be considered unsafe. (FSL)

Tuff Volcanic ash that has become stratified and consolidated. (SWT)

Tundra A treeless plain that is level or undulating, and that is characteristic of arctic regions. The tundra supports a type of ecosystem dominated by lichens, mosses, grasses, and woody plants. Tundra is found at high latitudes (arctic tundra) and high altitudes (alpine tundra). Arctic tundra is underlain by permanently frozen subsoil (permafrost) and is usually very wet. (FSL)

Turbidimeter A device that measures the amount of suspended solids in a liquid. (FSL)

Turbidity The presence of suspended material, such as clay, silt, finely divided organic material, plankton, and other inorganic material, in water. Turbidities in excess of 5 units are easily detectable in a glass of drinking water, and are usually objectionable for aesthetic reasons. (FSL)

Two-tailed test A statistical test in which the values of the parameter being investigated under the alternative hypothesis are allowed to be either greater than or less than the values of the parameters under the null hypothesis. (VWP)

Typhoid Mary Name given to Mary Mallon (1892–1932), a typhoid carrier who caused 10 outbreaks involving 51 cases and 3 deaths. She worked as a cook in New York City and was known to be a carrier as early as 1907. *See also* Carrier; salmonella. (FSL)

Tyrotoxism Poisoning due to the presence of a toxin in milk, cheese, or other dairy products. (FSL)

Ultisols Mineral soils that have an argillic horizon with a base saturation of less than 35% when measured at pH 8.2. Ultisols have a mean annual soil temperature of 8°C or higher. (SWT)

Ultra-clean coal (UCC) Coal that has been washed, ground into fine particles, chemically treated to remove sulfur, ash, silicone and other substances, and usually briquetted and coated with a sealant made from coal. (FSL)

Ultrasonic noise Noise characterized by a frequency greater than audible noise, typically greater than 20,000 Hz. (DEJ)

Ultraviolet rays Radiation, from the sun or other sources, that can be useful or potentially harmful. UV rays from one part of the spectrum enhance plant life and are useful in some medical and dental procedures. UV rays from other parts of the spectrum to which humans are exposed (e.g., while getting a tan) can cause skin cancer or other tissue damage. The ozone layer in the atmosphere provides a protective shield that limits the amount of ultraviolet rays that reach the earth's surface. (FSL)

Unclean-hands principle Legal principle that one who has "unclean hands" is guilty of some wrongdoing (in a controversy) and is not entitled to relief in a court of equity. (DEW)

Underground injection Under the Safe Drinking Water Act, the subsurface placement of fluid by well injection. 42 USC §300h(d). *See also* Safe Drinking Water Act; primary drinking water regulation. (DEW)

Underground storage tank (UST) Under the Resource Conservation and Recovery Act, any tank (or combination of tanks), including underground pipes connected thereto, that is used to contain an accumulation of regulated substances, and whose volume (including the volume of the underground pipes connected thereto) is 10% or more beneath the surface of the ground. Subject to specific exceptions. 42 USC §6991. *See also* Resource Conservation and Recovery Act. (DEW)

Uniformity coefficient As applied to soil percolation in subsurface sewage disposal systems, the coefficient obtained by dividing that size of sand of which 60% by weight of the soil particles are smaller, by that size of sand of which 10% by weight of the particles are smaller. *See also* Percolation test. (FSL)

Unimodal distribution A set of data with a single mode. (VWP)

Union of two events (biostatistics) The situation whereby one or both of two events occur. (VWP)

United States Code *See* Statute. (DEW)

Universal precautions An approach to infection control. According to this concept, all human blood and certain human body fluids are treated as if known to be infectious for HIV, HBV, and other bloodborne pathogens. *See also* Bloodborne pathogens; exposure control plan. (FSL)

Unsaturated flow Water movement in a soil that is not completely filled with water. (SWT)

Unsaturated zone The area above the water table, in which the soil pores are not fully saturated, although some water may be present. *See also* Aquifer. (FSL)

Unstable material A chemical substance that, in the pure state or as produced or transported, will vigorously polymerize, decompose, condense, or become self-reactive upon exposure to air, water, shock, pressure, or temperature fluctuation. *See also* Picric acid; shock-sensitive chemicals. (FSL)

Upper confidence limit (UCL) The maximum value expected for a parameter, using a given level of statistical confidence. (VWP)

Upper explosive limit (UEL) *See* Upper flammable limit. (DEJ)

Upper flammable limit (UFL) The concentration of a substance in air, usually expressed as a volume percent, above which combustion cannot be supported at normal room temperature because the mixture is too ''rich'' and there is insufficient oxygen. Combustion in air of a flammable material can occur only at concentrations between the lower and upper flammable limits. Each substance has distinct upper and lower flammable limits. *See also* Lower flammable limit. (DEJ)

Uranium A radioactive heavy-metal element used in nuclear reactors and in the production of nuclear weapons. The term refers usually to ^{238}U, the most abundant uranium isotope, although a small percentage of naturally occurring uranium is made up of other isotopes. (RMB)

Urban land A land area that has been so obstructed by urban development that soil feature identification is not feasible. (SWT)

Urban runoff Storm water, from city streets and adjacent domestic or commercial properties, that may carry pollutants of various kinds into sewer systems and/or receiving waters. (FSL)

Urea **(1)** A soluble breakdown product of proteins, excreted by many vertebrates. **(2)** As applied to herbicides, substituted urea compounds such as diuron and monuron, which control weeds by inhibiting the photosynthetic process. Urea compounds are very persistent in the soil, but they are not highly toxic to mammals—e.g., the acute oral LD_{50} for monuron in rats is 3600 mg/kg. *See also* Triazine herbicides. (FSL)

Vaccine A preparation containing killed or living whole or parts of microorganisms having antigenic properties. Vaccines are employed to induce in the recipient a specific active immunity against an infectious agent. *See also* Resistance; Sabin; Salk. (FSL)

Vacuum breaker A type of backflow prevention device. An atmospheric-vacuum breaker has an inside moving element that prevents water from discharging through the top of the breaker during flow and drops down to provide a vent opening when flow stops. When flow stops, the element drops onto the valve seat by the force of gravity. This prevents back-siphonage by allowing air to enter and break the siphon in the elevated pipe loop. A pressure breaker contains an assembly consisting of a spring-loaded check valve and a spring-loaded air valve. The check valve prevents backflow and the air valve opens to admit air when the pressure within the body of the breaker approaches atmospheric pressure. If the check valve does not close tightly because of interference from foreign matter, air is drawn in through the automatically operating vent valve to preclude backflow. In either case, a vacuum breaker can be used to prevent contamination of a potable water supply by water of unknown or questionable safety. *See also* Backflow; back-siphonage. (HMB)

Vapor A term used for a substance that, although present in the gaseous phase, usually exists as a liquid or solid at room temperature and pressure. *Vapor* is often used interchangeably with *gas*, which usually defines a substance that is in the gaseous phase at room temperature and pressure. (DEJ)

Vapor-capture system Any combination of hoods and ventilation systems that captures or contains organic vapors in order that they may be directed to an abatement or recovery device. (FSL)

Vapor dispersion The movement of vapor clouds in air, due to wind, gravity spreading, and mixing. (FSL)

Vaporization Conversion of a solid or liquid into a vapor without change in the chemical composition of the material involved. (FSL)

Vapor plumes Flue gases that are visible because they contain water droplets. (FSL)

Vapor pressure The pressure, usually expressed in millimeters of mercury or pounds per square inch, exerted when a solid or liquid is in equilibrium with its own vapor. Vapor pressure is a function of the substance and the temperature, and is often used as a measure of how rapidly a liquid will evaporate. (DEJ)

Variable A symbol that may represent any one of a given set of values. (VWP)

Variance **(1)** (Legal) Government permission for a delay or exception in the application of a given law, ordinance, or regulation. **(2)** (Biostatistics) *Population variance:* A quantity symbolized by σ^2 and given by the equation

$$\sigma^2 = \sum_{i=1}^{k} [(x_i - \mu)^2 \cdot p(x_i)]$$

where x_i is a population data value, μ is the population mean, $p(x_i)$ is the probability of x_i occurring, and $i = 1, 2, \ldots, k$.
Sample variance: A quantity symbolized by s^2 and given by the equation

$$s^2 = \sum_{i=1}^{n} \frac{(x_i - \bar{x})^2}{n - 1}$$

where x_i is a sample data value, $\bar{x}$ is the sample mean, and n is the number of sample values. (BEW,VWP)

Vector An arthropod or other invertebrate that can transmit an infectious agent from a source of infection to a susceptible host. Transmission takes place by inoculation into or through the skin or mucous membrane by biting or by the deposition of infective materials on the skin. Vectors may be infective, as in the case of mosquitoes that transmit malaria and other diseases. They may also be mechanical carriers of the infectious agent, such as houseflies that carry bacteria that cause the disease shigellosis. The bacteria are in fecal material that may then be transferred to susceptible persons and to foodstuffs. (FSL)

Vehicles Object or substances that are a source of infection or intoxication. Contaminated materials or objects, such as cooking or eating utensils, water, food, biological products, including blood and serum, and soiled clothing, that serve as an intermediate means by which an infectious agent is transported and introduced into a susceptible host. The term *vehicleborne* is used. The infectious agent may or may not have developed or increased in numbers on or in the vehicle before being introduced into the susceptible host. This type of transmission of the infectious agent is indirect. Many bacterial and viral agents may be vehicleborne. (FSL,HMB)

Vein **(1)** Any blood vessel that carries blood from some part of the body toward the heart. **(2)** A more or less continuous body of minerals or igneous or sedimentary rock occupying a fissure or zone, and differing in nature and separated from the enclosing rock. *See also* Artery. (FSL)

Velocity pressure The kinetic pressure in the direction of flow necessary to cause a fluid at rest to flow at a given velocity, usually expressed in inches water gauge. Velocity pressure is directly related to air velocity and is always positive. In standard air, the relationship is: $VP = (v/4005)^2$, where VP is velocity pressure (in inches water gauge) and v is velocity (in feet per minute). *See also* Static pressure; total pressure. (DEJ)

Venn diagram A chart developed by John Venn (1834–1923), an English logician, that uses circles to show relationships between events. Those points, and only those points, belonging to a given event are included inside the event's circle. (VWP)

Venom A toxic substance normally secreted by certain insects, snakes, and other animals. *See also* Toxin. (FSL)

Ventilation One of the principal methods of controlling health hazards, ventilation can be described generally as the causing of fresh air to circulate to replace contaminated air that is simultaneously removed. Exhaust ventilation is typically of two types: local (which consists of a means of capturing the contamination at its source before it can enter the room air) and general (which consists of a means of diluting air contaminants by mixing the air with uncontaminated air). Ventilation is characterized by volumetric air flow (expressed in cubic feet per minute) or air velocity (expressed in feet per minute) or air changes per unit time. *See also* Dilution ventilation; local exhaust ventilation. (DEJ)

Venting The release of gas from a vessel or confining structure in a controlled manner. *See also* Pressure vessel. (FSL)

Venturi-type collectors A technology used to remove dust from exhaust air by employing a high-velocity airstream that is used to break up water fed into a venturi throat. The fine water droplets collide with the particulates, and are scrubbed from the exhaust air. Venturi dust collectors are typically high-efficiency air cleaners. They consume a great deal of power, due to the high pressure drop across the collector. (DEJ)

Vermination An infestation with vermin (e.g., parasitic insects such as lice or bedbugs) or with worms such as pinworms, roundworms, or tapeworms. (FSL)

Vernal Of or pertaining to the spring season; appearing in spring. (SMC)

Vernal pond A pond that contains standing water in spring from snow melt and spring rains, but that dries up by mid- to late summer in all but the wettest years. Vernal ponds are important in the life cycles of amphibians and insects that use the seasonal ponds for reproduction and juvenile growth. The population dynamics of these organisms are strongly influenced by the length of time the ponds may contain water in a given year. (SMC)

Vertisols Mineral soils that have deep, wide slickensides or wedge-shaped structural aggregates tilted at an angle from the horizon. Vertisols contain 30% or more clay. (SWT)

Vesicant An agent capable of producing small circumscribed elevations of the skin that contain fluid; a bleb or blister. Vesicant gases such as mustard gas (dichlorodiethyldisulfide) have been used as chemical warfare agents. (FSL)

Vibration syndrome *See* Raynaud's disease. (FSL)

Vicarious liability Liability that falls on a person but is due to the actions of another. For example, an employer may be found liable for the actions of his or her employee. *See also* Respondeat superior. (DEW)

Vicia A plant whose beans or pollen contains a toxic compound capable of causing poisoning (favism) in humans. (FSL)

Vinyl chloride (CH_2:CHCl) A compressed gas that is easily liquefied and usually handled as a liquid. It is an extremely toxic and hazardous material by all routes of exposure and is carcinogenic. The TLV-TWA is 5 ppm in air. Vinyl chloride is used in organic synthesis, as an adhesive for plastic, and in many other applications. (FSL)

Viroid The smallest known agent of infectious disease; a part of a virus. (SMC)

Virulence The degree of pathogenicity of an infectious agent. (FSL)

Virus A submicroscopic, noncellular particle, composed of a nucleic acid core and a protein shell; viruses reproduce only within host cells and are the cause of many diseases, including measles, mumps, poliomyelitis, and acquired immunodeficiency syndrome (AIDS). *See also* Acquired immunodeficiency syndrome. (SMC)

Visibility The quality or state of being perceivable by the eye; often defined in terms of the distance at which an object can be seen by the eye or in terms of the contrast or size of the object. (DEJ)

Visual field The locus of objects or points in space that can be perceived when the head and eyes are kept fixed (either monocular or binocular). (DEJ)

Visual performance The quantitative assessment of the performance of a task, taking into consideration speed and accuracy. (DEJ)

Vital statistics Data relating to morbidity, mortality, death, marriage, divorce, and illnesses in a given geographic area for a specified period of time. *See also* Rate. (FSL)

Void ratio The ratio of the volume of void space to the volume of solid particles in soil. (SWT)

Volatile Describes chemicals that tend to evaporate rapidly, such as ether, chloroform, and benzene. *See also* Solvent. (FSL)

Volatile organic compound (VOC) Any organic compound that is involved in atmospheric photochemical reactions, except for those designated by the EPA as having negligible photochemical reactivity. (FSL)

Volatile synthetic organic chemicals Synthetic chemicals that tend to volatilize or evaporate. *See also* Solvent. (FSL)

Volatility The measure of the tendency of a chemical to vaporize or evaporate at ambient conditions. (FSL)

Vulnerability analysis Assessment of elements in the community that are susceptible to damage should a release of hazardous materials occur. (FSL)

Vulnerable zone An area over which the airborne concentration of a chemical involved in an accidental release could reach the level of concern. (FSL)

Walking and working surfaces Terms used by OSHA to categorize the area of safety that addresses the condition, design, and use of stairs, ramps, ladders, dock plates, gangplanks, and powered industrial platforms. OSHA has specific requirements in each case, including specifications for fall protection for work from heights. (SAN)

Walsh-Healey Public Contracts Act A federal law, passed by Congress in 1936, that established certain minimum occupational safety and health standards that must be observed by all contractors working on federal projects. The Walsh-Healey Act is often regarded as the precursor to the Occupational Safety and Health Act of 1970. *See also* Occupational Safety and Health Act. (DEJ)

Warfarin ($C_{19}H_{16}O_4$) A rodenticide, originally developed by the Wisconsin Alumni Research Foundation, that prevents the coagulation of blood in target animals such as rats and mice. Several doses must be ingested by the target pest over a period of a few days in order for it to die from internal hemorrhages. The active ingredient is coumarin or bis-hydroxycoumarin. *See also* Red squill; rodenticide; strychnine. (FSL)

Waste Unwanted materials remaining from a manufacturing process; refuse from places of human or animal habitation. *See also* Garbage; refuse. (FSL)

Waste codes EPA identifiers, consisting of one letter (D, F, P, U, or K) and three numbers, that must be indicated on hazardous-waste manifests; for example, D-001 refers to flammable liquids such as alcohol and indicates that the waste exhibits the characteristics of ignitability. *See also* Manifest. (FSL)

Wasteland Land with characteristics that are not suitable for, or capable of, producing materials or services of value. (SWT)

Waste load allocation The maximum load of pollutants each discharger of waste is allowed to release into a particular waterway. Discharge limits are usually required for each specific water quality criterion that is being, or is expected to be, violated. (FSL)

Waste minimization A program to reduce the volume or quantity and toxicity of hazardous waste by a variety of methods, such as altering production processes or recycling. It includes any source reduction or recycling activity that is undertaken by a generator and that results in the reduction of total volume or quantity of hazardous waste, the reduction of toxicity of hazardous waste, or both, as long as the reduction is consistent with the goal of minimizing present and future threats to human health and the environment. (DEW, FSL)

Waste reduction Reduction of the environmental and health hazards associated with waste, by means of a practice, other than dewatering, dilution, or evaporation, and including changes in production technology, materials, processes, operations, procedures, or use of in-process, in-line, or closed-loop recycling practices according to standard engineering practices, and that occurs without diluting or concentrating the waste before release, handling, storage, transport, treatment, or disposal. (FSL)

Waste stabilization pond Ponds, referred to as sewage lagoons or oxidation ponds, that have been used in various locations throughout the United States for the treatment of sewage. Treatment is accomplished by aerobic bacteria, which oxidize the organic carbon in the sewage to carbon dioxide and encourage the growth of algal cells that release oxygen. The oxygen is utilized by the bacteria and the cycle continues, causing degradation of the sewage in the process. This type of treatment has fallen into disfavor in recent years because of the high organic load in the effluent, which can have a deleterious impact on the receiving stream. *See also* Oxidation pond. (FSL)

Waste treatment plant A facility containing a series of tanks, screens, filters, and other devices by which pollutants are removed from water. (FSL)

Waste treatment stream The continuous movement of waste from generator to treater and disposer. (FSL)

Wastewater The spent or used water from individual homes, communities, farms, or industries that contains dissolved or suspended matter. *See also* Sewage. (FSL)

Wastewater operations and maintenance Actions taken after construction to assure that facilities constructed to treat wastewater will be properly operated, maintained, and managed to achieve efficiency levels and prescribed effluent levels in an optimum manner. (FSL)

Water activity The measurement of the availability of water for the growth of microorganisms. It is designated by A_w. The A_w of an organism's environment is defined as the ratio of the water vapor pressure of the menstruum (P) to that of pure water (P_0) at the same temperature, and is represented by the formula $A_w = P/P_0$. Microorganisms vary as to their water activity temperatures. Most bacteria, including pathogens, grow best at levels of A_w in the range of 0.995–0.980. The growth of most bacteria and fungi is confined to the A_w range above 0.90. Certain unusual microorganisms can survive an A_w level that approaches 0.68. The growth of most bacteria and fungi at levels above 0.90 is most critical to control in a food sanitation program in either food service or food processing. (HMB)

Water hardness Hard waters are generally considered to be those that require considerable amounts of soap to produce a lather or foam, and may also produce scale in pipes, heaters, or boilers. The hardness of water varies from place to place. In general, ground water is harder than surface water. The hardness of water reflects the nature of the geologic formations with which the water has been in contact, and is derived largely from contact with soil or rock formations. Hardness is caused by divalent metallic cations. These ions are capable of reacting with soap to form precipitates and with certain anions present in water to form scale. The main hardness-causing cations in order of relative importance are: calcium, magnesium, strontium, ferrous, and manganous ions. These cations are associated with several anions of varying abundance in natural waters, namely: HCO_3^-, SO_4^{2-}, Cl^-, and NO_3^-. With the advent of synthetic detergents, many of the disadvantages of hard water have been diminished. Nonetheless, hard water makes cleaning and, therefore, a balanced sanitation program more difficult to implement. *See also* Hard water; water softening; zeolite. (HMB)

Water pollution The presence in water of enough harmful or objectionable material to damage the water's quality. (FSL)

Water quality criteria Specific levels of water quality that, if reached, are expected to render a body of water suitable for its designated use. The criteria are based on specific levels of pollutants that would make water harmful if used for drinking, swimming, farming, fish production, or industrial processes. *See also* Clean Water Act. (FSL)

Water quality standards Under the Clean Water Act, standards that define the water quality goals for a water body or portion thereof, by designating the uses to be made of the water and by setting criteria necessary to protect the uses. States adopt water quality standards to protect public health or welfare, enhance the quality of water, and thus serve the purposes of the Clean Water Act. The standards should, wherever attainable, provide water quality for the protection and propagation of fish, shellfish, and wildlife, recreation in and on the water, and agricultural, industrial, and other purposes, including navigation. These standards serve the dual purpose of establishing the water quality goals for a specific water body and serving as the regulatory bases for the establishment of water-quality-based treatment controls and strategies beyond technology-based levels of treatment required under the Clean Water Act. 33 USC §1313, 40 CFR Part 131. *See also* Clean Water Act. (DEW)

Water-reactive Describes chemicals, such as sodium or potassium, that react with water to release a gas that may be flammable and/or may present a health hazard because of the toxic properties of the end product. (FSL)

Water rights The rights of individuals to use water for domestic, irrigation, or other purposes. Water rights vary from state to state. Some rights stem from ownership of the land bordering or overlying the source, whereas others are acquired by performance of certain acts required by law. The three basic types of water rights are as follows. **(1)** *Riparian:* Rights that are acquired together with title to the land bordering or overlying the source of water. **(2)** *Appropriative:* Rights that are acquired by following a specific legal procedure. **(3)** *Prescriptive:* Rights that are acquired by diverting and putting to use, for a period set by a statute, water to which other parties may or may not have prior claims. The procedure necessary to obtain prescriptive rights must conform with the conditions established by water rights laws of individual states. (FSL)

Watershed The area drained by a particular stream or river. Most water that falls within a watershed flows out (by surface or ground water) through the stream or river that drains the watershed. The characteristics of a watershed (vegetation type and age, percent cover, topography, etc.) strongly influence the characteristics of its stream or river (flow rate, constancy of flow, sedimentation, etc.). Watersheds are often used for ecosystem studies because the boundaries can be fairly and accurately delimited, inputs can be identified and measured, and, if the watershed is underlain by an impermeable layer, all outputs will be through evaporation or by the stream or river. (SMC)

Water softening A treatment process designed to remove, totally, or in part, hardness-producing ions. Two primary methods are used. In the lime–soda process, calcium and magnesium ions are precipitated as calcium carbonate and magnesium hydroxide, respectively, which are ultimately removed by sedimentation and filtration. In the second method, the base-exchange or zeolite method, calcium and magnesium ions are replaced by sodium ions from a zeolite. Zeolites are compounds that have the capability of exchanging ions as solutions are passed over them. The U.S. Geological

Survey has developed categories of hardness—e.g., water with less than 60 ppm of calcium or magnesium is considered soft water. *See also* Hard water; zeolite. (FSL)

Water solubility The maximum concentration that can result when a substance is dissolved in water. If a substance is water soluble, it can very readily disperse through the environment. (FSL)

Water supply system System for the collection, treatment, storage, and distribution of potable water from source to consumer. (DEW)

Water table The upper surface of ground water or that level in the ground where the water is at atmospheric pressure. (SWT)

Water table, perched The water table of a saturated layer of soil that is separated from an underlying saturated layer by an unsaturated layer. *See also* Aquifer. (SWT)

Water treatment *See* Primary wastewater treatment; secondary wastewater treatment. (FSL)

Water treatment disinfection The chemical process used to destroy pathogenic bacteria and other microorganisms present in water supplies. If all organisms are to be destroyed, the water to which the disinfectant is added must be low in turbidity. The desirable qualities of a chemical disinfectant are high germicidal power, stability, solubility, lack of toxicity to people and animals, economy, dependability, residual effect, and availability. Certain chlorine compounds meet these criteria and as a result are the most commonly used disinfectants in water treatment. The following terms are applicable. **(1)** *Chlorine feed (dosage):* The actual amount fed into the water system. **(2)** *Chlorine concentration:* Expressed in milligrams chlorine per liter of water. **(3)** *Chlorine demand:* The portion of the chlorine that combines with impurities in the water and therefore is not available for disinfectant action. **(4)** *Free and combined chlorine:* Chlorine can link with ammonia and nitrogen to form compounds that have some biocidal properties. These compounds are defined as *combined*. If no ammonia is present in the water, the chlorine that remains once the demand has been satisfied is the chlorine residual. **(5)** *Chlorine contact time:* The period of time that lapses between the addition of the chlorine and the use of the water. *See also* Chlorine disinfectants; chlorine disinfection. (FSL)

Weathering The physical and chemical changes that occur in rocks because of atmospheric agents. (SWT)

Web of causation The complex set of interrelationships of causal factors that, for example, affect the transmission of a foodborne disease. A ''web'' flowchart can be developed for each foodborne disease organism and each food. (HMB)

Weed A plant considered to be growing in the wrong place or thought to be valueless, troublesome, or noxious. (FSL)

Weir A device used to measure stream flow rate and volume. A weir resembles a dam with a notched top. Weirs also allow the collection of stream sediments and are often used in stream ecology and watershed studies. (SMC)

Welding curtains Barriers that are generally yellow and that attenuate the ultraviolet radiation given off by welding arcs and protect bystanders in a shop from flash burn. *See also* Flash burn. (SAN)

Well An excavation that is drilled, driven, dug, bored, or otherwise constructed for the purpose of locating, acquiring, or artificially recharging ground water or extracting oil or natural gas. (FSL)

Well injection The subsurface emplacement of fluids in a well. (FSL)

Well monitoring The measurement, by on-site instruments or laboratory methods, of the quality of water in a well. (FSL)

Well plug A watertight and gastight seal installed in a bore hole or well to prevent movement of liquids. (FSL)

Well seal An approved arrangement or device used to cap a well or to establish and maintain a junction between the casing or curbing of a well and the piping or equipment installed therein, the purpose or function of which is to prevent pollutants from entering the well at the upper terminal. (FSL)

Wet collector A dust collector operating through impaction of the dust onto liquid droplets, which are then separated by centrifugal force, through contact beds, through sheets of water, or through a venturi-shaped constriction. (DEJ)

Wetland **(1)** An area that is partially or completely saturated or occasionally inundated with water. Wetlands often have characteristic or unique communities or assemblages of organisms (e.g., pitcher plants, mosquitoes) that are adapted to the very moist conditions. Many wetlands filter water, moderate climate, and moderate the flow of water through nearby terrestrial and aquatic systems. Swamps, marshes, bogs, fens, estuaries, intertidal mud flats, and river deltas are examples of wetland ecosystems. **(2)** For purposes of the Clean Water Act, areas that are inundated or saturated by surface or ground water at a frequency or duration sufficient to support, and that under normal circumstances do support, a prevalence of vegetation typically adapted for life in saturated soil conditions. 33 CFR §328.3(b), 40 CFR §122.2. *See also* Clean Water Act. (SMC, DEW)

Wetting agent A chemical or a synthetic detergent that serves essentially the same function as soap: emulsifying fats and oils. Wetting agents have been described as excellent surface-active agents because their addition lowers the surface tension of the solution, promotes wetting of particles, and deflocculates and suspends soil particles. They are used in food sanitation as components in cleaning compounds and also serve as bactericides. Wetting agents may be divided into three major categories: cationic wetting agents, anionic wetting agents, and nonionic wetting agents. *See also* Anionic wetting agents; cationic wetting agents; nonionic wetting agents; quaternary ammonium compound. (HMB)

What if A hazard analysis technique used in process safety. An entire process is reviewed from the beginning (raw materials) to the end (product). At each handling or processing step, "what if" questions are formulated and answered to evaluate the effects of component failures or procedural errors in the process. Once the consequences of the answer to a particular question are determined, a discussion of the hazard follows, with emphasis on identifying process modifications to reduce or eliminate the potential hazard. *See also* Event-tree analysis; hazard analysis; failure mode and effects analysis; fault-tree analysis. (SAN)

Wheatstone bridge An electrical circuit commonly used in combustible-gas meters and other instrumentation. The air to be sampled is passed over a heated element, which causes combustion and additional heat release if a combustible gas is present. Another sealed part of the circuit contains identical heated elements that are not exposed to the air. The two portions of the circuit are connected so that the net effect or "imbalance" is related solely to the presence of combustible gas. *See also* Combustible-gas meter. (DEJ)

Whistleblowers Individuals who complain to official agencies about working conditions or practices related to safety, health, environmental problems, or illegal practices of any nature. Both OSHA and EPA have regulations that forbid reprisals against employees who report unsatisfactory situations. *See also* Worker complaint system. (SAN)

Wilcoxon rank sum test A nonparametric alternative to the two-sample t test for independent data samples. In the Wilcoxon test, the actual values are replaced by rank scores. (VWP)

Wildlife refuge An area designated for the protection of wild animals, within which hunting and fishing are either prohibited or strictly controlled. (FSL)

Willful violation In OSHA proceedings, a defiance or such reckless disregard of consequences as to be equivalent to a knowing, conscious, and deliberate flaunting of the Occupational Safety and Health Act or the regulations promulgated thereunder. An obstinate refusal to comply; or a repeated instance of an earlier violation. *See also* Occupational Safety and Health Act; serious violation. (DEJ)

Williams-Steiger Occupational Safety and Health Act *See* Occupational Safety and Health Act. (DEJ)

Windbreak Trees, shrubs, or other vegetation that are planted to protect against the effects of wind, such as erosion and snow or soil drift. Windbreaks are usually perpendicular to prevailing winds. (SWT)

Winter annual A herbaceous annual plant that grows through the winter and produces flowers and fruit in late winter and early spring. (SMC)

Wood-burning-stove pollution Air pollution caused by emissions of particulate matter, carbon monoxide, suspended particulates, and polycyclic organic matter from wood-burning stoves. (FSL)

Wood preservative A chemical, such as pentachlorophenol (C_6Cl_5OH), that is used to prevent the deterioration of telephone poles, fence posts, and pilings by invasive fungi or insects such as termites or powder post beetles. (FSL)

Worker complaint system A system that exists for the purpose of allowing employees to report hazards without fear of reprisal and to notify management of conditions that appear hazardous. Procedures and policies for handling worker complaints and subsequent allegations of reprisal must be developed and implemented. Employees should know procedures for reporting hazards and should receive appropriate and timely responses. Employee complaints must be evaluated by management to further control hazards and to improve the safety and health program as appropriate. Employees must be aware that they are protected from reprisal for exercising their rights. Allegations of reprisal must be investigated promptly and disposed of properly. *See also* Whistleblowers. (SAN)

Worker's compensation Mandatory insurance coverage provided to all workers in the event of occupational injury or illness that requires medical care (outside of that provided at the work site) and may involve loss of work time. The payment for disabilities that prevent the worker from returning to work is generally a percentage of the worker's pay. Disabilities are categorized as permanent partial, permanent total, temporary partial, and temporary total. *See also* Disability. (SAN)

Working level (WL) A unit of measure for documenting exposure to radon decay products. One working level is equal to approximately 200 picocuries per liter. *See also* Radon. (FSL)

Working Level Month (WLM) A unit of measure used to determine cumulative exposure to radon. *See also* Radon. (FSL)

Workplace environmental exposure limits (WEEL) Recommended exposure limits to certain chemicals, as developed by the American Industrial Hygiene Association. *See also* Occupational exposure limits; PEL; TLV. (DEJ)

Workplace Hazardous Materials Information System (WHIS) System mandated by Canadian legislation to provide employees, employers, and suppliers nationwide with specific vital information about hazardous materials. There are key requirements in the legislation, including: controlled product labeling, which alerts workers to the identity and dangers of products and to basic safety precautions; material safety data sheets (MSDS), technical bulletins that provide detailed hazard and precautionary information; and worker education and training programs, which inform employees of potential chemical hazards. (FSL)

Xenobiotic Describes nonnaturally occurring synthetic materials found in the environment (e.g., synthetic material, industrial or organic solvents, plastics). (FSL)

Xenopsylla cheopis The rat flea that transmits the organism *Yersina pestis*, the etiologic agent of plague. (FSL)

Xeric Dry; describes systems in which water availability is limited. A soil moisture regime common to Mediterranean climates, which have moist cool winters and warm dry summers. A limited amount of moisture is present but does not occur at optimum periods for plant growth. Irrigation or summer fallow is commonly necessary for crop production. (SMC)

Xerophytes Plants that are able to grow in or on extremely dry soils or soil materials. (SWT)

Xylem One of the two kinds of vascular tissues in plants. Xylem conducts water and most minerals. The fluids carried by xylem and phloem are commonly called sap. *See also* Phloem; sap. (SMC)

Xylene [$C_6H_4(CH_3)_2$] A commercial mixture of three isomers, also known as dimethylbenzene. Xylene is a clear liquid that is soluble in alcohol and ether and insoluble in water. It is flammable and toxic by ingestion and inhalation. TLV is 100 ppm in air. It is used in protective coatings, lacquers, enamels, and the synthesis of organic chemicals. It is an important laboratory chemical in that it is used in the preparation of slides of pathological or histological material. *See also* Solvent. (FSL)

Y

Yard trimmings Vegetative matter resulting from landscaping maintenance and land-clearing operations other than mining, agricultural, and silvacultural operations. *See also* Inert-waste landfill. (FSL)

Yersiniosis An illness caused by the bacterium *Yersinia enterocolitica*. The organism, a member of the family Enterobacteriaceae, is a non-spore-forming, nonmotile, facultatively anaerobic, Gram-negative straight rod or coccobacillus. There is a geographic partitioning of *Yersinia* serotypes, which may be the cause of a greater incidence of yersiniosis in European and Scandinavian countries than in the U.S. The symptoms of the disease are those of enterocolitis and include diarrhea, fever, and abdominal pain, especially in the lower right quadrant and often misdiagnosed as acute appendicitis. The incubation period for gastroenteritis may range from 3 to 7 days. The organism produces a heat-stable enterotoxin that is not formed at temperatures above 25°C. *Yersinia* is a psychrotroph and grows at refrigeration temperatures, making conventional cold storage an ineffective means of control. The organism is widely distributed in nature, with many animal reservoirs. Swine are implicated as a major reservoir for the serotypes associated with human yersiniosis. Foods that have been implicated in outbreaks in the United States include raw milk, tofu, turkey, and pasteurized milk and milk products. (HMB)

Yield The amount of a specified substance produced—e.g., crop yield. (SWT)

Z

Zeolite A natural hydrated silicate of aluminum and either sodium or calcium or both, of the type $Na_2O \cdot Al_2O_3 \cdot (SiO_2)_x \cdot (H_2O)_x$. Both natural and artificial zeolites are used extensively for water softening. When the zeolite in a water-softening unit becomes saturated with calcium or magnesium ions, it is flooded with a concentrated salt solution, and a reversal of cations takes place and the material is "recharged." Zeolites are also used as adsorbents and desiccants, and in solar collectors. *See also* Water softening. (FSL)

Zero point of charge The pH value for a solution in equilibrium with a particle whose net charge from all sources is zero. (SWT)

Zinc (Zn) A hard, blue-white metallic element with an atomic weight 65.37 and an atomic number of 30. It is used in alloys, particularly brass, and in the galvanization of iron. (FSL)

Zineb [$Zn(CS_2NHCH_2)_2$] A carbamate fungicide used for the control of blight and mildews on crops. It is toxic by inhalation and ingestion and an irritant to eyes and mucous membranes. (FSL)

Zoology The study of animals (Kingdom Animalia). (SMC)

Zoonosis An infection or an infectious disease transmissible under natural conditions between vertebrate animals and humans. In many cases, transmission of foodborne pathogens occurs indirectly from food animals or animal products. *See also* Animal products. (HMB, FSL)

Zooplankton Heterotrophic plankton; plankton that consume other organisms (usually phytoplankton); "animallike" plankton. *See also* Plankton. (SMC)

Z-score The position a particular data value has in terms of the number of standard deviations it is from the mean of the set of data to which it is being compared. The relationship is given by

$$z = \frac{x - \bar{x}}{s}$$

where x is a data value, $\bar{x}$ is the sample mean, and s is the sample standard deviation. (VWP)

Z value Heat resistance of a bacterium at a given temperature is represented by its D value. Points on a thermal-death-time curve indicate the relative resistance at different temperatures, thus permitting calculation of effects of differing temperatures. The slope of the curve is the Z value, and is the number of degrees Celsius required for the thermal-death-time curve to traverse one log cycle. Z values and D values are used to determine exact heat-processing times for foods. *See also* D value; thermal death time. (HMB)

Zymogenous flora Organisms that flourish in soils immediately after the addition of readily decomposable organic materials. (SWT)

Appendix I. Acronyms

AA. Atomic absorption

AAALAC. American Association for Accreditation of Laboratory Animal Care

AAAS. American Association for the Advancement of Science

AAEE. American Academy of Environmental Engineers

AAIH. American Academy of Industrial Hygienists

AALAS. American Association for Laboratory Animal Science

ABES. Alliance for Balanced Environmental Solutions

ABIH. American Board of Industrial Hygiene

ABSA. American Biological Safety Association

ACBM. Asbestos-containing building material

ACGIH. American Conference of Governmental Industrial Hygienists

ACLAM. American College of Laboratory Animal Medicine

ACM. Asbestos-containing material

ACS. American Chemical Society

ACSH. American Council on Science and Health

ADA. American Dental Association

ADI. Acceptable daily intake

AEA. Atomic Energy Act

AEC. Atomic Energy Commission

AERE. Association of Environmental and Resource Economists

AFB. Acid-fast bacillus

AFFF. Aqueous film-forming foam

AHERA. Asbestos Hazard Emergency Response Act (1986)

AICE. American Institute of Chemical Engineers

AIDS. Acquired immunodeficiency syndrome

AIHA. American Industrial Hygiene Association

AIHC. American Industrial Health Council

AIHF. American Industrial Hygiene Foundation

ALD. Average lethal dose

ALJ. Administrative law judge

AMA. American Medical Association

AMCA. Air Moving and Conditioning Association

ANPR. Advanced notice of proposed rulemaking under the Administrative Procedure Act

ANS. Autonomic nervous system

ANSI. American National Standards Institute

APA. Administrative Procedure Act

APCA. Air Pollution Control Association

APHA. American Public Health Association

APhA. American Pharmaceutical Association

APHIS. Animal and Plant Health Inspection Service (USDA)

ARARs. Applicable or relevant and appropriate requirements

ARCS. Alternative remedial contracting system

ASHAA. Asbestos in Schools Hazard Abatement Act

ASHRAE. American Society of Heating, Refrigeration, and Air Conditioning Engineers

ASLAP. American Society of Laboratory Animal Practitioners

ASM. American Society of Microbiology

ASSE. American Society of Safety Engineers

ASTM. American Society for Testing and Materials

ASTSWMO. Association of State and Territorial Solid Waste Management Officials

ATSDR. Agency for Toxic Substances and Disease Registry

AVMA. American Veterinary Medical Association

AWWARF. American Water Works Association Research Foundation

BACT. Best available control technology

BAT. Best available technology

BATF. Bureau of Alcohol, Tobacco, and Firearms

BCSP. Board of Certified Safety Professionals

BCT. Best conventional pollutant control technology

BDAT. Best demonstrated available technology

BEI. Biological exposure index

BLEVE. Boiling-liquid–expanding-vapor explosion

BLM. Bureau of Land Management

BLS. Bureau of Labor Statistics

BMP. Best management practices

BMR. Basal metabolic rate

BOD. Biochemical oxygen demand

BOM. Bureau of Mines

BP. Boiling point

BPJ. Best professional judgment

BPT. Best practicable technology; best practicable treatment

BSC. Biological safety cabinet

Btu. British thermal unit

BTZ. Below the treatment zone

C. Celsius; Centigrade

CA. Corrective action; on material safety data sheets: ''approximately''

CAA. Clean Air Act (1955, 1977)

CAAA. Clean Air Act Amendments

CAALR. Canadian Association of Administrators of Labor Relations

CAD. Computer-assisted design

CAER. Community awareness and emergency response

CAFE. Corporate average fuel economy standard

CAG. Carcinogenic assessment group

CAM. Computer-assisted manufacturing

CAS. Chemical abstracts service

CBA. Cost–benefit analysis

CCHW. Citizens Clearinghouse For Hazardous Wastes

CDC. Centers for Disease Control and Prevention (DHHS)

CEC. Council of European Communities

CELRF. Canadian Environmental Law Research Foundation

CEQ. Council on Environmental Quality

CERCLA. Comprehensive Environmental Responsibility, Compensation, and Liability Act (1980) (Superfund Act)

CERCLIS. Comprehensive Environmental Responsibility, Compensation, and Liability Information System

CERI. Center for Environmental Research Information

CFC. Chlorofluorocarbon

CFM. Cubic feet per minute

CFR. Code of Federal Regulations

CGL. Comprehensive general liability

CHEMTREC. Chemical transportation emergency center

CHESS. Community health and environmental surveillance system

CHMM. Certified hazardous-materials manager

CHP. Certified health physicist

CIH. Certified industrial hygienist

CITIES. Convention on International Trade in Endangered Species of Wild Fauna and Flora of 1973

CNS. Central nervous system

COCO. Contractor-owned/contractor-operated facility

COD. Chemical oxygen demand

COE. U.S. Army Corps of Engineers

COH. Coefficient of haze

COHN. Certified occupational health nurse

CPR. Cardiopulmonary resuscitation

CPSC. Consumer Product Safety Commission

CRAVE. Carcinogen Risk Assessment Verification Endeavor

CSF. Cerebrospinal fluid

CSP. Certified safety professional

CWA. Clean Water Act (1972)

CWTC. Chemical Waste Transportation Council

DASHO. Designated agency safety and health official

DCS. Decompression sickness

DDT. Dichlorodiphenyltrichloroethane

DMSO. Dimethylsulfoxide

DNA. Deoxyribonucleic acid

DOD. Department of Defense

DOE. Department of Energy

DOI. Department of Interior

DOJ. Department of Justice

DOL. Department of Labor

DOS. Department of State; in personal computers: disk operating system

DOT. Department of Transportation

DRE. Destruction or Removal Efficiency

DWS. Drinking-water standard

EA. Endangerment assessment; enforcement agreement; environmental action; environment assessment (NEPA); environmental audit

ECD. Electron capture detector

EDA. Emergency declaration area

EDF. Environmental Defense Fund

EDZ. Environment-driven zone

EIS. Environmental impact statement (NEPA)

EL. Exposure level

ELI. Environmental Law Institute

EM. Electron microscope

EMC. Emergency response commission

EMP. Emergency response policy

EPA. Environmental Protection Agency

EPAA. Environmental Programs Assistance Act

EPD. Emergency planning district

EPI. Environmental Policy Institute

EP Test. Extraction procedure test

ERT. Emergency response team

ERV. Expiratory reserve volume

ESA. Endangered Species Act; environmentally sensitive area

ESH. Environmental safety and health

ET. Emissions testing

F. Fahrenheit (degrees)

FACM. Friable asbestos-containing material

FACOSH. Federal Advisory Council for Occupational Safety and Health

FAM. Friable asbestos material

FAO. Food and Agriculture Organization (United Nations)

FDA. Food and Drug Administration (DHHS)

FDCA. Food, Drug, and Cosmetic Act

FEMA. Federal Emergency Management Agency

FEMA-REP-1. Response Plans and Preparedness in Support of Nuclear Power Plants

FEMA-REP-2. Guidance for Developing State and Local Radiological Emergency Response Plan and Preparedness for Transportation Actions

FEPCA. Federal Energy Policy and Conservation Act

FERC. Federal Energy Regulatory Commission

FFDCA. Federal Food, Drug, and Cosmetic Act

FIFRA. Federal Insecticide, Fungicide, and Rodenticide Act

FLPMA. Federal Land Policy and Management Act

FMC. Federal Maritime Commission

FOIA. Freedom of Information Act

FONSI. Finding of no significant impact

FPA. Federal Pesticide Act

FPD. Flame phosphorus detector

FR. Federal Register

FTC. Federal Trade Commission

FUA. Fuel Use Act

FWCA. Fish and Wildlife Coordination Act

FWPCA. Federal Water Pollution Control Act

FWS. Fish and Wildlife Service

GC/MS. Gas chromatograph/mass spectrograph

GEMS. Global environmental monitoring system

GLC. Gas–liquid chromatography

GLP. Good laboratory practice

GMP. Good manufacturing practice

GOCO. Government-owned/contractor-operated

GOGO. Government-owned/government-operated

GOPO. Government-owned/privately-operated

GRAS. Generally regarded as safe

HACCP. Hazard Analysis Critical Control Point

HAV. Hepatitis A virus

HAZMAT. Hazardous materials response team

HAZWOPER. Hazardous Waste Operations and Emergency Response (OSHA Standard, 29 CFR 1910.120)

HBV. Hepatitis B virus

HCFA. Health Care Finance Administration (DHHS)

HCS. Hazard communication standard

HEPA. High-efficiency particulate air (filter)

HHS. Department of Health and Human Services (formerly HEW)

HIV. Human immunodeficiency virus

HMCRI. Hazardous Materials Control Research Institute

HMIS. Hazardous Materials Information System

HMTA. Hazardous Materials Transportation Act

HPLC. High-performance liquid chromatography; high-pressure liquid chromatography

HPS. Health Physics Society

HRS. Hazard ranking system

HSWA. Hazardous and Solid Waste Amendments

HVAC. Heating, ventilation, and air conditioning (systems)

IACUC. Institutional Animal Care and Use Committee

IAG. Interagency agreement

IARC. International Association of Cancer Registries; International Agency for Research on Cancer

IBC. Institutional biosafety committee

IBRD. International Bank for Reconstruction and Development

ICAD. International Civil Aviation Organization

ICP, ICPES, ICP-OES. Inductively coupled plasma emission spectroscopy

ICRP. International Commission on Radiological Protection

ICRU. International Commission on Radiological Units and Measurements

ICS. Incident command system

IDLH. Immediately dangerous to life or health

IHMM. Institute of Hazardous Materials Management

IME. Institute of Makers of Explosives

IMO. International Maritime Organization

IPM. Integrated pest management

IR. Infrared spectroscopy

IRV. Inspiratory reserve volume

ITC. Interagency testing committee

I.U. International units (vitamins)

JCAH. Joint Commission on the Accreditation of Hospitals

JCAHCF. Joint Commission for Accreditation of Health Care Facilitators

LC_{50}. Lethal concentration at which 50% of test animals died

LCL. Lower confidence limit

LD_{50}. Lethal dose at which 50% of test animals died

LEL. Lower explosive limit

LEPC. Local emergency planning committee

LET. Linear energy transfer

LFL. Lower flammable limit

LLRW. Low-level radioactive waste

LLRWPA. Low-Level Radioactive Waste Policy Act

LOC. Level of concern

LPG. Liquefied petroleum gas

LUST. Leaking underground storage tank

MAC. Maximum allowable concentration

MCA. Manufacturing Chemists Association

MCL. Maximum containment level; maximum containment laboratory

MCLG. Maximum contaminant level goal

MEDLARS. Medical Literature Analysis and Retrieval System

MGD. Million gallons per day

MLD. Mean lethal dose; median lethal dose

MLSS. Mixed liquor suspended solids

MLVSS. Mixed liquor volatile suspended solids

MMWR. Morbidity and Mortality Weekly Report

MOA. Memorandum of agreement

MOU. Memorandum of understanding

MP. Melting point

MPC. Maximum permissible concentration

MPD. Maximum permissible dose

MPE. Maximum permissible exposure

MSDS. Material safety data sheets

MSHA. Mine Safety and Health Administration

MW. Molecular weight

NAAQS. National ambient air-quality standards

NAC. National Academy of Science

NACA. National Agricultural Chemicals Association

NAMRU. Naval Medical Research Unit (DOD)

NATICH. National Air Toxics Information Clearinghouse

NCCLS. National Committee for Clinical Laboratory Standards

NCHS. National Center for Health Statistics

NCI. National Cancer Institute (NIH)

NCP. National Contingency Plan

NCSBCS. National Conference of States on Building Codes and Standards

NCTR. National Center for Toxicological Research

NEC. National Electrical Code

NEHA. National Environmental Health Association

NEPA. National Environmental Policy Act

NESHAPS. National Emission Standard for Hazardous Air Pollutants

NF. National Formulary

NFPA. National Fire Protection Association

NFWS. National Fish and Wildlife Service (Department of the Interior)

NHTSA. National Highway Transportation Safety Administration

NIBS. National Institute of Building Sciences

NIEHS. National Institute of Environmental Health Sciences

NIH. National Institutes of Health (DHHS)

NIMBY. Not in my backyard

NIOSH. National Institute for Occupational Safety and Health

NIST. National Institute of Standards and Technology

NLM. National Library of Medicine

NOAA. National Oceanic and Atmospheric Administration

NOAEL. No observable adverse effect level

NOEL. No observable effect level

NOHSCP. National Oil and Hazardous Substances Contingency Plan

NPD. Nitrogen phosphorus detector

NPDES. National pollutant discharge elimination system

NPIRS. National Pesticide Information Retrieval System

NPL. National priorities list

NPRM. Notice of proposed rulemaking

NRA. Nuclear Regulatory Agency

NRC. National Response Center

NRC. Nuclear Regulatory Commission

NRDC. Natural Resources Defense Council

NRT. National response team

NSC. National Safety Council

NSF. National Sanitation Foundation; National Science Foundation

NSPS. New-source performance standard

NTP. National Toxicology Program

NTSB. National Transportation Safety Board

OD. Optical density

OFAP. Office of Federal Agency Programs

OMB. Office of Management and Budget

OPRR. Office of Protection from Research Risks (NIH)

ORDA. Office of Recombinant DNA Activities (NIH)

OSC. On-scene coordinator

OSHA. Occupational Safety and Health Act; Occupational Safety and Health Administration

OSHRC. Occupational Safety and Health Review Commission

PAHO. Pan-American Health Organization

PAPR. Powered air-purifying respirator

PCB. Polychlorinated biphenyl

PCC. Poison Control Center

PCP. *Pneumocystis pneumoniae* pneumonia; the illicit drug Phencyclidine (phenylcyclohexylpiperidine)

PDR. Physicians Desk Reference

PEL. Permissible exposure limit

PHS. Public Health Service

PIC. Partially incinerated compounds

PLIRRA. Pollution Liability Insurance and Risk Retention Act

PMA. Pharmaceutical Manufacturers Association

POE. Point of exposure

POGO. Privately-owned/government-operated

POTW. Publicly owned treatment works

PPB. Parts per billion

PPE. Personal protective equipment

PPM. Parts per million

PPT. Parts per trillion

PRP. Potentially responsible party

PZ. Prescriptive zone

RA. Remedial action

RAC. Recombinant Advisory Committee (NIH)

RACT. Reasonably available control technology

RAD. Radiation absorbed dose

RBE. Relative biological effectiveness

RCG. Radioactivity concentration guide

RCRA. Resource Conservation and Recovery Act

rDNA. Recombinant DNA (deoxyribonucleic acid)

REL. Recommended exposure limit

REM. Roentgen equivalent man (or mammal)

REP. Roentgen equivalent physical

RFD. Reference dose

RI/FS. Remedial investigation/feasibility study

RIFS. Remedial Investigation and Feasibility Study

RNA. Ribonucleic acid

RPAR. Rebuttal presumption against registration

RPG. Radiation protection guide

ROD. Record of decision under CERCLA

RPG. Radioactive protection guide

RPM. Remedial project manager

RQ. Reportable quantity

RS. Registered sanitarian

RSPA. Research and Special Programs Administrator (Department of Transportation)

RTECS. Registry of Toxic Effects of Chemical Substances

SAR. Sodium adsorption ratio

SARA. Superfund Amendments and Reauthorization Act of 1986

SCBA. Self-Contained breathing apparatus

SCUBA. Self-contained underwater breathing apparatus

SDWA. Safe Drinking Water Act

SENIC. Study of the Efficacy of Nosocomial Infection Control

SERC. State Emergency Response Commission

SG. Specific gravity

SHPDA. State health planning and development agency

SIP. State implementation plan

SMERA. Surface Mining Control and Reclamation Act of 1977

SMR. Standardized mortality ratio

SNUR. Significant-new-use rule

SPCC. Spill prevention control and countermeasures plan

SQG. Small-quantity generator

SSHO. Site safety and health officer

STEL. Short-term exposure limit

SUS. Saybolt universal seconds

SWDA. Solid Waste Disposal Act

SWMU. Solid-waste management unit

TAG. Technical assistance grant

TCDD. Tetrachlorodibenzodioxin

TCDF. Tetrachlorodibenzofuran

TCRI. Toxic chemical release inventory

TD_1. Tumor dose for 1% of exposed groups

TD_{50}. Tumor dose for 50% of exposed groups

THM. Trihalomethane

TLV. Threshold limit value

TRO. Temporary restraining order

TSCA. Toxic Substances Control Act

TSD. Treatment, storage, or disposal facility

TSS. Total suspended solids (under the CWA)

TWA. Time-weighted average

UAQI. Uniform Air Quality Index

UCL. Upper confidence limit

UEL. Upper explosive limit

UFL. Upper flammable limit

UL. Upper limit

UNHCR. United Nations High Commissioner for Refugees

UNICEF. United Nations International Children's Emergency Fund

USAID. United States Agency for International Development (Department of State)

USAMRIID. United States Army Medical Research Institute for Infectious Diseases (DOD)

USC. United States Code

USCG. United States Coast Guard

USDA. United States Department of Agriculture

USFS. United States Forest Service

USGS. United States Geological Survey

USP. United States Pharmacopeia

UST. Underground storage tank

USTR. United States Transuranic Registry

VDT. Video display terminal

VOC. Volatile organic compound

VP. Vapor pressure

WEEL. Workplace environmental exposure limits

WHO. World Health Organization

WPCF. Water Pollution Control Federation

WRC. Water Resources Council

WWEMA. Waste and Wastewater Equipment Manufacturers' Association

Appendix II. Chemical Abstracts Registry Numbers

Abscisic acid	Not listed	Benzene	71-43-2
Acetic acid	64-19-7	Benzopyrene	50-32-8
Acetone	67-64-1	Beryllium	7440-41-7
Acetylene	74-86-2	Bromine	7726-95-6
Activated alumina	Not listed	Cadmium	7440-43-9
Activated carbon	Not listed	Caffeine	58-08-2
Agar	Not listed	Calcium hydroxide	1305-62-0
Aldehydes	Not listed	Calcium oxide	1305-78-8
Aldrin	309-00-2	Cannabis	8063-14-7
Alkaloid	Not listed	Carbamate	Not listed
Ammonia	7664-41-7	Carbon dioxide	124-38-9
Anthracite	Not listed	Carbon monoxide	630-08-0
Antimony	7440-36-0	Carbon tetrachloride	56-23-5
Arsenic	7440-38-2	Cesium	7440-46-2
Asbestos	1332-21-4	Cetalkonium chloride	Not listed
Asphalt	8052-42-4	Chloramine	Not listed
Atrazine	Not listed	Chlordane	57-74-9
Barium	7440-39-3	Chlorine	7782-50-5

Chromic acid	7738-94-5	Graphite	Not listed
Chromium	7440-47-3	Graticle	7782-42-5
Coal tar	8007-45-2	Hexachlorophene	70-30-4
Cobalt	7440-48-4	Hydrazine	302-01-2
Coumarin	91-64-5	Hydrofluoric acid	7664-39-3
Creosote	Not listed	Hydrogen	1333-74-0
Cyanide	74-90-8	Hydrogen cyanide	74-90-8
Deoxyribonucleic acid	Not listed	Hydrogen peroxide	7722-84-1
Deuterium	Not listed	Hydrogen sulfide	7783-06-4
Dichlorodiphenyltrichloroethane (DDT)	50-29-3	Hypochlorous acid	7790-92-3
Dichlorophenoxyacetic acid (2,4-D)	94-75-7	Iodine	7553-56-2
Dichlorvos	62-73-7	Iron	7439-89-6
Dicumerol	Not listed	Iron pyrite	1309-36-0
Dieldrin	60-57-1	Kepone	143-50-0
Diethyl-m-toluamide	134-62-3	Kerosene	8008-20-6
Diethylstilbestrol	56-53-1	Lead	7439-92-1
Dimethoate	60-51-5	Lithium	7439-93-2
Dimethylsulfoxide	67-68-5	Lysergic acid diethylamide	50-37-3
Dinocap	6119-92-2	Maleic hydrazine	108-31-6
Diquat	85-00-7	Maneb	12427-38-2
Disulfiram	97-77-8	Manganese	7439-97-6
Diuron	330-54-1	Mercaptan	Not listed
Endrin	72-20-8	Mercury	7439-97-6
Ether	Not listed	Metaldehyde	9002-91-9
Ethyl acrylate	140-88-5	Methane	74-82-8
Ethylenediaminetetraacetic acid	64-02-8	Methanol	67-56-1
Ethylene oxide	75-21-8	Molybdenum	7439-98-7
Fluorine	7782-41-4	Nickel carbonyl	13463-39-3
Freon	Not listed	Nicotine	54-11-5
Fulvic acid	Not listed	Nitric acid	7697-37-2
Gold	7440-57-5	Nitrous oxide	10024-97-2

Oleic acid	112-80-1	Rosaniline	10043-92-2
Orthotolidine	119-93-7	Rotenone	83-79-4
Oxalic acid	144-62-7	Sarin	107-44-8
Oxygen	7782-44-7	Sodium	7440-23-05
Ozone	10028-15-6	Sodium chlorate	7775-09-9
Palmitic acid	57-10-3	Sodium hydroxide	Not listed
Paraldehyde	123-63-7	Sodium monofluoroacetate (1080)	Not listed
Paraquat	4685-14-7	Sodium nitrate	7631-99-4
Parathion	4685-14-7	Sodium sulfate	7757-82-6
Peracetic acid	79-21-0	Sodium sulfite	7757-83-7
Phenol	108-95-2	Soman	96-64-0
Phosgene	75-44-5	Strontium	7440-24-6
Phosphoric acid	7664-38-2	Strychnine	57-24-9
Phosphorus	4423-14-0	Sulfur	7704-34-9
Picric acid	88-89-1	Sulfur dioxide	7749-09-5
Pitchblende	13117-99-3	Sulfuric acid	7664-93-9
Plutonium	7440-07-5	Sulfur trioxide	7449-11-9
Polybrominated biphenyl	36355-01-8	Tabun	77-81-6
Polycarbonate	Not listed	Tetradotoxin	Not listed
Polychlorinated biphenyl	1336-36-3	Thallium	7740-28-0
Polyvinyl chloride	9002-86-2	Trichloroethylene	79-01-6
Potassium	7440-09-7	Vinyl chloride	75-01-4
Propylene oxide	75-56-9	Warfarin	Not listed
Pyridine	110-86-1	Xylene	106-42-3
Pyrite	1309-36-0	Zineb	12122-67-7
Radon	10043-92-2		

References

Ackerman, V. W. E. *Essentials of Human Physiology*. St. Louis, MO: Mosby Year Book, Inc., 1992.

Administrative Conference of the United States, Office of the Chairman. *Federal Administrative Handbook*. Washington, D.C.: Government Printing Office, 1985.

Agency for Toxic Substances and Disease Registry, Public Health Service, U.S. Department of Health and Human Services. *Managing Hazardous Materials Incidents, Vol. I: Emergency Medical Systems: A Planning Guide for the Management of Contaminated Patients*. Atlanta, GA: Division of Health Assessment and Consultation, Emergency Response and Consultation Branch (E57), 1992.

Agency for Toxic Substances and Disease Registry, Public Health Service, U.S. Department of Health and Human Services. *Managing Hazardous Materials Incidents, Vol. II: Hospital Emergency Departments: A Planning Guide for the Management of Contaminated Patients*. Atlanta, GA: Division of Health Assessment and Consultation, Emergency Response and Consultation Branch (E57), 1992.

Albert, Adrien. *Selective Toxicity and Related Topics*, 4th ed. London, England: Methuen & Company, Ltd., 1968.

Alexander, M. *Soil Microbiology*, 2nd ed. New York: John Wiley, 1977.

Allaby, Michael. *A Dictionary of the Environment*. New York: Van Nostrand Reinhold, 1977.

Allegri, T. H. *Handling and Management of Hazardous Materials*. New York: Chapman and Hall, 1986.

American Conference of Governmental Industrial Hygienists. *1991–1992 Threshold Limit Values for Chemical Substances and Physical Agents and Biological Exposure Indices*. Cincinnati: American Conference of Governmental Industrial Hygienists, 1991.

American Conference of Governmental Industrial Hygienists, Committee on Industrial Ventilation. *Industrial Ventilation: A Manual of Recommended Practice*. 20th ed. Cincinnati: American Conference of Governmental Industrial Hygienists, 1988.

American Conference of Government Industrial Hygienists. *Threshold Limit Values for Chemical Substances and Physical Agents and Biological Exposure Indices (1990–1991)*. Cincinnati, OH: American Conference of Governmental Industrial Hygienists, 1990.

American National Standards Institute. *Practices for Respiratory Protection*. ANSI Publication Z88.2. New York: American National Standards Institute, 1980.

Anderson, Ronald A. *Business Law*, 6th ed. Cincinnati, OH: Southwestern Publishing Company, 1961.

Archer, D. L., and F. E. Young. Contemporary issues: diseases with a food vector. *Clin. Microbiol. Rev.* 1:377–398, 1988.

Armour, M. A. *Hazardous Laboratory Chemicals Disposal Guide*. Boca Raton, FL: CRC Press, 1990.

Armour, Margaret-Ann, *et al. Hazardous Chemicals Information and Disposal Guide*. Edmonton, Alberta: University of Alberta Press, 1987.

Assar, M. *Guide to Sanitation in Natural Disasters*. Geneva, Switzerland: World Health Organization, 1971.

Baum, Janet, *et al. Guidelines for Laboratory Design: Health and Safety Considerations*. New York: John Wiley and Sons, 1987.

Beaudette, F. R. *Psittacosis—Diagnosis, Epidemiology and Control*. New Brunswick, NJ: Rutgers University Press, 1955.

Beaudette, F. R. *Progress in Psittacosis Research and Control*. New Brunswick, NJ: Rutgers University Press, 1958.

Benenson, A. S. *Control of Communicable Diseases in Man*, 15th ed. Washington, D.C.: American Public Health Association, 1990.

Birnbaum, S. L., and R. J. Phelan. *Complex Toxic Chemical or Hazardous Waste Cases*. New York: U.S. Practising Law Institute, 1984.

Black, H. C. *Black's Law Dictionary*. St. Paul, MN: West Publishing, 1968.

Block, S. S. (ed.). *Disinfection, Sterilization, and Preservation*, 3rd ed. Philadelphia, PA: Lea and Febiger, 1983.

Bollinger, N. J., *et al.* (eds.). *NIOSH Guide to Industrial Respiratory Protection*. DHHS (NIOSH) Pub 87-116. Washington, D.C.: U.S. Government Printing Office, 1987.

Boseleywoolf, Henry (ed.) *Webster's New Collegiate Dictionary*. New York: G & C Merriam Company, 1981.

Brady, N. C. *The Nature and Properties of Soils*, 8th ed. New York: Macmillan Publishing, 1974.

Brauer, R. L. *Safety and Health for Engineers*. New York: Van Nostrand Reinhold, 1990.

Bretherick, L. *Handbook of Reactive Chemical Hazards*. 4th ed. London, England: Butterworth Publishers, 1990.

Bretherick, L. *Hazards in the Chemical Laboratory*, 3rd ed. London, England: Royal Society of Chemistry, 1981.

Brewer, J. H. *Lectures on Sterilization*. Durham, NC: Duke University Press, 1973.

Brown, Stanley S., and Donald S. Davies (eds.). *Organ-Directed Toxicity—Chemical Indices and Mechanisms*. Oxford, England: Pergamon Press, 1981.

Bryan, F. L., Epidemiology of foodborne diseases. *Foodborne Infections and Intoxications*, 2nd ed., H. Riemann and F. L. Bryan (eds.). New York: Academic Press, 1979.

Bryan, F. L. Hazard analysis critical control point: What the system is and what it is not. *J. Environ. Health*, 50:400–401, 1988.

Bryan, F. L. Hazard analysis critical control point (HACCP) concept. *Dairy Food Environ. Sanit.* 10:416–418, 1990.

Buol, S. W., F. D. Hole, and R. J. McCraken. *Soil Genesis and Classification*, 2nd ed., Ames, IA: Iowa State University Press, 1980.

Bureau of National Affairs. *U.S. Environmental Laws*. 3rd Ed. Washington, D.C.: Bureau of National Affairs, 1991.

Burgess, W. A. *Recognition of Health Hazards in Industry—A Review of Materials and Processes*. New York: John Wiley and Sons, 1981.

Campbell, Reginold L., and Roland E. Langford. *Fundamentals of Hazardous Chemicals Incidents*. Chelsea, MI: Lewis Publishers, 1991.

Cano, R. J., and J. S. Colome. *Microbiology*. New York: West Publishing Company, 1986.

Cate, A. E., and J. L. Linville. *Fire Protection Handbook*, 16th ed. Quincy, MA: National Fire Protection Association, 1986.

Claus, Edward P., and Varro E. Tyler, *Pharmacognosy*, 5th ed. Philadelphia, PA: Lea & Febiger, 1965.

Committee on Hazardous Biological Substances in the Laboratory. National Research Council. *Biosafety in the Laboratory. Prudent Practices for the Handling and Disposal of Infectious Materials*. Washington, D.C.: National Academy Press, 1989.

Compressed Gas Association. *Handbook of Compound Gases*, 3rd ed. New York: Van Nostrand Reinhold, 1990.

Condensed 1988/1989 Hazardous Materials, Substances, and Waste Compliance Guide. Kutztown, PA: Transportation Skills Program, 1988.

Cralley, L. J., and L. V. Cralley (eds.). *Patty's Industrial Hygiene and Toxicology. Theory and Rationale of Industrial Hygiene Practice—The Work Environment*. Vols. 3A & 3B, 2nd ed. New York: Wiley Interscience, 1985.

Cross-Connection Control Manual. Washington, D.C.: Office of Water Programs, U.S. Environmental Protection Agency, 1973.

Davis, B. D., R. Dulbecco, H. N. Eisen, H. S. Ginsberg, and B. W. Wood. *Microbiology*. New York: Harper and Row, Haeber Medical Division, 1969.

Department of Labor. 29 CFR 1910. Occupational Safety and Health Standards for General Industry.

Department of Labor. 29 CFR 1926. Safety and Health Regulations for Construction.

Department of Labor. OSHA Instruction PUB 8-1.5. Subject: Guidelines for Pressure Vessel Safety Assessment. Directorate of Technical Support, Occupational Safety and Health Administration, Washington, D.C., 1989.

Department of Labor. OSHA Notice CPL 2, March 9, 1992. Subject: Special Emphasis Program in Petrochemical Industries, Standard Industrial Classification (SIC) Codes 2821, 2869, and 2911, 1992.

Design Manual: Onsite Wastewater Treatment and Disposal Systems. Washington, D.C.: Office of Water Program Operation, U.S. Environmental Protection Agency, 1980.

Devita, V. T., S. Hellman, S. A. Rosenberg. *AIDS—Etiology, Diagnosis, Treatment, and Prevention*. Philadelphia, PA: J. B. Lippincott Co., 1985.

Diberardinis, Louis J., *et al. Guidelines for Laboratory Design: Health and Safety Considerations*. New York: John Wiley and Sons, 1987.

Dornhoffer, Mary K. *Handling Chemical Carcinogens: A Safety Guide*. Kutztown, PA: Transportation Skills Programs, Inc., 1986.

Doyle, M. P. Food-borne pathogens of recent concern. *Ann. Rev. Nutr.* 5:25–41, 1985.

Dreisbach, Robert H. *Handbook of Poisoning Prevention, Diagnosis and Treatment*. Los Altos, CA: Lange Medical Publications, 1983.

Dunsmore, Donald J. *Safety Measures for Use in Outbreaks of Communicable Disease*. Geneva, Switzerland: World Health Organization, 1986.

Ehlers, V. M., and E. W. Steel. *Municipal and Rural Sanitation*, 6th ed. New York: McGraw-Hill Book Company, 1965.

Emergency Response Guidebook. Washington, D.C.: Office of Hazardous Materials Transportation, U.S. Department of Transportation, 1988.

Environmental Protection Agency. *Manual of Water Well Consumption Practices*. Washington, D.C.: Office of Water Supply, U.S. Environmental Protection Agency, EPA Publication 570/9 75-001, 1975.

Environmental Protection Agency. *National Priorities List Fact Book*. Washington, D.C.: U.S. Government Printing Office, 1986.

Environmental Protection Agency, Office of Communications and Public Affairs (A-107). *Glossary of Environmental Terms and Acronym List*. Washington, D.C.: Environmental Protection Agency, 1986.

Fawcett, Howard H., and William S. Wood (eds.). *Safety and Accident Prevention in Chemical Operations*, 2nd ed. New York: John Wiley and Sons, 1982.

Federal Environment Laws. St. Paul, MN: West Publishing Company, 1991.

Fire Protection Guide and Hazardous Materials, 9th ed. Quincy, MA: National Fire Protection Association, 1986.

Fischer, P. M., L. A. Addison, P. Curtis, J. M. Mitchell. *The Office Laboratory*. Norwalk, CT: Appleton-Century-Crofts, 1983.

Fisk, W. G., *et al. Indoor Air Quality Control Techniques*. Park Ridge, NJ: Noyes Data Corporation, 1987.

Food and Drug Administration, Department of Health, Education, and Welfare. *Food Service Sanitation Manual*. DHEW Publication #(FDA) 78-2081. Washington, D.C.: U.S. Government Printing Office, 1978.

Freedman, B. *Sanitarian's Handbook: Theory and Administrative Practice for Environmental Health*. New Orleans, LA: Peerless Publishing Co., 1977.

Ganong, W. F. *Review of Medical Physiology*, 15th ed. Norwalk, CT: Appleton & Lange Publications, 1991.

Gasch, M. *Toxic Tort Litigation*. New York: Practising Law Institute, United States, 1984.

Gifis, S. H. *Law Dictionary*. Woodbury, NY: Barron's Educational Series, Inc., 1975.

Gilpin, A. *Dictionary of Environmental Terms*. London and Henley, England: Routledge and Kegan Paul, 1976.

Goth, Andres. *Medical Pharmacology*, 4th ed. St. Louis, MO: C. V. Mosby Company, 1968.

Greenberg, M. I., and J. R. Roberts. *Emergency Medicine—A Clinical Approach to Challenging Problems*. Philadelphia, PA: F. A. Davis Co., 1982.

Gressel, M. G., and J. A. Gideon. An overview of process hazard evaluation techniques. *AIHA J.* 52(4):158–163, 1991.

Gudiksen, P. H., T. F. Harvey, and R. Lange. Chernobyl Source Term, Atmospheric Dispersion, and Dose Estimation. *Health Physics* 57(5):697–706, 1989.

Guyton, A. C. *Textbook of Medical Physiology*, 8th ed. Philadelphia, PA: W. B. Saunders Company, 1991.

Hammer, M. J. *Water and Wastewater Technology*, 2nd ed. New York: John Wiley and Sons, 1986.

Hauk, C. C., and R. Post. *Chemistry: Concepts and Problems*. New York: John Wiley and Sons, 1977.

Hillel, D. *Introduction to Soil Physics*. New York: Academic Press, 1982.

Hinkle, Lawrence E., and William C. Loring (eds.). *The Effect of the Man-Made Environment on*

Health and Behavior. DHEW publication #(CDC) 77-8318. Atlanta, GA: Centers for Disease Control, 1977.

Hoeprick, P. D. *Infectious Diseases*, 1st ed. Hagerstown, MD: Harper and Row, 1972.

Howard, B. G., J. Klass, S. J. Rubin, A. S. Weissfeld, and R. C. Telton. *Clinical and Pathogenic Microbiology*. Washington, D.C.: C. V. Mosby Co., 1987.

Hubbert, W. T., and H. V. Hagstad. *Food Safety and Quality Assurance: Foods of Animal Origin*. Ames, IA: Iowa State University Press, 1991.

Hubbert, W. T., W. F. McCullouch, and P. R. Schnurenberger (eds.). *Diseases Transmitted from Animals to Man*, 6th ed. Springfield, IL: Charles C. Thomas, 1975.

Inhorn, S. L. *Quality Assurance Practices for Health Laboratories*. Washington, D.C.: American Public Health Association, 1978.

Institute of Medicine, National Academy of Sciences. *Confronting AIDS: Directions for Public Health, Health Care, and Research*. Washington, D.C.: National Academy Press, 1986.

International Commission on Microbiological Specifications for Foods (ICMSP) of the International Union of Microbiological Societies. *Microorganisms in Foods 4: Application of the Hazard Analysis Critical Control Point (HACCP) System to Ensure Microbiological Safety and Quality*. Boston, MA: Blackwell Scientific Publications, 1988.

Johnson, Robert. *Elementary Statistics*. 5th ed. Boston, MA: PWS-Kent Publishing Company, 1988.

Kamrin, Michael A. *Toxicology*. Chelsea, MI: Lewis Publishers, 1988.

Katz, John C., and Keith F. Purcell. *Chemistry and Chemical Reactivity*. Philadelphia, PA: Saunders College Publishing, 1987.

Kaufman, J. A. (ed.). *Waste Disposal in Academic Institutions*. Chelsea, MI: Lewis Publishers, Inc., 1990.

Keller's Chemical Regulatory Cross-References. Neenah, WI: J. J. Keller and Associates, Inc., 1987.

Key, Marcus M. (ed.). *Occupational Diseases—A Guide to Their Recognition*. Washington, D.C.: U.S. Dept. of Health, Education, and Welfare. Gov. Printing Office, Stock # 017-033-00266-5, 1977.

Kilgore, Wendell W., and Richard L. I. Doutt. 1967. *Pest Control: Biological, Physical and Selected Chemical Methods*. New York: Academic Press, 1967.

Kimbrough, R. D., and A. A. Jensen (eds.). *Halogenated Biphenyls, Terphenyls, Naphthalenes, Dibenzodioxins, and Related Products*, 2nd ed. Amsterdam: Elsevier Press, 1989.

Kingsbury, John M. *Poisonous Plants of the United States and Canada*. Englewood Cliffs, NJ: Prentice-Hall, Inc., 1964.

Klaasen, Curtis D., Mary O. Amdur, and John Doull (eds.). *Casarett & Doull's Toxicology*, 3rd ed. New York: Macmillan Publishing Company, 1986.

Klein, B. R. (ed.). *Health Care Facilities Handbook*, 3rd ed. Quincy, MA: National Fire Protection Association, 1990.

Kolb, J., and S. S. Ross. *Product Safety and Liability: A Desk Reference*. New York: McGraw-Hill, 1980.

Kormondy, Edward J. *Concepts of Ecology*, 2nd ed. Englewood Cliffs, NJ: Prentice-Hall, Inc., 1976.

Krebs, Charles J. *Ecology: The Experimental Analysis of Distribution and Abundance*, 3rd ed. New York: Harper & Row, 1985.

Laboratory Biosafety Guidelines. Medical Research Council of Canada. Ottawa, Ontario: Laboratory Centre for Disease Control, Health and Welfare, 1990.

Landau, S. I. *International Dictionary of Medicine and Biology*. Vols. I, II, III. New York: John Wiley and Sons, 1986.

Lawrence, Carl A., and Seymour S. Block. *Disinfection, Sterilization, and Preservation*. Philadelphia, PA: Lea & Febiger, 1968.

Legal Education Institute. *The Dynamics of Environmental Law*. Washington, D.C.: The Department of Justice, 1988.

Lennette, E. H., A. Ballows, W. J. Hausler, and J. P. Truant. *Manual of Clinical Microbiology*, 3rd ed. Washington, D.C.: American Society for Microbiology, 1980.

Lennette, E. H., A. Ballows, W. J. Hausler, and J. H. Shadomy. *Manual of Clinical Microbiology*, 3rd ed. Washington, D.C.: American Society of Microbiology, 1980.

Lennette, E. H., and N. J. Schmidt. *Diagnostic Procedures for Viral, Rickettsial, and Chlamydial Infections*, 5th ed. Washington, D.C.: American Public Health Association, 1979.

Levine, Steven P., and William F. Martin. *Protecting Personnel at Hazardous Waste Sites*. Boston, MA: Butterworth Publishers, 1985.

Lewis, Richard J. *Carcinogenically Active Chemicals*. New York: Van Nostrand Reinhold, 1991.

Lisella, F. S., Epidemiology of poisonings by chemicals. *J. Environ. Health*, 34(6):603–612, 1972.

Lisella, F. S., and L. W. Herring. Promising biotechnology industry raises unique safety, health issues. *Occupational Health and Safety*, 57(6):62–81, 1988.

Lisella, F. S., W. Johnson, and K. Holt. Mortality from carbon monoxide in Georgia 1961–1973. *J. Med. Assoc. GA*. Feb., 1976, pp. 98–100.

Lisella, F. S., W. Johnson, and C. Lewis. Health aspects of Organophosphate insecticide usage. *J. Environ. Health*, 33(3), 1971.

Lisella, F. S., and K. R. Long. Accidental poisonings in the Cedar Rapids–Linn County, Iowa area. *Bull. Inst. Agric. Med.*, Univ. of Iowa, Iowa City, IA, 1972.

Lisella, F. S., and K. R. Long. Epidemiologic aspects of self-induced poisonings in the Cedar Rapids–Linn County, Iowa area. *Bull. Inst. Agric. Med.*, Univ. of Iowa, Iowa City, IA, 1972

Lisella, F. S., K. R. Long, and H. G. Scott. Health aspects of arsenicals in the environment. *J. Environ. Health*, 33(5), 1972.

Lisella, F. S., K. R. Long, and H. G. Scott. Toxicology of rodenticides and their relation to human health (Parts I and II) Epidemiology of poisonings by chemicals. *J. Environ. Health*, 33(6):231–237, 1971, and 33(4):362–365, 1972.

Lisella, F. S., E. P. Savage, and H. G. Scott. Pesticides in the institutional environment. *J. Environ. Health*, 33(3), 1971.

Loomis, Ted A. *Essentials of Toxicology*. Philadelphia, PA: Lea and Febiger, 1968.

Lundberg, George, and J. C. Segen. *The Dictionary of Modern Medicine*. Park Ridge, NJ: The Parthenon Publishing Group, 1992.

Lunn, George, and Eric B. Samsone. 1989. *Destruction of Hazardous Chemicals in the Laboratory*. New York: John Wiley & Sons, Inc., 1989.

MacMahon, Brian, *et al. Epidemiologic Methods*. Boston, MA: Little, Brown, & Company, 1960.

Maletskos, C. J., and C. C. Lushbaugh. The Goiania radiation incident: editorial. *Health Physics*, 60(1):1, 1991.

Manahan, Stanley E. *Hazardous Waste Chemistry, Toxicology, and Treatment*. Chelsea, MI: Lewis Publishers, 1990.

Marriott, N. G. *Principles of Food Sanitation*, 2nd ed. New York: Van Nostrand Reinhold, 1989.

Martin, A., and S. A. Harbison. *An Introduction to Radiation Protection*. Oxford: Methuen, Inc., 1986.

Martin, William F., *et al. Hazardous Waste Handbook for Health and Safety*. Boston, MA: Butterworth Publishers, 1987.

Masterson, W. L., and E. J. Slowinski. *Chemical Principles*. Philadelphia, PA:W. B. Saunders Co., 1973.

Maxy-Ronseau. *Public Health and Preventive Medicine*, 12th ed. J. M. Last (ed.). Norwalk, CT: Appleton-Century Crofts, 1986.

McEwen, F. L., and G. R. Stephenson. *The Use and Significance of Pesticides in the Environment*. New York: John Wiley and Sons, 1979.

McIntyre, C. R. Hazard analysis critical control point (HACCP) identification. *Dairy Food and Environ. Sanit*. 7:357–358, 1991.

McKechnie, Jean L. (ed.) *Webster's New Universal Unabridged Dictionary*, 2nd ed. New York: Simon and Schuster, 1983.

Metry, Amis A. *The Handbook of Hazardous Waste Management*. Westport, CT: Technomic Publishing Company, Inc., 1980.

Meyer, Rudolph. *Explosives*. New York: VCH Publishers, 1987.

Miller, Brinton M. (ed.). *Laboratory Safety: Principles and Practices*. Washington, D.C.: American Society for Microbiology, 1986.

Miller, K. L., and W. A. Weidner (eds.). *CRC Handbook of Management of Radiation Protection Programs*. Boca Raton: CRC Press, 1986.

Morris, William (ed.). *The American Heritage Dictionary of the English Language*. Boston, MA: Houghton Mifflin Company, 1979.

Morrison, R. T., and R. N. Boyd. *Organic Chemistry*. Boston, MA: Allyn and Bacon, Inc., 1973.

National Fire Protection Association. *Flammable and Combustible Liquids Code*. Publication 30. Quincy, MA: National Fire Protection Association, 1980.

National Fire Protection Association. *Fire Protection Handbook*. Quincy, MA: National Fire Protection Association, 1981.

National Fire Protection Association. *Fire Protection for Laboratories Using Chemicals*. Publication 45. Quincy, MA: National Fire Protection Association, 1982.

National Institute of Occupational Safety and Health. *The Industrial Environment—Its Evaluation and Control*. NIOSH pub. S/N 017-001-00396-4. Cincinnati, OH: National Institute of Occupational Safety and Health, 1973.

National Institute of Occupational Safety and Health. *Work Practices Guide for Manual Lifting (Rep 81-122)*. Cincinnati, OH: National Institute of Occupational Safety and Health, 1981.

National Research Council. *AIDS: The Second Decade*. Washington, D.C.: National Academy Press, 1990.

National Research Council, Committee on Risk Perception and Communication. *Improving Risk Communication*. Washington, D.C.: National Academy Press, 1989.

National Research Council. *Prudent Practices for Handling Hazardous Chemicals in Laboratories*. Washington, D.C.: National Academy Press, 1981.

National Sanitation Foundation. Food Service Equipment Standards (#2, 3, 4, 7–1990, 26) Ann Arbor, MI: National Sanitation Foundation, various years.

Newton, Jim. *Environmental Auditing*. Northbrook, IL: Pudvan Publishing Company, Inc., 1989.

O'Brien, R. D. *Insecticides—Action and Metabolism*. New York: Academic Press, 1967.

Olishifski, J. B., and E. R. Harford. *Industrial Noise and Hearing Conservation*. Chicago, IL: National Safety Council, 1975.

Olishifski, J. B., and F. E. McElroy (eds.). *Fundamentals of Industrial Hygiene*, 2nd ed. Chicago, IL: National Safety Council, 1982.

Okun, D. A., and G. Ponghis. *Community Wastewater Collection and Disposal*. Geneva, Switzerland: World Health Organization, 1975.

Parish, H. J. *Antisera, Toxoids, Vaccines, and Tuberculines in Prophylaxis and Treatment*, 3rd ed. Baltimore, MD: Williams & Wilkins Company, 1954.

Parker, S. P. (ed.). *Concise Encyclopedia of Science and Technology*. New York: McGraw-Hill, 1984.

Pelczar, M. J., R. D. Reed, and E. C. S. Chan. *Microbiology*, 4th ed. New York: McGraw-Hill, 1977.

Phifer, Russell W., and William R. McTigue. *Handbook of Hazardous Waste Management for Small Quantity Generators*. Chelsea, MI: Lewis Publishers, Inc., 1989.

Phillips, G. B., and W. S. Miller. *Industrial Sterilization*. Durham, NC: Duke University Press, 1972.

Pipitone, David A. *Safe Storage of Laboratory Chemicals*. New York: John Wiley and Sons, 1984.

Planka, Eric R. *Evolutionary Ecology*, 4th ed. New York: Harper & Row, 1988.

Pollitzer, R. *Chlorera*. Geneva, Switzerland: World Health Organization, 1959.

Pontius, F. W. (ed.). *Water Quality and Treatment, A Handbook of Community Water Supplies*. New York: McGraw-Hill, 1990.

Purdom, P. W. (ed.). *Environmental Health*, 2nd ed. New York: Academic Press, 1980.

Radiological Health Handbook. Bureau of Radiological Health. U.S. Department of Health, Education, and Welfare. Washington, D.C.: U.S. Government Printing Office, 1970.

Raven, Peter H., Ray R. Evert, and Helena Curtis. *Biology of Plants*, 3rd ed. New York: Worth Publishers, Inc., 1981.

Reddish, G. F. (ed.). *Antiseptics, Disinfectants, Fungicides, and Chemical and Physical Sterilization*. Philadelphia, PA: Lea and Febiger, 1975.

Richardson, John H., and Emmett W. Barkley (eds.). *Biosafety in Microbiological and Biomedical Laboratories*, 2nd ed. Washington, D.C.: U.S. Government Printing Office, Stock # 17-40-508-3, 1988.

Roberts, J. M., Sr. *OSHA Compliance Manual*. Reston, VA: Reston Publishing Company, 1976.

Robinson, J. S. (ed.). *Hazardous Chemical Spill Cleanup*. Park Ridge, NJ: Noyes Data Corporation, 1979.

Rogers, Fred B. *Studies in Epidemiology: Selected Papers of Morris Greenberg*. New York: G. P. Putnam's Sons, 1965.

Roland, H. E., and B. Moriarty. *System Safety Engineering and Management*, 2nd ed. New York: John Wiley & Sons, 1990.

Rose, Susan L. *Chemical Laboratory Safety*. Philadelphia, PA: J. B. Lippincott Company, 1984.

Rosehlund, S. J. *The Chemical Laboratory: Its Design and Operation*,. Park Ridge, NJ: Noyes Publishers, 1987.

Roser, Bernard. *Fundamentals of Biostatistics*. Boston, MA: PWS-Kent Publishing Company, 1990.

Rossnagel, W. E., L. R. Higgins, and J. A. MacDonald. *Handbook of Rigging*, 4th ed. New York: McGraw-Hill, 1988.

Ryser, E. T., and E. J. Marth. *Listeria, Listeriosis and Food Safety*. New York: Marcel Dekker, 1991.

Salvato, J. A. *Guide to Sanitation in Tourist Establishments*. Geneva, Switzerland: World Health Organization, 1976.

Salvato, J. A. *Environmental Engineering*, 3rd ed. New York: John Wiley & Sons, 1982.

Sawyer, Clair N., and M. C. Carty. *Chemistry for Sanitary Engineers*. New York: McGraw-Hill Book Company, 1967.

Sax, Irving N., and Richard J. Lewis (eds.). *Rapid Guide to Hazardous Chemicals in the Workplace*. New York: Van Nostrand Reinhold, 1986.

Sax, Irving N., and Richard J. Lewis. *Hazardous Chemicals Desk Reference*. New York: Van Nostrand Reinhold, 1987.

Sax, Irving N., and Richard J. Lewis. *Hawley's Condensed Chemical Dictionary*, 11th ed. New York: Van Nostrand Reinhold, 1987.

Sherman, Janette D. *Chemical Exposure and Disease: Diagnostic and Investigative Techniques*. New York: Van Nostrand Reinhold, 1988.

Sittig, Marshall. *Handbook of Toxic and Hazardous Chemicals and Carcinogens*, 2nd ed. Park Ridge, NJ: Noyes Publications, 1985.

Sive, D., and F. Friedman. *A Practical Guide to Environmental Law*. Philadelphia, PA: American Law Institute, American Bar Association Committee on Continuing Professional Education, 1987.

Skoog, Douglas A., and Donald M. West (eds.). *Fundamentals of Analytical Chemistry*, 3rd ed. New York: Holt, Rinehart, and Winston, 1976.

Smith, Robert Leo. *Ecology and Field Biology*, 3rd ed. New York: Harper & Row, 1980.

Soil Science Society of America. *Glossary of Soil Science Terms*. Madison, WI: Science Society of America, 1984.

Soil Taxonomy, Soil Survey Staff, Soil Conservation Service, United States Department of Agriculture. *Agricultural Handbook No. 436*, Washington, D.C.: 1975.

Spiegel, Murray R. *Schaum's Outline Series—Theory and Problems of Statistics*. New York: Schaum Publishing Company, 1961.

Stedman's Medical Dictionary, 25th ed. Baltimore, MD: Williams & Williams, 1990.

Surveillance of Drinking Water Quality. Geneva, Switzerland: World Health Organization, 1976.

Tan, K. H. *Principles of Soil Chemistry*, New York: Marcel Dekker, Inc., 1982.

Taylor, Ian, and John Knowelden. *Principles of Epidemiology*. Boston, MA: Little, Brown & Company, 1958.

The Merck Veterinary Manual, 5th ed. Rahway, NJ: Merck and Co., Inc., 1979.

The United States Government Manual 1988/89. Office of the Federal Register, National Archive and Records Administration. Washington, D.C.: U.S. Government Printing Office, 1988.

Theiler, M., and W. G. Downs. *The Arthropod-Borne Viruses of Vertebrates*. New Haven, CT: Yale University Press, 1973.

Tisdale, S. L., and W. L. Nelson. *Soil Fertility and Fertilizers*, 3rd ed. New York: Macmillan Publishing, 1975.

The Soap and Detergent Association. *A Handbook of Industry Terms*, 2nd ed. New York, 1981.

Top, Franklin H. *Communicable and Infectious Diseases*, 6th ed. St. Louis, MO: C. V. Mosby Company, 1960.

Troeh, F. R., J. A. Hobbs, and R. L. Donahue. *Soil and Water Conservation*. Englewood Cliffs, NJ: Prentice-Hall Publishing, 1980.

Velz, Clarence J. *Applied Stream Sanitation*. New York: Wiley Interscience, 1970.

Wade, L. G., Jr. *Organic Chemistry*. Englewood Cliffs, NJ: Prentice-Hall, Inc., 1987.

Wagner, Edmund G., and J. N. Lanoix. *Excreta Disposal for Rural Areas and Small Communities*. Geneva, Switzerland: World Health Organization, 1958.

Wagner, Edmund G., and J. N. Lanoix. *Water Supply for Rural Areas and Small Communities*. Geneva, Switzerland: World Health Organization, 1959.

Wang, C. H., and David L. Willis. *Radiotracer Methodology in Biological Sciences*. Englewood Cliffs, NJ: Prentice-Hall, Inc., 1965.

Ware, G. W. (ed.). *Review of Environmental Contamination and Toxicology*. U.S. Environmental Protection Agency, Office of Drinking Water Health Advisories, Vol. 107. New York: Springer-Verlag, 1990.

Weast, Robert C., and Melvin J. Astle (eds.). *Handbook of Chemistry and Physics*, 61st ed. Boca Raton, FL: CRC Press, Inc., 1980–81.

Weiss, G. *Hazardous Chemicals Data Book*, 2nd ed. Park Ridge, NJ: Noyes Data Corporation, 1986.

Weiss, Neil, and Matthew Hasset. *Introductory Statistics*. Addison-Wesley Publishing Company, 1982.

Whittaker, Robert H. *Communities and Ecosystems*, 2nd ed. New York: Macmillan Publishing Co., Inc., 1975.

Whitten, K. W., and K. D. Gailey. *General Chemistry with Qualitative Analysis*, 2nd ed. New York: Saunders College Publishing, 1984.

Williams, Philip L., and James L. Burson (eds.). *Industrial Toxicology—Safety and Health Applications in the Workplace*. New York: Van Nostrand Reinhold, 1985.

Winter, F. H., and M. L. Shourd. *Review of Human Physiology*. Philadelphia, PA: W. B. Saunders Company, 1987.

Wylie, Lawrence G. *Hazardous and Infectious Waste Management for Health Care Facilities*. New York: Fred S. James & Company, Inc., 1984.

Young, Jay A., *et al*. *Improving Safety in the Chemical Laboratory*. New York: John Wiley & Sons, Inc., 1987.

Zinsser, Hans. *Rats, Lice, and History*, 31st printing. New York: Bantam Books, Inc., 1965.

About the Editor

Frank S. Lisella, Ph.D., M.P.H., is Director of the Environmental Health and Safety Office of the School of Medicine at Emory University in Atlanta, Georgia. Formerly, he was Director of the Office of Biosafety for the Centers for Disease Control in Atlanta, where he worked for over 20 years. He has received the U.S. Public Health Service Outstanding Service Award and numerous other honors, and is the author of over 40 technical articles. Dr. Lisella has also been a consultant to the World Health Organization. He earned his Ph.D. in Preventive Medicine and Environmental Health from the University of Iowa, his M.P.H. in Environmental Health and Epidemiology from Tulane University, and his bachelor's degree from Millersville University.